NANOPHOTONICS

NANOPHOTONICS

PARAS N. PRASAD

A JOHN WILEY & SONS, INC., PUBLICATION

Copyright © 2004 by John Wiley & Sons, Inc. All rights reserved.

Published by John Wiley & Sons, Inc., Hoboken, New Jersey.
Published simultaneously in Canada.

No part of this publication may be reproduced, stored in a retrieval system or transmitted in any form or by any means, electronic, mechanical, photocopying, recording, scanning or otherwise, except as permitted under Section 107 or 108 of the 1976 United States Copyright Act, without either the prior written permission of the Publisher, or authorization through payment of the appropriate per-copy fee to the Copyright Clearance Center, Inc., 222 Rosewood Drive, Danvers, MA 01923, (978) 750-8400, fax (978) 646-8600, or on the web at www.copyright.com. Requests to the Publisher for permission should be addressed to the Permissions Department, John Wiley & Sons, Inc., 111 River Street, Hoboken, NJ 07030, (201) 748-6011, fax (201) 748-6008.

Limit of Liability/Disclaimer of Warranty: While the publisher and author have used their best efforts in preparing this book, they make no representation or warranties with respect to the accuracy or completeness of the contents of this book and specifically disclaim any implied warranties of merchantability or fitness for a particular purpose. No warranty may be created or extended by sales representatives or written sales materials. The advice and strategies contained herein may not be suitable for your situation. You should consult with a professional where appropriate. Neither the publisher nor author shall be liable for any loss of profit or any other commercial damages, including but not limited to special, incidental, consequential, or other damages.

For general information on our other products and services please contact our Customer Care Department within the U.S. at 877-762-2974, outside the U.S. at 317-572-3993 or fax 317-572-4002.

Wiley also publishes its books in a variety of electronic formats. Some content that appears in print, however, may not be available in electronic format.

Library of Congress Cataloging-in-Publication Data:

Prasad, Paras N.
   Nanophotonics / Paras N. Prasad.
     p. cm.
   Includes bibliographical references and index.
   ISBN 0-471-64988-0 (cloth)
    1. Photonics. 2. Nonotechnology. I. Title.
TA1520.P73 2004
621.36—dc22   2004001186

Printed in the United States of America.

10 9 8 7 6 5 4 3 2 1

Summary of Contents

1. Introduction
2. Foundations for Nanophotonics
3. Near-Field Interaction and Microscopy
4. Quantum-Confined Materials
5. Plasmonics
6. Nanocontrol of Excitation Dynamics
7. Growth and Characterization of Nanomaterials
8. Nanostructured Molecular Architectures
9. Photonic Crystals
10. Nanocomposites
11. Nanolithography
12. Biomaterials and Nanophotonics
13. Nanophotonics for Biotechnology and Nanomedicine
14. Nanophotonics and the Marketplace

Contents

Preface	xiii
Acknowledgments	xv
1. Introduction	**1**
1.1 Nanophotonics—An Exciting Frontier in Nanotechnology	1
1.2 Nanophotonics at a Glance	1
1.3 Multidisciplinary Education, Training, and Research	3
1.4. Rationale for this Book	4
1.5 Opportunities for Basic Research and Development of New Technologies	5
1.6 Scope of this Book	6
References	8
2. Foundations for Nanophotonics	**9**
2.1 Photons and Electrons: Similarities and Differences	10
2.1.1 Free-Space Propagation	12
2.1.2 Confinement of Photons and Electrons	14
2.1.3 Propagation Through a Classically Forbidden Zone: Tunneling	19
2.1.4 Localization Under a Periodic Potential: Bandgap	21
2.1.5 Cooperative Effects for Photons and Electrons	24
2.2 Nanoscale Optical Interactions	28
2.2.1 Axial Nanoscopic Localization	29
2.2.2 Lateral Nanoscopic Localization	32
2.3 Nanoscale Confinement of Electronic Interactions	33
2.3.1 Quantum Confinement Effects	34
2.3.2 Nanoscopic Interaction Dynamics	34
2.3.3 New Cooperative Transitions	34
2.3.4 Nanoscale Electronic Energy Transfer	35
2.3.5 Cooperative Emission	36
2.4 Highlights of the Chapter	37
References	38

3. Near-Field Interaction and Microscopy — 41

- 3.1 Near-Field Optics — 42
- 3.2 Theoretical Modeling of Near-Field Nanoscopic Interactions — 44
- 3.3 Near-Field Microscopy — 48
- 3.4 Examples of Near-Field Studies — 51
  - 3.4.1 Study of Quantum Dots — 51
  - 3.4.2 Single-Molecule Spectroscopy — 53
  - 3.4.3 Study of Nonlinear Optical Processes — 55
- 3.5 Apertureless Near-Field Spectroscopy and Microscopy — 62
- 3.6 Nanoscale Enhancement of Optical Interactions — 65
- 3.7 Time- and Space-Resolved Studies of Nanoscale Dynamics — 69
- 3.8 Commercially Available Sources for Near-Field Microscope — 73
- 3.9 Highlights of the Chapter — 73
- References — 75

4. Quantum-Confined Materials — 79

- 4.1 Inorganic Semiconductors — 80
  - 4.1.1 Quantum Wells — 81
  - 4.1.2 Quantum Wires — 85
  - 4.1.3 Quantum Dots — 86
  - 4.1.4 Quantum Rings — 88
- 4.2 Manifestations of Quantum Confinement — 88
  - 4.2.1 Optical Properties — 88
  - 4.2.2 Examples — 91
  - 4.2.3 Nonlinear Optical Properties — 95
  - 4.2.4 Quantum-Confined Stark Effect — 96
- 4.3 Dielectric Confinement Effect — 99
- 4.4 Superlattices — 100
- 4.5 Core-Shell Quantum Dots and Quantum Dot-Quantum Wells — 104
- 4.6 Quantum-Confined Structures as Lasing Media — 106
- 4.7 Organic Quantum-Confined Structures — 115
- 4.8 Highlights of the Chapter — 120
- References — 122

5. Plasmonics — 129

- 5.1 Metallic Nanoparticles and Nanorods — 130
- 5.2 Metallic Nanoshells — 135
- 5.3 Local Field Enhancement — 137
- 5.4 Subwavelength Aperture Plasmonics — 138
- 5.5 Plasmonic Wave Guiding — 139
- 5.6 Applications of Metallic Nanostructures — 141
- 5.7 Radiative Decay Engineering — 142
- 5.8 Highlights of the Chapter — 147
- References — 149

6. Nanocontrol of Excitation Dynamics — 153

    6.1 Nanostructure and Excited States — 154
    6.2 Rare-Earth Doped Nanostructures — 158
    6.3 Up-Converting Nanophores — 161
    6.4 Photon Avalanche — 165
    6.5 Quantum Cutting — 166
    6.6 Site Isolating Nanoparticles — 171
    6.7 Highlights of the Chapter — 171
          References — 173

7. Growth and Characterization of Nanomaterials — 177

    7.1 Growth Methods for Nanomaterials — 178
          7.1.1 Epitaxial Growth — 179
          7.1.2 Laser-Assisted Vapor Deposition (LAVD) — 183
          7.1.3 Nanochemistry — 185
    7.2 Characterization of Nanomaterials — 189
          7.2.1 X-Ray Characterization — 190
                7.2.1.1 X-Ray Diffraction — 190
                7.2.1.2 X-Ray Photoelectron Spectroscopy — 192
          7.2.2 Electron Microscopy — 194
                7.2.2.1 Transmission Electron Microscopy (TEM) — 195
                7.2.2.2 Scanning Electron Microscopy (SEM) — 195
          7.2.3 Other Electron Beam Techniques — 197
          7.2.4 Scanning Probe Microscopy (SPM) — 199
    7.3 Highlights of the Chapter — 204
          References — 206

8. Nanostructured Molecular Architectures — 209

    8.1 Noncovalent Interactions — 210
    8.2 Nanostructured Polymeric Media — 212
    8.3 Molecular Machines — 215
    8.4 Dendrimers — 217
    8.5 Supramolecular Structures — 225
    8.6 Monolayer and Multilayer Molecular Assemblies — 229
    8.7 Highlights of the Chapter — 233
          References — 235

9. Photonic Crystals — 239

    9.1 Basics Concepts — 240
    9.2 Theoretical Modeling of Photonic Crystals — 242
    9.3 Features of Photonic Crystals — 246
    9.4 Methods of Fabrication — 252
    9.5 Photonic Crystal Optical Circuitry — 259

x CONTENTS

9.6	Nonlinear Photonic Crystals	260
9.7	Photonic Crystal Fibers (PCF)	264
9.8	Photonic Crystals and Optical Communications	266
9.9	Photonic Crystal Sensors	267
9.10	Highlights of the Chapter	270
	References	272

10. Nanocomposites — 277

10.1	Nanocomposites as Photonic Media	278
10.2	Nanocomposite Waveguides	280
10.3	Random Lasers: Laser Paints	283
10.4	Local Field Enhancement	284
10.5	Multiphasic Nanocomposites	286
10.6	Nanocomposites for Optoelectronics	290
10.7	Polymer-Dispersed Liquid Crystals (PDLC)	297
10.8	Nanocomposite Metamaterials	301
10.9	Highlights of the Chapter	302
	References	304

11. Nanolithography — 309

11.1	Two-Photon Lithography	311
11.2	Near-Field Lithography	317
11.3	Near-Field Phase-Mask Soft Lithography	322
11.4	Plasmon Printing	324
11.5	Nanosphere Lithography	325
11.6	Dip-Pen Nanolithography	328
11.7	Nanoimprint Lithography	330
11.8	Photonically Aligned Nanoarrays	331
11.9	Highlights of the Chapter	332
	References	334

12. Biomaterials and Nanophotonics — 337

12.1	Bioderived Materials	338
12.2	Bioinspired Materials	344
12.3	Biotemplates	346
12.4	Bacteria as Biosynthesizers	347
12.5	Highlights of the Chapter	350
	References	350

13. Nanophotonics for Biotechnology and Nanomedicine — 355

13.1	Near-Field Bioimaging	356
13.2	Nanoparticles for Optical Diagnostics and Targeted Therapy	357
13.3	Semiconductor Quantum Dots for Bioimaging	358

13.4	Up-Converting Nanophores for Bioimaging	359
13.5	Biosensing	360
13.6	Nanoclinics for Optical Diagnostics and Targeted Therapy	365
13.7	Nanoclinic Gene Delivery	367
13.8	Nanoclinics for Photodynamic Therapy	371
13.9	Highlights of the Chapter	375
	References	376

14. Nanophotonics and the Marketplace — 381

- 14.1 Nanotechnology, Lasers, and Photonics — 382
  - 14.1.1 Nanonetchnology — 382
  - 14.1.2 Worldwide Laser Sales — 383
  - 14.1.3 Photonics — 383
  - 14.1.4 Nanophotonics — 386
- 14.2 Optical Nanomaterials — 386
  - 14.2.1 Nanoparticle Coatings — 387
  - 14.2.2 Sunscreen Nanoparticles — 389
  - 14.2.3 Self-Cleaning Glass — 389
  - 14.2.4 Fluorescent Quantum Dots — 390
  - 14.2.5 Nanobarcodes — 391
  - 14.2.6 Photonic Crystals — 391
  - 14.2.7 Photonic Crystal Fibers — 391
- 14.3 Quantum-Confined Lasers — 392
- 14.4 Near-Field Microscopy — 392
- 14.5 Nanolithography — 393
- 14.6 Future Outlook for Nanophotonics — 394
  - 14.6.1 Power Generation and Conversion — 394
  - 14.6.2 Information Technology — 395
  - 14.6.3 Sensor Technology — 395
  - 14.6.4 Nanomedicine — 395
- 14.7 Highlights of the Chapter — 396
  - References — 397

Index — 399

Preface

Nanophotonics, defined by the fusion of nanotechnology and photonics, is an emerging frontier providing challenges for fundamental research and opportunities for new technologies. Nanophotonics has already made its impact in the marketplace. It is a multidisciplinary field, creating opportunities in physics, chemistry, applied sciences, engineering, and biology, as well as in biomedical technology.

Nanophotonics has meant different things to different people, in each case being defined with a narrow focus. Several books and reviews exist that cover selective aspects of nanophotonics. However, there is a need for an up-to-date monograph that provides a unified synthesis of this subject. This book fills this need by providing a unifying, multifaceted description of nanophotonics to benefit a multidisciplinary readership. The objective is to provide a basic knowledge of a broad range of topics so that individuals in all disciplines can rapidly acquire the minimal necessary background for research and development in nanophotonics. The author intends this book to serve both as a textbook for education and training as well as a reference book that aids research and development in those areas integrating light, photonics, and nanotechnology. Another aim of the book is to stimulate the interest of researchers, industries, and businesses to foster collaboration through multidisciplinary programs in this frontier science, leading to development and transition of the resulting technology.

This book encompasses the fundamentals and various applications involving the integration of nanotechnology, photonics, and biology. Each chapter begins with an introduction describing what a reader will find in that chapter. Each chapter ends with highlights that are basically the take-home message and may serve as a review of the materials presented.

In writing this book, which covers a very broad range of topics, I received help from a large number of individuals at the Institute for Lasers, Photonics, and Biophotonics at the State University of New York–Buffalo and from elsewhere. This help has consisted of furnishing technical information, creating illustrations, providing critiques, and preparing the manuscript. A separate Acknowledgement recognizes these individuals.

Here I would like to acknowledge the individuals whose broad-based support has been of paramount value in completing the book. I owe a great deal of sincere gratitude to my wife, Nadia Shahram. She has been a constant source of inspiration, providing support and encouragement for this writing, in spite of her own very busy professional schedule. I am also indebted to our daughters and our princesses, Melanie and Natasha, for showing their love and understanding by sacrificing their quality time with me.

I express my sincere appreciation to my colleague, Professor Stanley Bruckenstein, for his endless support and encouragement. I thank Dr. Marek Samoc, Professor Joseph Haus, and Dr. Andrey Kuzmin for their valuable general support and technical help. I owe thanks to my administrative assistant, Ms. Margie Weber, for assuming responsibility for many of the noncritical administrative issues at the Institute. Finally, I thank Ms. Theresa Skurzewski and Ms. Barbara Raff, whose clerical help in manuscript preparation was invaluable.

Acknowledgments

Technical Contents:

Mr. Martin Casstevens, Professor Joseph Haus, Dr. Andrey Kuzmin, Dr. Paul Markowicz, Dr. Tymish Ohulchanskyy, Dr. Yudhisthira Sahoo, Dr. Marek Samoc, Professor Wieslaw Strek, Professor Albert Titus

Technical Illustrations and References:

Dr. E. James Bergey, Professor Jean M.J. Frechet, Mr. Christopher Friend, Dr. Madalina Furis, Professor Bing Gong, Dr. James Grote, Dr. Aliaksandr Kachynski, Professor Raoul Kopelman, Professor Charles Lieber, Dr. Tzu Chau Lin, Dr. Derrick Lucey, Professor Hong Luo, Professor Tobin J. Marks, Professor Chad Mirkin, Dr. Haridas Pudavar, Dr. Kaushik RoyChoudhury, Dr. Yudhisthira Sahoo, Dr. Yuzchen Shen, Mr. Hanifi Tiryaki, Dr. Richard Vaia, Mr. QingDong Zheng

Chapter Critiques:

Dr. E. James Bergey, Dr. Jeet Bhatia, Professor Robert W. Boyd, Professor Stanley Bruckenstein, Dr. Timothy Bunning, Professor Alexander N. Cartwright, Professor Cid de Araújo, Dr. Edward Furlani, Professor Sergey Gaponenko, Dr. Kathleen Havelka, Professor Alex Jen, Professor Iam Choon Khoo, Professor Kwang-Sup Lee, Dr. Nick Lepinski, Professor Hong Luo, Dr. Glauco Maciel, Professor Seth Marder, Professor Bruce McCombe, Professor Vladimir Mitin, Dr. Robert Nelson, Professor Lucas Novotny, Dr. Amitava Patra, Professor Andre Persoons, Dr. Corey Radloff, Professor George Schatz, Professor George Stegeman, Dr. Richard Vaia

Manuscript Preparation:

Michelle Murray, Barbara Raff, Theresa Skurzewski, Marjorie Weber

CHAPTER 1

Introduction

1.1 NANOPHOTONICS—AN EXCITING FRONTIER IN NANOTECHNOLOGY

Nanophotonics is an exciting new frontier that has captured the imaginations of people worldwide. It deals with the interaction of light with matter on a nanometer size scale. By adding a new dimension to nanoscale science and technology, nanophotonics provides challenges for fundamental research and creates opportunities for new technologies. The interest in nanoscience is a realization of a famous statement by Feynman that "There's Plenty of Room at the Bottom" (Feynman, 1961). He was pointing out that if one takes a length scale of one micrometer and divides it in nanometer segments, which are a billionth of a meter, one can imagine how many segments and compartments become available to manipulate.

We are living in an age of "nano-mania." Everything nano is considered to be exciting and worthwhile. Many countries have started Nanotechnology Initiatives. A detailed report for the U.S. National Nanotechnology Initiative has been published by the National Research Council (NRC Report, 2002). While nanotechnology can't claim to provide a better solution for every problem, nanophotonics does create exciting opportunities and enables new technologies. The key fact is that nanophotonics deals with interactions between light and matter at a scale shorter than the wavelength of light itself. This book covers interactions and materials that constitute nanophotonics, and it also describes their applications. Its goal is to present nanophotonics in a way to entice one into this new and exciting area. Purely for the sake of convenience, the examples presented are selected wherever possible from the work conducted at our Institute for Lasers, Photonics, and Biophotonics, which has a comprehensive program in nanophotonics.

As a supplemental reference, a CD-ROM of the author's SPIE short course on nanophotonics, produced by SPIE, is recommended. This CD-ROM (CDV 497) provides numerous technical illustrations in color in the PowerPoint format.

1.2 NANOPHOTONICS AT A GLANCE

Nanophotonics can conceptually be divided into three parts as shown in Table 1.1 (Shen et al., 2000). One way to induce interactions between light and matter on a

Table 1.1. Nanophotonics

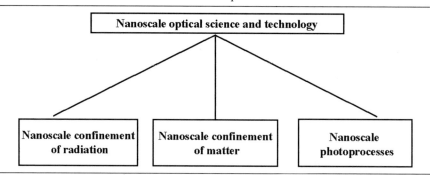

nanometer size scale is to confine light to nanoscale dimensions that are much smaller than the wavelength of light. The second approach is to confine matter to nanoscale dimensions, thereby limiting interactions between light and matter to nanoscopic dimensions. This defines the field of nanomaterials. The last way is nanoscale confinement of a photoprocess where we induce photochemistry or a light-induced phase change. This approach provides methods for nanofabrication of photonic structures and functional units.

Let's look at nanoscale confinement of radiation. There are a number of ways in which one can confine the light to a nanometer size scale. One of them is using near-field optical propagation, which we discuss in detail in Chapter 3 of this book. One example is light squeezed through a metal-coated and tapered optical fiber where the light emanates through a tip opening that is much smaller than the wavelength of light.

The nanoscale confinement of matter to make nanomaterials for photonics involves various ways of confining the dimensions of matter to produce nanostructures. For example, one can utilize nanoparticles that exhibit unique electronic and photonic properties. It is gratifying to find that these nanoparticles are already being used for various applications of nanophotonics such as UV absorbers in sunscreen lotions. Nanoparticles can be made of either inorganic or organic materials. Nanomers, which are nanometer size oligomers (a small number of repeat units) of monomeric organic structures, are organic analogues of nanoparticles. In contrast, polymers are long chain structures involving a large number of repeat units. These nanomers exhibit size-dependent optical properties. Metallic nanoparticles exhibit unique optical response and enhanced electromagnetic field and constitute the area of "plasmonics." Then there are nanoparticles which up-convert two absorbed IR photons into a photon in the visible UV range; conversely, there are nanoparticles, called quantum cutters, that down-convert an absorbed vacuum UV photon to two photons in the visible range. A hot area of nanomaterials is a photonic crystal that represents a periodic dielectric structure with a repeat unit of the order of wavelength of light. Nanocomposites comprise nanodomains of two or more dissimilar materials that are phase-separated on a nanometer size scale. Each nanodomain in the nanocomposite can impart a particular optical property to the bulk media. Flow of optical energy by energy transfer (optical communications) between different domains can also be controlled.

Nanoscale photoprocesses can be used for nanolithography to fabricate nanostructures. These nanostructures can be used to form nanoscale sensors and actuators. A nanoscale optical memory is one of exciting concepts of nanofabrication. An important feature of nanofabrication is that the photoprocesses can be confined to well-defined nanoregions so that structures can be fabricated in a precise geometry and arrangement.

1.3 MULTIDISCIPLINARY EDUCATION, TRAINING, AND RESEARCH

We live in a complex world where revolutionary progress has been and continues to be made in communications, computer memory, and data processing. There is a growing need for new technologies that rapidly detect and treat diseases at an early stage or even pre-stage. As we get accustomed to these advances, our expectations will demand more compact, energy-efficient, rapidly responding, and environmentally safe technologies. Photonic-based technology, coupled with nanotechnology, can meet many of these challenges. In the medical area, new modes of photonic diagnostics, which are noninvasive and molecular-based, may recognize the pre-stages and onset of a disease such as cancer and thus provide a major leap (Prasad, 2003). Nanomedicine, combined with light-guided and activated therapy, will advance individualized therapy that is based on molecular recognition and thus have minimal side effects.

The past several decades have witnessed major technological breakthroughs produced by fusion of different disciplines. This trend is even more likely in this millenium. Nanophotonics in its broader vision offers opportunities for interactions among many traditionally disparate disciplines of science, technology, and medicine. As shall be illustrated in this book, nanophotonics is an interdisciplinary field that comprises physics, chemistry, applied sciences and engineering, biology, and biomedical technology.

A significant multidisciplinary challenge lies ahead for the broader nanophotonics visions to become reality. These challenges require a significant increase in the number of knowledgeable researchers and trained personnel in this field. This need can be met by providing a multidisciplinary training for a future generation of researchers at both undergraduate and graduate levels, worldwide. A worldwide recognition of this vital need is evident from the growing number of conferences and workshops being held on this topic, as well as from the education and training programs being offered or contemplated at various institutions. For example, the author has offered a multidisciplinary course in nanophotonics at Buffalo as well as a short course in this subject at the SPIE professional society meetings. Much of the material covered in this book was developed during the teaching of these courses and was refined by valuable feedback from these course participants.

It is hoped that this book will serve both as an education and training text and as a reference book for research and development. Also, this book should be of value to industries and businesses, because the last chapter attempts to provide a critical evaluation of the current status of nanophotonic-based technologies.

1.4 RATIONALE FOR THIS BOOK

Naturally, for a hot area such as nanophotonics, many excellent reviews and edited books exist. Nanophotonics has meant different things to different people. Some have considered near-field interactions and near-field microscopy as the major thrust of nanophotonics, while others have considered it to be focused in photonic crystals. Another major direction has been nanomaterials, particularly the ones exhibiting size dependence of their optical properties; these are the quantum-confined structures. For engineers, nanoscale optical devices and nanolithography are the most relevant aspects of nanophotonics.

In terms of optical materials, the scientific community is often divided in two traditional groups: inorganic and organic, with very little cross-fertilization. The physics community focuses on inorganic semiconductors and metals, while shying away from complex organic structures. The chemical community, on the other hand, deals traditionally with organic structures and biomaterials and feels less comfortable with inorganic semiconductors, particularly with concepts defining their electronic and optical properties. Importantly, a new generation of hybrid nanomaterials, which involve different levels of integration of organic and inorganic structures, holds considerable promise for new fundamental science and novel technologies. For example, novel chemical routes can be utilized to prepare inorganic semiconductor nanostructures for nanophotonics. Engineers, who could exploit these new materials' flexibility for fabrication of components with diverse functionalities and their heterogeneous integration, often lack experience in dealing with these materials. Biologists have a great deal to offer by providing biomaterials for nanophotonics. At the same time, biological and biomedical researchers can utilize nanophotonics to study cellular processes and use nano-optical probes for diagnosis and to effect light-guided and activated therapy.

Often, a major hurdle is the lack of a common language to foster effective communication across disciplines. Therefore, much is to be gained by creating an environment that includes these disciplines and facilitates their interactions.

This book will address all these issues. It proposes to fill the existing void by providing the following features:

- A unifying, multifaceted description of nanophotonics that includes near-field interactions, nanomaterials, photonic crystals, and nanofabrication
- A focus on nanoscale optical interactions, nanostructured optical materials and applications of nanophotonics
- A coverage of inorganic, organic materials, and biomaterials as well as their hybrids

- A broad view of nanolithography for nanofabrication
- A coverage of nanophotonics for biomedical research and nanomedicine
- A critical assessment of nanophotonics in the market place, with future forecasts

1.5 OPPORTUNITIES FOR BASIC RESEARCH AND DEVELOPMENT OF NEW TECHNOLOGIES

Nanophotonics integrates a number of major technology thrust areas: lasers, photonics, photovoltaics, nanotechnology, and biotechnology. Each of these technologies either already generates or shows the potential to generate over $100 billion per year of sales revenue. Nanophotonics also offers numerous opportunities for multidisciplinary research. Provided below is a glimpse of these opportunities, categorized by disciplines.

Chemists and Chemical Engineers
- Novel synthetic routes and processing of nanomaterials
- New types of molecular nanostructures and supramolecular assemblies with varied nanoarchitectures
- Self-assembled periodic and aperiodic nanostructures to induce multifunctionality and cooperative effects
- Chemistry for surface modifications to produce nanotemplates
- One-pot syntheses that do not require changing reaction vessels
- Scalable production to make large quantities economically

Physicists
- Quantum electrodynamics to study novel optical phenomena in nanocavities
- Single photon source for quantum information processing
- Nanoscale nonlinear optical processes
- Nanocontrol of interactions between electrons, phonons and photons
- Time-resolved and spectrally resolved studies of nanoscopic excitation dynamics

Device Engineers
- Nanolithography for nanofabrication of emitters, detectors, and couplers
- Nanoscale integration of emitters, transmission channels, signal processors, and detectors, coupled with power generators
- Photonic crystal circuits and microcavity-based devices
- Combination of photonic crystals and plasmonics to enhance various linear and nonlinear optical functions
- Quantum dot and quantum wire lasers

- Highly efficient broadband and lightweight solar panels that can be packaged as rolls
- Quantum cutters to split vacuum UV photons into two visible photons for new-generation fluorescent lamps and lighting

Biologists
- Genetic manipulation of biomaterials for photonics
- Biological principles to guide development of bio-inspired photonic materials
- Novel biocolloids and biotemplates for photonic structures
- Bacterial synthesis of photonic materials

Biomedical Researchers
- Novel optical nanoprobes for diagnostics
- Targeted therapy using light-guided nanomedicine
- New modalities of light-activated therapy using nanoparticles
- Nanotechnology for biosensors

1.6 SCOPE OF THIS BOOK

This book is written for a multidisciplinary readership with the goal to provide introduction to a wide range of topics encompassing nanophotonics. A major emphasis is placed on elucidating concepts with minimal mathematical details; examples are provided to illustrate principles and applications. The book can readily enable a newcomer to this field to acquire the minimum necessary background to undertake research and development.

A major challenge for researchers working in a multidisciplinary area is the need to learn relevant concepts outside of their expertise. This may require searching through a vast amount of literature, often leading to frustrations of not being able to extract pertinent information quickly. By providing a multifaceted description of nanophotonics, it is hoped that the book will mitigate this problem and serve as a reference source.

The book is structured so that it can also be of value to educators teaching undergraduate and graduate courses in multiple departments. For them, it will serve as a textbook that elucidates basic principles and multidisciplinary approaches. Most chapters are essentially independent of each other, providing flexibility in choice of topics to be covered. Thus, the book can also readily be adopted for training and tutorial short courses at universities as well as at various professional society meetings.

Each chapter begins with an introduction describing its contents. This introduction also provides a guide to what may be omitted by a reader familiar with the specific content or by someone who is less inclined to go through details. Each chapter ends with highlights of the content covered in it. The highlights provide the take-home message from the chapter and serve to review the materials learned. Also, for researchers interested in a cursory glimpse of a chapter, the highlights provide an overview of topics covered. For an instructor, the highlights may also be useful in the preparation of lecture notes or PowerPoint presentations.

Chapter 2 provides an introduction to the foundations of nanophotonics. The nanoscale interactions are defined by discussing similarities and differences between photons and electrons. Spatial confinement effects on photons and electrons are presented. Other topics covered are photon and electron tunneling, effect of a periodic potential in producing a bandgap, and cooperative effects. Ways to localize optical interactions axially and laterally on a nanoscale are described.

Chapter 3 defines near-field interactions and describes near-field microscopy. A brief theoretical description of near-field interactions and the various experimental geometries used to effect them are introduced. The theoretical section may be skipped by those more experimentally oriented. Various optical and higher-order nonlinear optical interactions in nanoscopic domains are described. Applications to the highly active field of single molecule spectroscopy is described.

Chapter 4 covers quantum-confined materials whose optical properties are size-dependent. Described here are semiconductor quantum wells, quantum wires, quantum dots, and their organic analogues. A succinct description of manifestations of quantum confinement effects presented in this chapter should be of significant value to those (e.g., some chemists and life scientists) encountering this topic for the first time. The applications of these materials in semiconductor lasers, described here, exemplify the technological significance of this class of materials.

Chapter 5 covers the topic of metallic nanostructures, now with a new buzzword "plasmonics" describing the subject. Relevant concepts together with potential applications are introduced. Guiding of light through dimensions smaller than the wavelength of light by using plasmonic guiding is described. The applications of metallic nanostructures to chemical and biological sensing is presented.

Chapter 6 deals with nanoscale materials and nanoparticles, for which the electronic energy gap does not change with a change in size. However, the excitation dynamics—that is, emission properties, energy transfer, and cooperative optical transitions in these nanoparticles—are dependent on their nanostructures. Thus nanocontrol of excitation dynamics is introduced. Important processes described are (i) energy up-conversion acting as an optical transformer to convert two IR photons to a visible photon and (ii) quantum cutting, which causes the down-conversion of a vacuum UV photon to two visible photons.

Chapter 7 describes various methods of fabrications and characterization of nanomaterials. In addition to the traditional semiconductor processing methods such as molecular beam epitaxy (MBE) and metal–organic chemical vapor deposition (MOCVD), the use of nanochemistry, which utilizes wet chemical synthesis approach, is also described. Some characterization techniques introduced are specific to nanomaterials.

Chapter 8 introduces nanostructured molecular architectures that include a rich class of nanomaterials often unfamiliar to physicists and engineers. These nanostructures involve organic and inorganic–organic hybrid structures. They are stabilized in a three-dimensional architecture by both covalent bonds (chemical) and noncovalent interactions (e.g., hydrogen bond). This topic is presented with a minimum amount of chemical details so that nonchemists are not overburdened. Nanomaterials covered in this chapter are block copolymers, molecular motors, dendrimers, supramolecules, Langmuir–Blogdett films, and self-assembled structures.

Chapter 9 presents the subject of photonic crystals, which is another major thrust of nanophotonics that is receiving a great deal of worldwide attention. Photonic crystals are periodic nanostructures. The chapter covers concepts, methods of fabrication, theoretical methods to calculate their band structure, and applications of photonic crystals. One can easily omit the theory section and still appreciate the novel features of photonic crystals, which are clearly and concisely described.

Chapter 10 covers nanocomposites. A significant emphasis is placed on the nanocomposite materials that incorporate nanodomains of highly dissimilar materials such as inorganic semiconductors or inorganic glasses and plastics. Merits of nanocomposites are discussed together with illustrative examples of applications, such as to energy-efficient broadband solar cells and other optoelectric devices.

Chapter 11 introduces nanolithography, broadly defined, that is used to fabricate nanoscale optical structures. Both optical and nonoptical methods are described, and some illustrative examples of applications are presented. The use of direct two-photon absorption, a nonlinear optical process, provides improved resolution leading to smaller photoproduced nanostructures compared to those produced by linear absorption.

Chapter 12 deals with biomaterials that are emerging as an important class of materials for nanophotonic applications. Bioderived and bioinspired materials are described together with bioassemblies that can be used as templates. Applications discussed are energy-harvesting, low-threshold lasing, and high-density data storage.

Chapter 13 introduces the application of nanophotonics for optical diagnostics, as well as for light-guided and light-activated therapy. Use of nanoparticles for bioimaging and sensing, as well as for targeted drug delivery in the form of nanomedicine, is discussed.

Chapter 14 provides a critical assessment of the current status of nanophotonics in the marketplace. Current applications of near-field microscopy, nanomaterials, quantum-confined lasers, photonic crystals, and nanolithography are analyzed. The chapter concludes with future outlook for nanophotonics.

REFERENCES

Feynman, R. P., There's Plenty of Room at the Bottom, in *Miniaturization,* Horace D. Gilbert, ed., Reinhold, New York, 1961, pp. 282–296.

NRC Report, *Small Wonders, Endless Frontiers—A Review of the National Nanotechnology Initiative,* National Academy Press, Washington, D.C., 2002.

Prasad, P. N., *Introduction to Biophotonics,* Wiley-Interscience, New York, 2003.

Shen, Y., Friend, C. S., Jiang, Y., Jakubczyk, D., Swiatkiewicz, J., and Prasad, P. N., Nanophotonics: Interactions, Materials, and Applications, *J. Phys. Chem. B* **140,** 7577–7587 (2000).

CHAPTER 2
Foundations for Nanophotonics

The basic foundation of nanophotonics involves interaction between light and matter. This interaction is for most systems electronic, that is, it involves changes in the properties of the electrons present in the system. Hence, an important starting point is a parallel discussion of the nature of photons and electrons. Section 2.1 begins with a discussion of the similarities and differences between photons and electrons, including the nature of the propagation and interactions. This section is followed by a description of confinement effects. Photon and electron tunneling through a classically energetically forbidden zone are discussed, drawing similarities between them. Localization of photons and electrons in periodic structures—photonic crystals and electronic semiconductor crystals, respectively—is presented.

Cooperative effects dealing with electron–electron, electron–hole, and photon–photon interactions are described. Examples of photon–photon interactions are nonlinear optical effects. Examples of electron–electron interactions are formation of electron pairs in a superconducting medium. Electron–hole interactions produce excitons and biexcitons.

Section 2.2 discusses optical interactions localized on nanometer scale. Methods of evanescent wave and surface plasmon resonance to produce axial localization of electromagnetic field of light are presented. Lateral (in-plane) localization of light can be achieved by using a near-field geometry.

Section 2.3 provides examples of nanometer scale electronic interactions leading to major modifications or new manifestations in the optical properties. Examples given are of new cooperative transitions, nanoscale electronic energy transfer, and the phenomenon of cooperative emission. Other examples, quantum confinement effects and nanoscopic interaction dynamics, are deferred for Chapters 4 and 6.

Finally, Section 2.4 provides highlights of the chapter.

Some parts of the chapter utilize a more rigorous mathematical approach. The less mathematically inclined readers may skip the mathematical details, because they are not crucial for understanding the remainder of the book.

For further reading the suggested materials are:

Joannopoulos, Meade, and Winn (1995): *Photonic Crystals*
Kawata, Ohtsu, and Irie, eds. (2002): *Nano-Optics*
Saleh and Teich (1991): *Fundamentals of Photonics*

2.1 PHOTONS AND ELECTRONS: SIMILARITIES AND DIFFERENCES

In the language of physics, both photons and electrons are elementary particles that simultaneously exhibit particle and wave-type behavior (Born and Wolf, 1998; Feynman et al., 1963). Upon first comparing them, photons and electrons may appear to be quite different as described by classical physics, which defines photons as electromagnetic waves transporting energy and electrons as the fundamental charged particle (lowest mass) of matter. A quantum description, on the other hand, reveals that photons and electrons can be treated analogously and exhibit many similar characteristics. Table 2.1 summarizes some similarities in characteristics of photons and electrons. A detailed description of features listed in the table follows.

When both photons and electrons treated as waves, the wavelength associated with their properties is described, according to the famous de Broglie postulate, by

Table 2.1. Similarities in Characteristics of Photons and Electrons

Photons	Electrons
Wavelength	
$\lambda = \dfrac{h}{p} = \dfrac{c}{v}$	$\lambda = \dfrac{h}{p} = \dfrac{h}{mv}$
Eigenvalue (Wave) Equation	
$\left\{ \nabla \times \dfrac{1}{\varepsilon(r)} \nabla \times \right\} \mathbf{B}(r) = \left(\dfrac{\omega}{c}\right)^2 \mathbf{B}(r)$	$\hat{H}\psi(r) = -\dfrac{\hbar^2}{2m}(\nabla \cdot \nabla + V(r))\psi(r) = E\psi$
Free-Space Propagation	
Plane wave $\mathbf{E} = (\tfrac{1}{2})\mathbf{E}°(e^{i\mathbf{k}\cdot\mathbf{r}-\omega t} + e^{-i\mathbf{k}\cdot\mathbf{r}+\omega t})$ $\mathbf{k}$ = wavevector, a real quantity	Plane wave: $\Psi = c(e^{i\mathbf{k}\cdot\mathbf{r}-\omega t} + e^{-i\mathbf{k}\cdot\mathbf{r}+\omega t})$ $\mathbf{k}$ = wavevector, a real quantity
Interaction Potential in a Medium	
Dielectric constant (refractive index)	Coulomb interactions
Propagation Through a Classically Forbidden Zone	
Photon tunneling (evanescent wave) with wavevector, $\mathbf{k}$, imaginary and hence amplitude decaying exponentially in the forbidden zone	Electron-tunneling with the amplitude (probability) decaying exponentially in the forbidden zone
Localization	
Strong scattering derived from large variations in dielectric constant (e.g., in photonic crystals)	Strong scattering derived from a large variation in Coulomb interactions (e.g., in electronic semiconductor crystals)
Cooperative Effects	
Nonlinear optical interactions	Many-body correlation Superconducting Cooper pairs Biexciton formation

the same relation $\lambda = h/p$, where p is the particle momentum (Atkins and dePaula, 2002). One difference between them is on the length scale: The wavelengths associated with electrons are usually considerably shorter than those of photons. Under most circumstances the electrons will be characterized by relatively larger values of momentum (derived from orders of magnitude larger rest mass of an electron than the relativistic mass of a photon given by $m = h\nu/c^2$) than photons of similar energies. This is why electron microscopy (in which the electron energy and momentum are controlled by the value of the accelerating high voltage) provides a significantly improved resolution over optical (photon) microscopy, since the ultimate resolution of a microscope is diffraction-limited to the size of the wavelength. The values of momentum, which may be ascribed to electrons bound in atoms or molecules or the conduction electrons propagating in a solid, are also relatively high compared to those of photons, thus the characteristic lengths are shorter than wavelengths of light. An important consequence derived from this feature is that "size" or "confinement" effects for photons take place at larger size scales than those for electrons.

The propagation of photons as waves is described in form of an electromagnetic disturbance in a medium, which involves an electric field **E** (corresponding displacement **D**) and an orthogonal magnetic field **H** (corresponding displacement **B**), both being perpendicular to the direction of propagation in a free space (Born and Wolf, 1998). A set of Maxwell's equations describes these electric and magnetic fields. For any dielectric medium, the propagation of an electromagnetic wave of angular frequency ω is described by an eigenvalue equation shown in Table 2.1. An eigenvalue equation describes a mathematical operation $\hat{O}$ on a function **F** as

$$\hat{O}\mathbf{F} = C\mathbf{F} \tag{2.1}$$

to yield a product of a constant C, called eigenvalue, and the same function. Functions fulfilling the above condition are called eigenfunctions of the operator $\hat{O}$. In the eigenvalue equation in Table 2.1, $\varepsilon(r)$ is the dielectric constant of the medium at the frequency ω of the electromagnetic wave, which, in the optical region, is equal to $n^2(\omega)$, where n is the refractive index of the medium at angular frequency ω. The dielectric constant, and thereby the refractive index, describes the resistance of a medium to the propagation of an electromagnetic wave through it. Therefore, light propagation speed c in a medium is reduced from light propagation speed c_0 in vacuum by the relation

$$c = \frac{c_0}{n} = \frac{c_0}{\varepsilon^{1/2}} \tag{2.2}$$

The eigenvalue equation in Table 2.1 involves the magnetic displacement vector, **B**. Since the electric field **E** and the magnetic displacement **B** are related by the Maxwell's equations, an equivalent equation can be written in terms of **E**. However, this equation is often solved using **B**, because of its more suitable mathematical character. (The operator for **B** has a desirable character called Hermitian). The eigenvalue C in the equations for photon is $(\omega/c)^2$, which gives a set of allowed frequencies ω (thus energies) of photons in a medium of dielectric constant ε(r) and thus refractive index n(r). The dielectric constant may be either constant throughout the medium or dependent on spatial location **r** and the wavevector **k**.

The corresponding wave equation for electrons is the Schrödinger equation, and its time-independent form is often written in the form of an eigenvalue equation shown in Table 2.1 (Levine, 2000; Merzbacher, 1998). Here $\hat{H}$, called the *Hamiltonian operator,* consists of the sum of operator forms of the kinetic and the potential energies of an electron and is thus given as

$$\hat{H} = -\frac{\hbar^2}{2m}\left(\frac{\partial^2}{\partial x^2} + \frac{\partial^2}{\partial y^2} + \frac{\partial^2}{\partial z^2}\right) + V(r) = -\frac{\hbar^2}{2m}\nabla^2 + V(r) \qquad (2.3)$$

The first term (all in the parentheses) is derived from the kinetic energy; the second term, $V(\mathbf{r})$, is derived from the potential energy of the electron due to its interaction (Coulomb) with the surrounding medium.

The solution of the Schrödinger equation provides the allowed energy states, eigenvalues E of energy, of the electron, while wavefunction ψ yields a probabilistic description of the electron. The square of this function, $|\psi(\mathbf{r})|^2$, describes the probability density for the electron at a position **r**. Thus, the wavefunction ψ for an electron can be considered an amplitude of the electron wave, the counterpart of the electric field **E** for an electromagnetic wave.

The analogue of the interaction potential for electromagnetic wave propagation in a medium is the spatial variation of the dielectric constant (or refractive index). This variation causes a modification of the propagation characteristics as well as, in some cases, a modification of the allowed energy values for photons. The interaction potential V of Eq. (2.3) describes the Coulomb interactions (such as electrostatic attraction between an electron and a nucleus). This interaction modifies the nature of the electronic wavefunction (thus the probability distribution) as well as the allowed set of energy values E (eigenvalues) obtained by the solution of the Schrödinger equation listed in Table 2.1. Some examples of these wave propagation properties for photons and electrons under different interaction potentials are discussed below.

Despite the similarities described above, electrons and photons also have important differences. The electrons generate a scalar field while the photons are vector fields (light is polarized). Electrons possess spin, and thus their distribution is described by Fermi–Dirac statistics. For this reason, they are also called *fermions*. Photons have no spin, and their distribution is described by Bose–Einstein statistics. For this reason, photons are called *bosons*. Finally, since electrons bear a charge while the charge of photons is zero, there are principal differences in their interactions with external static electric and magnetic fields.

2.1.1 Free-Space Propagation

In a "free-space" propagation, there is no interaction potential or it is constant in space. For photons, it simply implies that no spatial variation of refractive index n

occurs. In such a case, the propagation of the electromagnetic wave is described by a plane wave for which the electric field described in the complex plane (using a real and an imaginary part) is shown in the Table 2.1. It is a propagating electromagnetic wave with oscillating (sinusoidal) electric field **E** (and the corresponding magnetic field **B**) (Born and Wolf, 1998). The amplitude of the field is described by **E°**. The direction of propagation is described by the propagation vector **k** whose magnitude relates to the momentum as

$$\mathbf{p} = \hbar \mathbf{k}$$

The wavevector **k** has the length equal to

$$\mathbf{k} = |k| = \frac{2\pi}{\lambda} \tag{2.4}$$

The positive **k** describes forward propagation (e.g., left to right), while the negative **k** describes the propagation in the backward direction (right to left). The corpuscular properties of the electromagnetic wave are described by the photon energy

$$E = h\nu = \hbar\omega = \frac{hc}{\lambda}$$

These relationships yield the dispersion relation

$$\omega = c|k| \tag{2.5}$$

which describes the dependence of the frequency (or energy) of a photon on its wavevector and is illustrated by the linear dispersion relation shown in Figure 2.1.

Similarly, for free-space propagation of an unbound electron, the wavefunction obtained by the solution of the Schrödinger equation is an oscillating (sinusoidal) plane wave, similar to that for a photon, and is characterized by a wavevector **k**. Therefore, the probability density described by the absolute value of the square of

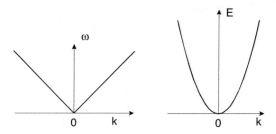

Figure 2.1. Dispersion relation showing the dependence of energy on the wavevector for a free-space propagation. (a) Dispersion for photons. (b) Dispersion for electrons.

the wavefunction is the same everywhere, which again conforms to the free state of the electron. The momentum of the electron similarly is defined by $\hbar \mathbf{k}$. The energy dispersion for a free electron is described by a parabolic relation (quadratic dependence on k) as

$$E = \frac{\hbar^2 k^2}{2m} \tag{2.6}$$

where m is the mass of the electron. A modification of the free electron theory is often used to describe the characteristics of electrons in metals (the so-called Drude model). One often refers to such behavior as delocalization of electron or electronic wavefunction. This dependence of the electronic energy on the wavevector is also represented in Figure 2.1. One can clearly see that, even though analogous descriptions can be used for photons and electrons, the wavevector dependence of energy is different for photons (linear dependence) and electrons (quadratic dependence).

For free-space propagation, all values of frequency ω for photons and energy E for electrons are permitted. This set of allowed continuous values of frequency (or energy) form together a band, and the band structure refers to the characteristics of the dependence of the frequency (or energy) on the wavevector $\mathbf{k}$.

2.1.2 Confinement of Photons and Electrons

The propagation of photons and electrons can be dimensionally confined by using areas of varying interaction potential in their propagation path to reflect or backscatter these particles, thus confining their propagation to a particular trajectory or a set of those.

In the case of photons, the confinement can be introduced by trapping light in a region of high refractive index or with high surface reflectivity (Saleh and Teich, 1991). This confining region can be a waveguide or a cavity resonator. The examples of various confinements are shown in Figure 2.2. The confinements can be produced in one dimension such as in a plane, as in the case of a planar optical waveguide. Here, the light propagation is confined in a layer (such as a thin film) of high refractive index, with the condition that the refractive index n_1 of the light-guiding layer is higher than the refractive index n_2 of the surrounding medium, as shown in Figure 2.2. The figure shows the classical optics picture using a ray path to describe light guiding (trapping) due to total internal reflection. In the case of a planar waveguide, the confinement is only in the vertical (x direction). The propagation direction is z. In the case of a fiber or a channel waveguide, the confinement is in the x and y directions. A microsphere is an example of an optical medium confining the light in all dimensions. The light is confined by the refractive index contrast between the guiding medium and the surrounding medium. Thus the contrast n_1/n_2 acts as a scattering potential creating barrier to light propagation (Joannopoulos et al., 1995).

In the direction of propagation (z axis), the light behaves as a plane wave with a propagation constant β, analogous to the propagation vector k for free space. In the

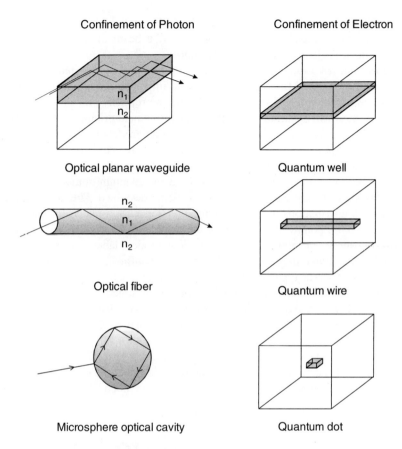

Figure 2.2. Confinements of photons and electrons in various dimensions and the configurations used for them. The propagation direction is z.

confining direction, the electric field distribution has a different spatial profile. Thus the envelope of the electric field **E** (only spatially dependent part), represented in Table 2.1 for free-space propagation, is modified for a waveguide as (Prasad and Williams, 1991; Saleh and Teich, 1991)

$$\mathbf{E} = \tfrac{1}{2} f(x, y) a(z)(e^{i\beta z} + e^{-i\beta z}) \qquad (2.7)$$

Equation (2.7) describes guiding in a fiber or a channel waveguide (a rectangular or square guiding channel) which has two-dimensional confinement. The function $a(z)$ is the electric amplitude in the z direction, which (in the absence of losses) is constant. The function $f(x, y)$ represents the electric field distribution in the confinement plane. In the case of a planar waveguide, producing confinement only in the x direction, only the x component shows a spatial distribution limited by the confining potential, while the y component of f is like that for a plane wave (free space) if a plane wave is used to excite the waveguide mode. On the other hand, it will show the characteristic spreading in the y direction if a beam of limited size (e.g., a Gaussian beam) is launched into the waveguide.

The field distribution and the corresponding propagation constant are obtained by the solution of the Maxwell's equation and imposing the boundary conditions (defining the boundaries of the waveguide and the refractive index contrast). The solution of the wave equation shows that the confinement produces certain discrete sets of field distributions called *eigenmodes*, which are labeled by quantum numbers (integer). For a one-dimensional confinement, there is only one quantum number n, which can assume values 0, 1, and so on. Unfortunately, the same letter n, used to represent refractive index, is also used for quantum number. The readers should distinguish them based on the context being described. The examples of the various field distributions in the confining direction x for a planar (or slab) waveguide using a TE polarized light (where the polarization of light is in the plane of the film) is shown in Figure 2.3 for the various modes (labeled as TE_0, TE_1, TE_2 etc.). From the above description, it is clear that confinement produces quantization—that is, discrete types of the field distributions, labeled by integral sets of quantum numbers used to represent the various eigenmodes.

We shall see now that confinement of electrons also leads to modification of their wave properties and produces quantization—that is, discrete values for the possible eigenmodes (Merzbacher, 1998; Levine, 2000). The corresponding one-, two-, and three-dimensional confinement of electrons is also exhibited in Figure 2.2 (Kelly, 1995). Here the potential confining the electron is the energy barrier—that is, regions where the potential energy V of Eq. (2.3) is much higher than the energy

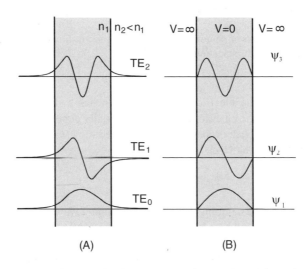

Figure 2.3. (A) Electric field distribution for TE modes $n = 0, 1, 2$ in a planar waveguide with one-dimensional confinement of photons. (B) Wavefunction ψ for quantum levels $n = 1, 2, 3$ for an electron in a one-dimensional box.

E of the electron. Classically, the electron will be completely confined within the potential energy barriers (walls). This is true if the potential barriers are infinite (as shown in the figure). However, for finite potential barriers the wavefunction does enter the region of the barrier and the pattern becomes similar to that for photons in Figure 2.2. The confinement for an electron is thus similar to that for a photon. However, the length scales are different. To produce confinement effect for photons, the dimensions of the confining regions are in micrometers. But, for electrons, which have a significantly shorter wavelength, the confining dimensions have to be in nanometers to produce a significant quantization effect.

The quantum confinement effect on electrons is discussed in more detail in Chapter 4. Here, only a general description of electronic confinement is given. A simple example that illustrates confinement effect is demonstrated by the model of an electron in a one-dimensional box, as depicted in Figure 2.4 (Levine, 2000).

The electron is trapped (confined) in a box of length l within which the potential energy is zero. The potential energy rises to infinity at the ends of the box and stays at infinity outside the box.

Inside the box, the Schrödinger equation is solved with the following conditions:

$$V(x) = 0$$
$$\Psi(x) = 0 \quad \text{at} \quad x = 0 \quad \text{and} \quad x = l \tag{2.8}$$

The solution yields sets of allowed values of E and the corresponding functions ψ, each defining a given energy state of the particle and labeled by a quantum number n which takes an integral value starting from 1 (Atkins and dePaula, 2002). These values are defined as

$$E_n = \frac{n^2 h^2}{8ml^2} \quad \text{where} \quad n = 1, 2, 3, \ldots \tag{2.9}$$

The lowest value of total energy is $E_1 = h^2/8ml^2$. Therefore, total energy E of an electron can never be zero when bound (or confined), even though its potential energy is zero. The discrete energy values are E_1, E_2, and so on, corresponding to quantum numbers $n = 1, 2, 3$, and so on. These values represent the various allowed

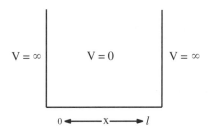

Figure 2.4. Schematics of a particle in a one-dimensional box.

energy levels of an electron confined in a one-dimensional box. The gap between two successive levels describes the effect of quantization (discreteness). If it were zero, we would have a continuous variation of the energy as for a free electron and there would be no quantization.

The gap ΔE between two successive levels E_n and E_{n+1} can be given as

$$\Delta E = (2n + 1)\frac{h^2}{8ml^2} \qquad (2.10)$$

This equation reveals that the gap between two successive levels decreases as l^2 when the length of the box increases. Thus the spacing between successive electronic levels decreases as the electron is spread (delocalized) over a longer confining distance, as in the case of the π electrons in a conjugated structure. In the limit of $l \to \infty$, there is no confinement and $\Delta E = 0$, indicating no quantization. The wavefunction for the various quantum levels is also modified from a plane wave, shown in Table 2.1.

The wavefunctions for the different quantum states n are given as (Levine, 2000)

$$\Psi_n(x) = \left(\frac{2}{l}\right)^{1/2} \sin\left(\frac{n\pi x}{l}\right) = \frac{1}{2i}\left(\frac{2}{l}\right)^{1/2}(e^{ikx} - e^{-ikx}) \qquad (2.11)$$

where $k = n\pi/l$.

Figure 2.3 also shows the wavefunction Ψ_n as a function of n for various quantum states $n = 1, 2, 3$. From a correlation between the field distribution for photons for $n = 0, 1, 2$ modes of a planar waveguide and the wavefunctions for $n = 1, 2, 3$ quantum states of electrons, it can be clearly seen that the one-dimensional confinement effects are quite analogous.

The probability density $|\Psi_n|^2$ varies with position within the box, and this variation is different for different quantum number states n. Again, this is a modification from the constant probability density for a free electron. For example, for $n = 1$ the maximum probability density is at the center of the box, in contrast to the plane wave picture showing equal probability at every place. The two-dimensional analogue will involve a rectangular box in which the potential barriers in the x and y directions are $V = \infty$ regions at distances l_1 and l_2. The solution of a two-dimensional Schrödinger equation now yields energy eigenvalues depending on two quantum numbers n_1 and n_2 as

$$E_{n_1,n_2} = \left(\frac{n_1^2}{l_1^2} + \frac{n_2^2}{l_2^2}\right)\frac{h^2}{8m} \qquad (2.12)$$

and the corresponding wavefunction is

$$\Psi_{n_1,n_2}(x,y) = \frac{2}{(l_1 l_2)^{1/2}} \sin\left(\frac{n_1 \pi x}{l_1}\right)\sin\left(\frac{n_2 \pi y}{l_2}\right) \qquad (2.13)$$

with quantum numbers n_1 and n_2 each having allowed values of 1, 2, 3, and so on. Similarly, three-dimensional confinement, like in a box of dimension l_1, l_2, and l_3, are characterized by three quantum numbers, n_1, n_2, and n_3, each assuming values 1, 2, 3, and so on. The eigenvalues E_{n_1,n_2,n_3} and wavefunctions ψ_{n_1,n_2,n_3} are simple extensions of Eqs. (2.12) and (2.13) to include a third term depending on n_3 and l_3.

2.1.3 Propagation Through a Classically Forbidden Zone: Tunneling

In a classical picture, the photons and electrons are completely confined in the regions of confinement. For photons, it is seen by the ray optics for the propagating wave as shown in Figure 2.2. Similarly, classical physics predicts that, once trapped within the potential energy barriers where the energy E of an electron is less than the potential energy V due to the barrier, the electron will remain completely confined within the walls. However, the wave picture does not predict so. As shown in Figure 2.3, the field distribution of light confined in a waveguide extends beyond the boundaries of the waveguide. This behavior is also shown in Figure 2.5.

Hence, light can leak into the region outside the waveguide, a classically forbidden region. This light leakage generates an electromagnetic field called evanescent wave (Courjon, 2003). The field distribution in the region outside the waveguide (classically forbidden region) does not behave like a plane wave having the wavevector **k**, as a real quantity (Saleh and Teich, 1991). The electric field amplitude extending into the classically forbidden region decays exponentially with distance x into the medium of lower refractive index, from the boundary of the guiding region, according to the equation

$$\mathbf{E}_x = \mathbf{E}_0 \exp(-x/d_p) \qquad (2.14)$$

$\mathbf{E}_0$ in this equation is the electric field at the boundary of the waveguide. The parameter d_p, also called the penetration depth, is defined as the distance at which the electric field amplitude reduces to 1/e of E_0. Comparing the field $\mathbf{E}_x$ of Eq. (2.14)

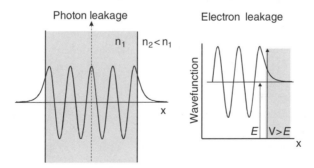

Figure 2.5. Schematic representation of leakage of photons and electrons into classically energetically forbidden regions.

with that for a plane wave shown in Table 2.1, it can be seen that Eq. (2.14) represents a case where the wavevector **k** is imaginary (as $k = i/d_p$, hence making e^{ikx} as e^{-x/d_p}). This exponentially decaying wave with an imaginary wavevector **k** is the evanescent wave. Typically, the penetration depths d_p for the visible light are 50–100 nm. Thus, this wave can be used for nanophotonics, because optical interactions manifested by the evanescent wave can be kept localized within nanometer range. Evanescent waves have been used for numerous surface selective excitations. The nanoscale optical interactions utilizing evanescent waves are further discussed in Section 2.2.

In an analogous fashion, an electron shows a leakage through regions where $E < V$, also shown in Figure 2.5. Again, this process results from the wave nature of the electron that is described by a wavefunction. Figure 2.5 shows a situation as that of a particle in a one-dimensional box (Figure 2.4). For Figure 2.4, the potential barrier represents $V = \infty$, but for Figure 2.5 the potential barrier simply represents the case where $V > E$. Within the box, again $V = 0$, where the wavefunctions are the ones represented by Eq. (2.11) for which $|\mathbf{k}|$ is a real quantity. In Figure 2.5 the represented wavefunction within the box corresponds to a high quantum number n (hence many oscillating cycles). But the wavefunction extending beyond the box into the region of $V > E$ decays exponentially, just like the evanescent wave for confined light.

Electron tunneling is defined as the passage of electrons from one allowed zone ($E > V$) through a classically forbidden zone ($V > E$), called a barrier layer, to another allowed zone ($E > V$) as shown in Figure 2.6 (Merzbacher, 1998). Similarly, photon tunneling is defined as passage of photon through a barrier layer of lower refractive index, also shown in Figure 2.6 (Gonokami et al., 2002; Fillard, 1996). The tunneling probability, often described by T, called the transmission probability, is given as

$$T = ae^{-2kl} \tag{2.15}$$

where a is a function of E/V and k is equal to $(2mE)^{1/2}/\hbar$. The latter parameter is equivalent to wavevector k of the plane wave for a free electron, which is now imaginary, thus producing an exponential decay.

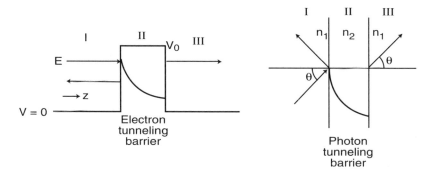

Figure 2.6. Schematics of electron and photon tunneling through a barrier.

2.1.4 Localization Under a Periodic Potential: Bandgap

Both photons and electrons show an analogous behavior when subjected to a periodic potential. The example of an electron subjected to a periodic potential is provided by a semiconductor crystal, which consists of a periodic arrangement of atoms. The electrons are free to move through a lattice (ordered arrangement) of atoms, but as they move, they experience a strong Coulomb (attractive) interaction by the nucleus of the atom at each lattice site. Figure 2.7 shows a schematic of a semiconductor crystal, which we can call an electronic crystal. A rapidly growing field in nanophotonics is that of photonic crystals, which will be discussed in detail in Chapter 9. Here a brief description is presented to draw an analogy between an electronic crystal and a photonic crystal. A photonic crystal represents an ordered arrangement of a dielectric lattice, which represents a periodic variation of the dielectric constant (Joannopoulos et al., 1995). An example presented in Figure 2.7 is that of close-packed, highly uniform colloidal particles such as silica or polystyrene spheres. The refractive index contrast (n_1/n_2), where n_1 is the refractive index of the packing spheres and n_2 is that of the interstitial medium between them (which can be air, a liquid, or more desirably, a very high refractive index material), acts as a periodic potential.

The periodicity in the two cases is of different length scale. In the case of an electronic (semiconductor) crystal, the atomic arrangement (lattice spacings) is on the subnanometer scale. This range of dimensions in the domain of electromagnetic waves corresponds to X rays, and X rays can be diffracted on crystal lattices to produce Bragg scattering of X-ray waves. The Bragg equation determining the directions in space at which the diffraction takes place, is given as

$$m\lambda = 2nd \sin\theta \qquad (2.16)$$

where d is the lattice spacing and λ is the wavelength of the wave, m is the order of diffraction, n is the refractive index, and θ is the ray incidence angle.

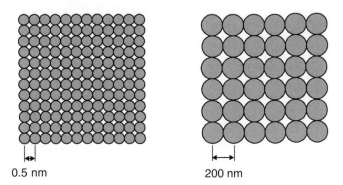

Figure 2.7. Schematic representation of an electronic crystal (*left*) and a photonic crystal (*right*).

22 FOUNDATIONS FOR NANOPHOTONICS

In the case of a photonic crystal, the same Bragg scattering produces diffraction of optical waves (Joannopoulos et al., 1995). Use of Eq. (2.16) suggests that the lattice spacing (distance between the centers of the packed spheres) should be, for example, 200 nm to produce Bragg scattering of light of wavelength 500 nm.

The solution of the Schrödinger equation for the energy of electrons, now subjected to the periodic potential V, produces a splitting of the electronic band, shown in Figure 2.8 for a free electron (Kittel, 2003). The lower energy band is called the *valence band,* and the higher energy band is called the *conduction band.* In the language of chemists, this situation can also be described as in the case of highly π-conjugated structures that contain alternate single and double bonds. The Hückel theory predicts a set of closely spaced, occupied bonding molecular orbitals (π), which are like the valence band, and a set of closely spaced, empty anti-bonding molecular orbitals (π^*) which are equivalent to the conduction band (Levine, 2000). The dispersion behavior for the valence and the conduction bonds, given by the plot of E versus the wavevector k for two possible cases, are shown in Figure 2.8. These two bands are separated by a "forbidden" energy gap, the width of which is called the *bandgap.* The bandgap energy is often labeled as E_g and plays an important role in determining the electrical and optical properties of the semiconductor (Cohen and Chalikovsky, 1988).

The dispersion relation for each band has a parabolic form, just as for free electrons. Under the lowest energy condition, all the valence bands are completely occupied and thus no flow of electrons can occur. The conduction band consequently is empty. If an electron is excited either thermally or optically, or an electron is injected to the conduction band by an impurity (n-doping), this electron can move in the conduction band, producing electronic conduction under an applied electric

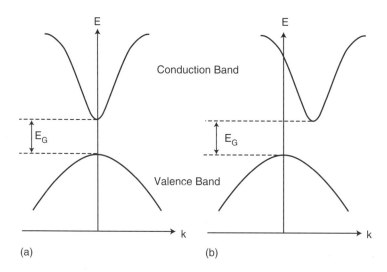

Figure 2.8. Schematics of electron energy in (a) direct bandgap (e.g., GaAs, InP, CdS) and (b) indirect bandgap (e.g., Si, Ge, GaP) semiconductors.

field. In the case of excitation of an electron to the conduction band, a vacancy is left in the valence band with a net positive charge, which is also treated as a particle of positive charge, called a *hole*. The hole (positive vacancy) can move through the valence band, providing conduction.

The energy, E_{CB}, of an electron near the bottom of the conduction band is given by the relation (Kittel, 2003)

$$E_{CB} = E_C^0 + \frac{\hbar^2 k^2}{2m_e^*} \qquad (2.17)$$

Here E_C^0 is the energy at the bottom of the conduction band, and m_e^* is the effective mass of the electron in the conduction band, which is modified from the real mass of the electron because of the periodic potential experienced by the electron. Similarly, near the top of the valence band, the energy is given as

$$E_{VB} = E_V^0 + \frac{\hbar^2 k^2}{2m_h^*} \qquad (2.18)$$

where $E_V^0 = E_C^0 - E_g$ is the energy at the top of the valence band and m_h^* is the effective mass of a hole in the valence band. Both Eqs. (2.17) and (2.18) predict a parabolic relation between E and k, if m_e^* and m_h^* are assumed to be independent of k. The effective masses m_e^* and m_h^* can be obtained from the curvature of the calculated band structure (E versus k).

Figure 2.8 shows that the two types of cases encountered are as follows: (a) Direct-gap materials for which the top of the valence band and the bottom of the conduction band are at the same value of k. An example is a binary semiconductor GaAs. (b) Indirect-gap materials for which the top of the valence band and the bottom of the conduction band are not at the same value of k. An example of this type of semiconductor is silicon. Hence in the case of an indirect-gap semiconductor, the transition of an electron between the valence band and the conduction band involves a substantial change in the momentum (given by $\hbar k$) of the electron. This has important consequences for optical transitions between these two bands, induced by absorption or emission of photons. For example, emission of a photon (luminescence) leading to transition of an electron from the conduction band to the valence band requires a conservation of momentum. In other words, the momentum of the electron in the conduction band should be the same as the sum of momenta of the electron in the valence band and the emitted photon. Since the photon has a very small momentum ($k \sim 0$) because of its long wavelength compared to that of electrons, the optically induced electronic transition between the conduction band and the valence band requires $\Delta k = 0$. Hence, emission in the case of an indirect-gap semiconductor, such as silicon, is essentially forbidden by this selection rule, and thus Si in the bulk form is not a luminescent light emitter. GaAs, on the other hand being a direct-gap material, is an efficient emitter. This discussion of electronic semiconductor bulk property will also be useful for Chapter 4, where we see how the bulk semiconductor properties are modified in confined semiconductor structures such as quantum wells, quantum wires, and quantum dots.

In the case of a photonic crystal, the eigenvalue equation for photons, as shown in Table 2.1, can be used to calculate the dispersion relation ω versus k. Figure 2.9 shows the dispersion curve calculated for a one-dimensional photonic crystal (also known as Bragg stack), which consists of alternating layers of two dielectric media of refractive indices n_1 and n_2 (Joannopoulos et al., 1995).

Again, a similar type of band splitting is observed for a photonic crystal, and a forbidden frequency region exists between the two bands, similar to that between the valence and the conduction band of an electronic crystal, which is often called the photonic bandgap. Similar to electronic bandgap, no photon frequencies in the gap region of the photonic crystal correspond to the allowed state for this medium. Hence, the photons in this frequency range cannot propagate through the photonic crystal. Using the Bragg-diffraction model, it can also be seen that the photons of the bandgap frequency meet the Bragg-diffraction conditions. Hence, another way to visualize the photon localization in the photonic bandgap region (not propagating) is that the photons of these frequencies are multiply scattered by the scattering potential produced by the large refractive index contrast n_1/n_2, resulting in their localization. In other words, if photons of frequencies corresponding to the bandgap region are incident on the photonic crystal, they will be reflected from the surface of the crystal and will not enter the crystal. If a photon in the bandgap region is generated inside the crystal by emission, it will not exit the crystal because of its localization (lack of propagation). These properties and their consequences are detailed in Chapter 9.

2.1.5 Cooperative Effects for Photons and Electrons

Cooperative effects refer to the interaction between more than one particle. Although conventionally the cooperative effects for photons and electrons have been

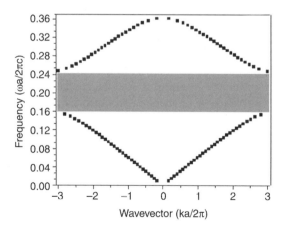

Figure 2.9. Dispersion curve for a one-dimensional photonic crystal showing the lowest energy bandgap.

described separately, an analogy can be drawn. However, it should be made clear that while electrons can interact directly, photons can interact only through the mediation of a material in which they propagate.

In the case of photons, an example of a cooperative effect can be a nonlinear optical effect produced in an optically nonlinear medium (Shen, 1984). In a linear medium, photons propagate as an electromagnetic wave without interacting with each other. As described above, the propagating electromagnetic wave senses the medium response in the form of its dielectric constant or refractive index. For a linear medium, the dielectric constant or the refractive index is related to the linear susceptibility $\chi^{(1)}$ of the medium by the following relation (Shen, 1984; Boyd, 1992)

$$\varepsilon = n^2 = 1 + \chi^{(1)} \tag{2.19}$$

where $\chi^{(1)}$, the linear susceptibility, is the coefficient that relates the polarization (charge distortion) of the medium, induced linearly by the electric field **E** of light to its magnitudes as

$$\mathbf{P} = \chi^{(1)}\mathbf{E} \tag{2.20}$$

Because $\chi^{(1)}$ relates two vectors, **P** and **E**, it actually is a second-rank tensor.

In a strong optical field such as a laser beam, the amplitude of the electric field is so large (comparable with electrical fields corresponding to electronic interactions) that in a highly polarizable nonlinear medium the linear polarization behavior [Eq. (2.20)] does not hold. The polarization **P** in this case also depends on higher powers of electric field, as below (Shen, 1984; Prasad and Williams, 1991; Boyd, 1992).

$$\mathbf{P} = \chi^{(1)}\mathbf{E} + \chi^{(2)}\mathbf{E}\mathbf{E} + \chi^{(3)}\mathbf{E}\mathbf{E}\mathbf{E} \ldots \tag{2.21}$$

The higher-order terms in the electric field **E** produce nonlinear optical interactions, whereby the photons interact with each other. Some of the nonlinear optical interactions discussed in this book are described below.

An important manifestation of photon–photon interaction is frequency conversion. The most important examples of such conversion processes are as follows:

- Interaction of two photons of frequency ω by the $\chi^{(2)}$ term to produce a photon of up-converted frequency 2ω. This process is referred to as *second harmonic generation* (SHG). For example, if the original photon is of wavelength 1.06 μm (IR), the new output of $\lambda/2$ is at 532 nm in the green.
- Interaction of two photons of different frequencies ω_1 and ω_2, again by the $\chi^{(2)}$ term, to produce a new photon at the sum frequency value $\omega_1 + \omega_2$ or difference frequency value $\omega_1 - \omega_2$. This process is called *parametric mixing* or *parametric generation*.

- Interaction of three photons of frequency ω through the third-order nonlinear optical susceptibility term $\chi^{(3)}$ to produce a new photon of frequency 3ω. This process is called *third harmonic generation* (THG).
- Simultaneous absorption of two photons (two-photon absorption), to produce an electronic excitation. This is another important $\chi^{(3)}$ process.

These frequency conversion processes are conceptually described by simple energy diagrams in the book *Introduction to Biophotonics* by this author (Prasad, 2003). Other important types of nonlinear optical interactions produce field dependence of the refractive index of an optically nonlinear medium. These are:

- Pockels effect, which describes linear dependence of the refractive index on the applied electric field. This linear electro-optic effect can be used to affect the propagation of photons by application of an electric field and thereby produce devices such as electro-optic modulators.
- Kerr effect (more precisely optical Kerr effect), which describes linear dependence of the refractive index on the intensity of light. This effect is thus all-optical, whereby varying the intensity of a controlling intense light beam can affect the propagation of another light beam (signal). This effect provides the basis for all optical signal processing.

An example of a cooperative process for electrons is electron–electron interaction to bind together and produce a Cooper pair in a superconducting medium, as proposed by Bardeen, Cooper, and Schrieffer (BCS) to explain superconductivity (Kittel, 2003). Two electrons, each carrying a negative charge, will be expected to repel each other electrostatically. However, an electron in a cation lattice distorts the lattice around it by so-called electron–phonon (lattice vibration) interactions, creating an area of increased positive charge density around itself, which can attract another electron. In other words, the two electrons are attracted toward each other by electron–phonon interaction, which acts as a spring, and they form what is called a *Cooper pair*. This pair formation is schematically represented in Figure 2.10.

The binding energy between an electron pair is of the order of milli-electron-volts, sufficient to keep them paired at extremely low temperatures (below a temperature called *critical temperature*, T_c). The Cooper pair experiences less resistance and leads to superconductivity at lower temperature, where current flows without resistance.

Another example of a cooperative effect is the binding between an electron and a hole to form an exciton, as well as the binding between two excitons to form a bound state called a *biexciton* (Kittel, 2003). Excitons are formed when an electron and the corresponding hole in the valence band are bound so that they cannot move independently. Thus, the electron and the hole move together as a bound particle called an *exciton*. In organic insulators, the electron and the hole are tightly bound at the same lattice site (i.e., within a small radius, usually within the same molecule). Such a tightly bound electron–hole pair is called a *Frenkel exciton*. In the

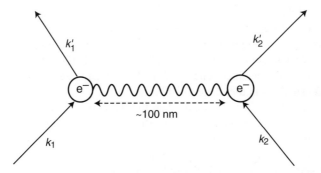

Figure 2.10. Schematic representation of a phonon-mediated Cooper-pair formation between two electrons.

case of a semiconductor, the electrons in the conduction band and the holes in the valence band are not independent and exhibit coupled behavior, giving rise to an exciton, which may exhibit a larger separation between the electron and the hole (spread over more than one lattice site). This type of exciton is called a *Wannier exciton*. The exciton, being composed of a negatively charged electron and a positively charged hole, is a neutral particle and has quantum properties. They are analogous to those of a hydrogen-like atom where an electron and a proton are bound by coulombic interactions. Just like in the case of a hydrogen atom, the energy of an exciton is described by a set of quantized energy levels, which is below the bandgap (E_g) and described by (Gaponenko, 1999)

$$E_n(k) = E_g - \frac{R_y}{n^2} + \frac{\hbar^2 k^2}{2m} \qquad (2.22)$$

In the above quotation, R_y is called the exciton Rydberg energy and is defined as

$$R_y = \frac{e^2}{2\varepsilon a_B} \qquad (2.23)$$

in which ε is the dielectric constant of the crystal. The term α_B, called the *exciton Bohr radius* or often simply the *Bohr radius of a specific semiconductor*, is defined as

$$a_B = \frac{\varepsilon \hbar^2}{\mu e^2} \qquad (2.24)$$

In the above equation, μ is the reduced mass of the electron–hole pair defined as

$$\mu^{-1} = m_e^{*-1} + m_h^{*-1} \qquad (2.25)$$

28 FOUNDATIONS FOR NANOPHOTONICS

The exciton Bohr radius gives an estimate of the size of the exciton (most probable distance of the electron from the hole) in a semiconductor. k is the wavevector for the exciton. For optically generated exciton, $k \cong 0$. Thus the lowest energy of excitonic transition corresponds to $E_1 = E_g - R_y$ [with $n = 1$ in Eq. (2.22)] and is below the bandgap E_g. The Rydberg energy R_y is thus an estimate of the exciton binding energy and usually is in the range 1–100 meV.

A bound exciton will be formed when the thermal energy $kT < R_y$. If $kT \gg R_y$, most excitons are ionized and behave like the separated electrons and holes.

Under high excitation density, two excitons can bind to form a biexciton (Klingshirn, 1995). The formation of biexcitons has been extensively investigated for a number of semiconductors such as CuCl. It has also been a subject of investigation in quantum-confined structures, such as quantum wells, quantum wires, and quantum dots, which are discussed in detail in Chapter 4.

2.2 NANOSCALE OPTICAL INTERACTIONS

The electric field associated with a photon can be confined by using a number of geometries to induce optical interactions on nanoscale. The optical field can be localized on nanoscale both axially and laterally. Table 2.2 lists the methods that can be used for such a purpose.

Table 2.2. Methods for Nanoscale Localization of Electromagnetic Field

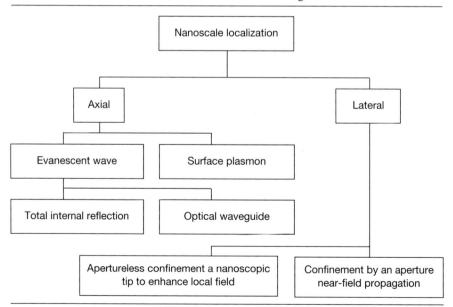

2.2.1 Axial Nanoscopic Localization

Evanescent Wave. The evanescent wave derived from photon tunneling, in the case of a waveguide, has been discussed in the previous section. The evanescent wave from a waveguide surface penetrates the surrounding medium of lower refractive index where it decays exponentially in the axial direction (away from the waveguide). This evanescent field extends about 50–100 nm and can be used to induce nanoscale optical interactions. This evanescent wave excitation has been used for fluorescence sensing with high near surface selectivity (Prasad, 2003). Another example of nanoscale optical interaction is coupling of two waveguides by the evanescent wave, which is schematically shown in Figure 2.11. Here, photon launched in one waveguide can tunnel from it to another waveguide (Saleh and Teich, 1991). The evanescent wave-coupled waveguides can be used as directional couplers for switching of signal in an optical communication network. Evanescent wave-coupled waveguides have also been proposed for sensor application, where sensing produces a change in photon tunneling from one waveguide channel to another (Prasad, 2003).

Another example of a geometry producing an evanescent wave is provided by total internal reflection involving the propagation of light through a prism of refractive index n_1 to an environment of a lower refractive index n_2 (Courjon, 2003; Prasad, 2003). At the interface, the light refracts and partially passes into the second medium at a sufficiently small incidence angle. But when the angle of incidence exceeds a value θ_c, called the *critical angle,* the light beam is reflected from the interface as shown in Figure 2.12. This process is called *total internal reflection* (TIR). The critical angle θ_c is given by the equation

$$\theta_c = \sin^{-1}(n_2/n_1) \tag{2.26}$$

As shown in Figure 2.12 for incidence angle $>\theta_c$, the light is totally internally reflected back to the prism from the prism/environment interface. The refractive index n_1 of a standard glass prism is about 1.52, while the refractive index n_2 of the surrounding environment, say an aqueous buffer, may be 1.33, yielding a critical angle of 61°.

Even under the condition of TIR, a portion of the incident energy penetrates the prism surface as an evanescent wave and enters the environment in contact with the prism surface (Courjon, 2003). As described earlier, its electric field amplitude E_z decays exponentially with distance z into the surrounding medium of lower refractive index n_2 as $\exp(-z/d_p)$.

Figure 2.11. Evanescent wave-coupled waveguides.

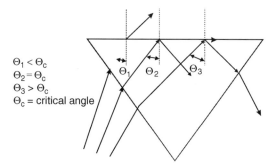

Figure 2.12. Principle of total internal reflection.

The term d_p for TIR can be shown to be given as

$$d_p = \lambda/[4\pi n_1\{\sin^2\theta - (n_2/n_1)^2\}^{1/2}] \qquad (2.27)$$

Typically, the penetration depths d_p for the visible light are 50–100 nm. The evanescent wave energy can be absorbed by an emitter, a fluorophore, to generate fluorescence emission that can be used to image fluorescently labeled biological targets (Prasad, 2003). However, because of the short-range exponentially decaying nature of the evanescent field, only the fluorescently labeled biological specimen near the substrate (prism) surface generates fluorescence and can be thus imaged. The fluorophores, which are further away in the bulk of the cellular medium, are not excited. This feature allows one to use TIR for microscopy and obtain a high-quality image of the fluorescently labeled biologic near the surface, with the following advantages (Axelrod, 2001):

- Very low background fluorescence
- No out-of-focus fluorescence
- Minimal exposure of cells to light in any other planes in the sample, except near the interface

Surface Plasmon Resonance (SPR). In principle, the SPR technique provides an extension of evanescent wave interaction, described above, except that a waveguide or a prism is replaced by a metal–dielectric interface. Surface plasmons are electromagnetic waves that propagate along the interface between a metal film and a dielectric material such as organic films (Wallis and Stegeman, 1986; Fillard, 1996). Since the surface plasmons propagate in a metal film in the frequency and wavevector ranges for which no light propagation is allowed in either of the two media, no direct excitation of surface plasmons is possible. The most commonly used method to generate a surface plasmon wave is attenuated total reflection (ATR).

The Kretschmann configuration of ATR is widely used to excite surface plasmons (Wallis and Stegeman, 1986). This configuration is shown in Figure 2.13. A

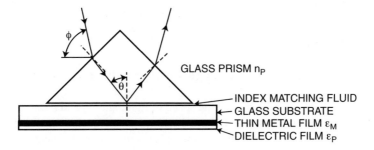

Figure 2.13. Kretschmann (ATR) geometry used to excite surface plasmons (Wallis and Stegeman, 1986).

microscopic slide is coated with a thin film of metal (usually a 40- to 50-nm-thick gold or silver film by vacuum deposition). The microscopic slide is now coupled to a prism through an index-matching fluid or a polymer layer. A p-polarized laser beam (or light from a light-emitting diode) is incident at the prism. The reflection of the laser beam is monitored. At a certain θ_{sp}, the electromagnetic wave couples to the interface as a surface plasmon. At the same time, an evanescent field propagates away from the interface, extending to about 100 nm above and below the metal surface. At this angle the intensity of the reflected light (the ATR signal) drops. This dip in reflectivity is shown by the left-hand curve in Figure 2.14.

The angle is determined by the relationship

$$k_{sp} = kn_p \sin \theta_{sp} \tag{2.28}$$

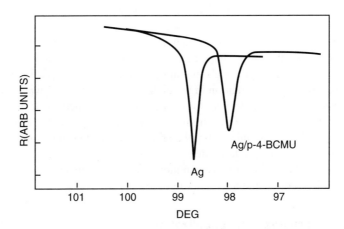

Figure 2.14. Surface plasmon resonance curves. The left-hand curve is for just the silver film (labeled Ag); the right-hand side shows the curve (labeled Ag/p-4-BCMU) shifted on the deposition of a monolayer Langmuir–Blodgett film of poly-4-BCMU on the silver film (Prasad, 1988).

where k_{sp} is the wavevector of the surface plasmon, k is the wavevector of the bulk electromagnetic wave, and n_p is the refractive index of the prism. The surface plasmon wavevector k_{sp} is given by

$$k_{sp} = (\omega/c)\,[(\varepsilon_m \varepsilon_d)/(\varepsilon_m + \varepsilon_d)]^{1/2} \tag{2.29}$$

where ω is the optical frequency, c the speed of light, and ε_m and ε_d are the relative dielectric constants of the metal and the dielectric, respectively, which are of opposite signs. In the case of a bare metal film, ε_d (or square of the refractive index for a dielectric) is the dielectric constant of air and the dip in reflectivity occurs at a certain angle. In the case of metal coated with another dielectric layer (which can be used for photonic processing or sensing), this angle shifts. Figure 2.14 shows as an illustration the shift in the coupling angle on deposition of a monolayer Langmuir–Blodgett film of a polydiacetylene, poly-4-BCMU. The shifted SPR curve is shown on the right-hand side in Figure 2.14 (Prasad, 1988).

In this experiment, one can measure the angle for the reflectivity minimum, the minimum value of reflectivity, and the width of the resonance curves. These observables are used to generate a computer fit of the resonance curve using a least-squares fitting procedure with the Fresnel reflection formulas yielding three parameters: the real and the imaginary parts of the refractive index and the thickness of the dielectric layer.

From the above equations, one can see that the change $\delta\theta$ in the surface plasmon resonance angle (the angle corresponding to minimum reflectivity; for simplicity the subscript sp is dropped) caused by changes $\delta\varepsilon_m$ and $\delta\varepsilon_d$ in the dielectric constants of the metal and covering film, respectively, is given by (Nunzi and Ricard, 1984)

$$\cot\theta\,\delta\theta = (2\varepsilon_m\varepsilon_d(\varepsilon_m + \varepsilon_d))^{-1}(\varepsilon_m^2 \delta\varepsilon_d + \varepsilon_d^2 \delta\varepsilon_m) \tag{2.30}$$

Since $|\varepsilon_m| \gg |\varepsilon_d|$, the change in θ is much more sensitive to a change in ε_d (i.e., of the dielectric layer) than to a change in ε_m. Therefore, this method appears to be ideally suited to obtain $\delta\varepsilon_d$ (or a change in the refractive index) as a function of interactions or structural perturbation in the dielectric layer. Another way to visualize the high sensitivity of SPR to variations in the optical properties of the dielectric above the metal is to consider the strength of the evanescent field in the dielectric, which is an order of magnitude higher than that in a typical evanescent wave source utilizing an optical waveguide as described above (see Chapter 5). This surface plasmon enhanced evanescent wave can more efficiently generate nonlinear optical processes, which require a much higher intensity.

2.2.2 Lateral Nanoscopic Localization

A lateral nanoscale confinement of light can be conveniently obtained using a near-field geometry in which the sample under optical illumination is within a fraction of the wavelength of light from the source or aperture (Fillard, 1996; Courjon, 2003;

Saiki and Narita, 2002). In the near-field geometry, an electric field distribution around a nanoscopic structure produces spatially localized optical interactions. Also, the spatially localized electric field distribution contains a significant evanescent character—that is, decaying exponential because of the imaginary wavevector character. A near-field geometry is conveniently realized using a near-field scanning optical microscope, abbreviated as NSOM or sometimes as SNOM (interchanging the terms near-field and scanning). This topic forms the content of Chapter 3. Here, only the NSOM is briefly described.

In a more commonly used NSOM approach, a submicron size 50 to 100-nm aperture, such as an opening tip of a tapered optical fiber, is used to confine light. For an apertureless NSOM arrangement, a nanoscopic metal tip (such as the ones used in scanning tunneling microscopes, STM) or a nanoparticle (e.g., metallic nanoparticle) is used in close proximity of the sample to enhance the local field. This field enhancement is discussed in Chapter 5.

2.3 NANOSCALE CONFINEMENT OF ELECTRONIC INTERACTIONS

This section provides some selected examples of nanoscale electronic interactions, which produce major modifications or new manifestations in the optical properties of a material. Table 2.3 lists these interactions. Brief discussions of these interactions then follow.

Table 2.3. Various Nanoscale Electronic Interactions Producing Important Consequences in the Optical Properties of Materials

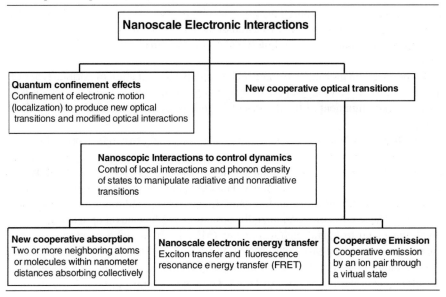

2.3.1 Quantum Confinement Effects

The quantum confinement effects are discussed in detail in Chapter 4.

2.3.2 Nanoscopic Interaction Dynamics

Chapter 6 covers examples of control of nanoscopic interactions, whereby a particular radiative transition (emission at a particular wavelength) is enhanced by local interactions. An example is the use of a nanocrystal host environment with low-frequency phonons (vibrations of lattice) so that multiphonon relaxation of excitation energy in a rare-earth ion is significantly reduced to enhance the emission efficiency. Because the electronic transitions in a rare-earth ion are very sensitive to nanoscale interactions, only a nanocrystal environment is sufficient to control the nature of electronic interactions. This provides an opportunity to use a glass or a plastic medium containing these nanocrystals for many device applications.

Nanoscale electronic interactions also produce new types of optical transitions and enhanced optical communications between two electronic centers. These interactions are described below.

2.3.3 New Cooperative Transitions

In a collection of ions, atoms, or molecules, two neighboring species can interact to produce new optical absorption bands or allow new multiphoton absorption processes. Some examples are provided here. An example already discussed in Section 2.1 is that of the formation of biexcitons in a semiconductor, such as CuCl, or in a quantum-confined structure to be covered in Chapter 4. This produces new optical absorption and emission from a biexcitonic state, whose energy is lower than that of two separate excitons. This difference in energy corresponds to the binding energy of the two excitons. An extension of the biexciton concept to multiexciton or exciton string, produced by binding (condensation) of many excitons, has also been proposed.

In the case of a molecular system, an analogy is the formation of various types of aggregates, such as a J-aggregate of dyes (Kobayashi, 1996). The J-aggregate is a head-to-head alignment of the dipoles of various dyes as schematically represented in Figure 2.15.

Another type of nanoscale electronic interaction giving rise to new optical transitions manifests when an electron-donating group (or molecule) is in the nearest neighboring proximity within nanoscopic distance of an electron withdrawing group or molecule (electron acceptor). The examples are organometallic structures involving a binding between an inorganic (metallic) ion and many organic groups (ligands). These types of organometallic structures produce novel optical transitions involving metal-to-ligand charge transfer (MLCT) or in some cases the reverse charge transfer induced by light absorption (Prasad, 2003). Another example is an organic donor (D)–acceptor (A) intermolecular complex, which in the excited state produces a charge transfer species D^+A^-. These charge-transfer complexes display

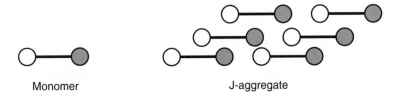

Figure 2.15. Schematics of a J-aggregate of a linear charge-transfer dye. The two circles represent the two ends of the dipole.

intense visible color derived from new charge-transfer transitions in the visible, even though the components D and A are colorless, thus having no absorption individually in the visible spectral range.

Yet another type of cooperative transition is provided by dimer formation between a species A in the excited electronic state (often labeled as A*) and another species B in the ground electronic state (Prasad, 2003). This excited-state dimer formation, produced by optical absorption, can be represented as

$$A \xrightarrow{h\nu} A^*, \quad A^* + B \longrightarrow (AB)^* \quad (2.31)$$

If A and B are the same, the resulting excited-state dimer is called an excimer. If A and B are different, the resulting heterodimer is called an exciplex. It should be emphasized that an exciplex does not involve any electron (charge) transfer between A and B. They both are still neutral species in the exciplex state, but bound together by favorable nanoscopic interactions. The optical emission from these excimeric or exciplex state is considerably red-shifted (toward longer wavelength or lower energy), compared to the emission from the monomeric excited form A*. Furthermore, the excimer or exciplex emission is fairly broad and featureless (no structures in the emission band). The excimer and exciplex emission is a very sensitive probe to the nanoscopic structure and orientation surrounding a molecule and has been extensively used to probe local environment and dynamical processes in biology.

Still another example of cooperative transition is shown by rare-earth ion pairs where one ion absorbs energy and transfers it to another ion, which then absorbs another photon to climb to yet another higher electronic level. The emission can then be up-converted in energy compared to the excitation.

2.3.4 Nanoscale Electronic Energy Transfer

The excess electronic energy supplied by an optical transition (or by a chemical reaction as in a chemical laser) can be transferred from one center (ion, atom, or molecule) to another, often on nanoscopic scale, although long-range energy transfer can also be achieved. This electronic energy transfer involves transfer of excess energy and not the transfer of electrons. Hence, in this process, one center has the excess energy (excited electronic state) and acts as an energy donor, which transfers the excitation to an energy acceptor. Consequently, the excited electron in the energy donor returns to the ground state while an electron in the energy acceptor group is promoted to an excited state. The interaction among energetically equivalent centers produces exciton migration either coherently (through many closely spaced levels forming an exciton band) or incoherently by hopping of an electron–hole pair from one center to another.

Another type of energy transfer is between two different types of molecules, a process often called *fluorescence resonance energy transfer* (FRET). This type of transfer, often used with two fluorescent centers within nanometers apart, is detected as fluorescence from the energy acceptor when the energy donor molecule is optically excited to a higher electronic level. FRET is a popular method in bioimaging to probe nanoscale interactions among cellular components, such as to monitor protein–protein interactions (Prasad, 2003). In this case, one protein may be labeled with a fluorescent dye which acts as the energy donor when electronically excited by light. The other protein is labeled by an energy acceptor, which can receive energy when the two proteins are within nanoscopic distances in the range of 1–10 nm. This energy transfer occurs often by dipole-dipole interaction between the energy donor and the energy acceptor, which shows a distance dependence of R^{-6}. To maximize the FRET process, there should be a significant spectral overlap between the emission spectrum of the donor and the absorption spectrum of the acceptor.

2.3.5 Cooperative Emission

Cooperative emission is another example of manifestation of electronic interactions. Here two neighboring centers within nanoscopic distances, when electronically excited, can emit a photon of higher energy through a virtual state of the pair centers, as shown in Figure 2.16.

This process exhibited by rare-earth ions produces up-converted emission of a higher energy photon than the energy of excitation of individual ions ($v_e > v_a, v_b$). The interaction is again manifested when the two neighboring ions are separated within nanometers. The interaction between the two ions may be of multipole–multipole or electron exchange type, depending on the nature of electronic excitation in

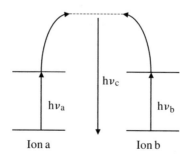

Figure 2.16. Cooperative emission from an ion pair.

individual ions. It should be pointed out that the emission is not from a real level of the ion pair, but from a virtual level, which is not an allowed electronic level either of the individual ion or of the ion pair.

2.4 HIGHLIGHTS OF THE CHAPTER

- Both photons and electrons simultaneously exhibit particle and wave-type behavior.
- For both photons and electrons as waves, the wavelength, λ, associated with their motion is given as $\lambda = h/p$. The difference is on the length scale, with electrons having wavelengths considerably smaller than that for photons.
- Equivalent eigenvalue wave equations describe the propagation of photons and electrons and their allowed energy values in a medium.
- Similar to the electrostatic interactions introducing resistance to the flow of electrons, the dielectric constant and the related refractive index describe the resistance of a medium to the propagation of photons.
- Photons differ from electrons in two ways: (i) Photons are vector field (light can be polarized) while the wavefunctions of electrons are scalar; (ii) photons have no spin and no charge, while electrons possess spin and charge.
- In a free space, both electrons and photons are described by a propagating plane wave whose amplitude is constant throughout the space, and a propagation vector, **k**, describes the direction of the propagation; the magnitude of **k** relates to the momentum.
- Confinement of electrons and photons to dimensions comparable to their wavelengths produces quantization where only certain discrete values of energies (for electrons) and field distribution (for photons) are permissible.
- Both electrons and photons have finite amplitude in a region, which classically is energetically not allowed for their propagation. For light, this electromagnetic field in the forbidden region is called an *evanescent wave,* which decays exponentially with the penetration depth.
- Like electron tunneling, photon tunneling describes passage of photons from an allowed zone to another through a barrier where its propagation is energetically forbidden.
- Electrons face a periodic electrostatic potential (due to nuclear attraction) in an electronic semiconductor crystalline structure, which produces the splitting of the conduction (high-energy) band from the valence band and thus creates a bandgap.
- A photonic crystal, analogous to an electronic crystal, describes a periodic dielectric domain (periodic modulation of refractive index), but now with the periodicity at a much longer scale, matched to the wavelength of photons. A photonic bandgap is produced, in analogy to the bandgap for electrons in semiconductors.

- Both photons and electrons exhibit cooperative effects. For photons, the cooperative effects are nonlinear optical effects produced at high field strength (optical intensities). Examples of cooperative effects for electrons are electron–electron interactions in superconductivity and formation of excitons and biexcitons in semiconductors.
- Nanoscale localization of optical interactions can be produced axially through the evanescent wave and surface plasmon waves, as well as laterally by using a near-field geometry. Surface plasmons are electromagnetic waves that propagate along interface between a metal film and a dielectric.
- Nanoscopic electronic interactions between two neighboring ions produce new optical absorption band or allow new multiphoton absorption processes.
- Formation of biexcitons resulting from binding of two excitons in a semiconductor is an example of producing new optical absorption and emission. In molecular systems, the analogous effect is intermolecular interactions producing various types of aggregates (such as J-aggregates) with different spectral characteristics.
- Another type of nanoscopic electronic interaction giving rise to new optical transitions (called charge-transfer bands) manifests when an electron-donating specie is within nanoscopic distance from an electron-withdrawing group.
- Yet another nanoscopic interaction produces excited-state dimers called excimers (in case of the same molecules) and exciplexes (in case the dimer consists of two different molecules).
- Nanoscopic interactions also produce excited-state energy transfer whereby the absorbing molecule (energy donor) returns to ground state by transferring its excitation energy to a different type of molecule of lower-energy excited state (energy acceptor). If the acceptor fluoresces, the process is referred to as *fluorescence resonance energy transfer* (FRET).
- Two neighboring centers within nanoscopic distances, when electronically excited, can emit a photon of higher energy through a virtual state of the pair centers. This process is called *cooperative emission*.

REFERENCES

Atkins, P., and dePaula, J., *Physical Chemistry,* 7th edition, W. H. Freeman, New York, 2002.

Axelrod, D., Total Internal Reflection Fluorescence Microscopy, in *Methods in Cellular Imaging,* A. Periasamy, ed., Oxford University Press, Hong Kong, 2001, pp. 362–380.

Born, M., and Wolf, E., *Principles of Optics,* 7th edition, Pergamon Press, Oxford, 1998.

Boyd, R. W., *Nonlinear Optics,* Academic Press, New York, 1992.

Cohen, M. L., and Chelikowsky, J. R., *Electronic Structure and Optical Properties of Semiconductors,* Springer-Verlag, Berlin, 1988.

Courjon, D., *Near-Field Microscopy and Near-Field Optics,* Imperial College Press, Singapore, 2003.

Feynman, R. P., Leighton, R. B., and Sands, M., *The Feynman Lectures of Physics,* Vol. 1, Addison-Wesley, Reading, MA, 1963.

Fillard, J. P, *Near Field Optics and Nanoscopy,* World Scientific, Singapore, 1996.

Gaponenko, S. V., *Optical Properties of Semiconductor Nanocrystals,* Cambridge University Press, Cambridge, 1999.

Gonokami, M., Akiyama, H., and Fukui, M., Near-Field Imaging of Quantum Devices and Photonic Structures, in *Nano-Optics,* S. Kawata, M. Ohtsu, and M. Irie, eds., Springer-Verlag, Berlin, 2002, pp. 237–286.

Joannopoulos, J. D., Meade, R. D., and Winn, J. N., *Photonic Crystals,* Princeton University Press, Singapore, 1995.

Kelly, M. J., *Low-Dimensional Semiconductors,* Clarendon Press, Oxford, 1995.

Kittel, C., *Introduction to Solid State Physics,* 7th edition, John Wiley & Sons, New York, 2003.

Klingshirn, C. F., *Semiconductor Optics,* Springer-Verlag, Berlin, 1995.

Kobayashi, T., *J-Aggregates,* World Scientific, Japan, 1996.

Levine, I. N., *Quantum Chemistry,* 5th edition, Prentice-Hall, Upper Saddle River, NJ, 2000.

Merzbacher, E., *Quantum Mechanics,* 3rd edition, John Wiley & Sons, New York, 1998.

Nunzi, J. M., and Ricard, D., Optical Phase Conjugation and Related Experiments with Surface Plasmon Waves, *Appl. Phys. B* **35,** 209–216 (1984).

Prasad, P. N., Design, Ultrastructure, and Dynamics of Nonlinear Optical Effects in Polymeric Thin Films, in *Nonlinear Optical and Electroactive Polymers,* P. N. Prasad and D. R. Ulrich, eds., Plenum Press, New York, 1988, pp. 41–67.

Prasad, P. N., *Introduction to Biophotonics,* Wiley-Interscience, New York, 2003.

Prasad, P. N., and Williams, D. J., *Introduction to Nonlinear Optical Effects in Molecules and Polymers,* Wiley-Interscience, New York, 1991.

Saiki, T., and Narita, Y., Recent Advances in Near-Field Scanning Optical Microscopy, *JSAP Int.,* no. 5, 22–29 (January 2002).

Saleh, B. E. A., and Teich, M. C., *Fundamentals of Photonics,* Wiley-Interscience, New York, 1991.

Shen, Y. R., *The Principles of Nonlinear Optics,* John Wiley & Sons, New York, 1984.

Wallis, R. F., and Stegeman, G. I., eds., *Electromagnetic Surface Excitations,* Springer-Verlag, Berlin, 1986.

CHAPTER 3
Near-Field Interaction and Microscopy

This chapter discusses the principles and applications of near-field optics where a near-field geometry is utilized to confine light on nanometer scale. These principles form the basis of near-field scanning optical microscopy (NSOM), which provides a resolution of ≤ 100 nm, significantly better than the diffraction limit imposed on far-field microscopy. NSOM is emerging as a powerful technique for studying optical interactions in nanodomains as well as for nanoscopic imaging. The applications of NSOM have ranged from single-molecule detection to bioimaging of viruses and bacteria. Bioimaging using NSOM is described separately in Chapter 13.

After a general description of near-field optics in Section 3.1, theoretical modeling of near-field nanoscopic interactions is presented in Section 3.2. Readers less theoretically inclined may skip this section. Section 3.3 presents various approaches used for near-field microscopy. Some illustrative examples of optical interactions and dynamics utilizing NSOM are presented in various sections. Section 3.4 discusses spectroscopy of quantum dots and single molecules, as well as studies of nonlinear optical processes in nanoscopic domains. Section 3.5 introduces apertureless NSOM that utilizes a metallic tip to enhance the local field. Applications of this approach are peresented. Section 3.6 discusses enhancement of optical interactions using a surface plasmon geometry incorporated in an NSOM assembly. Section 3.7 describes time- and space-resolved studies of nanoscale dynamics.

Section 3.8 lists some of the commercial manufacturers of near-field microscopes. Section 3.9 provides highlights of the chapter.

For further reading, the following books and reviews are recommended:

Courjon (2003): *Near-Field Microscopy and Near-Field Optics*

Fillard (1997): *Near-Field Optics and Nanoscopy*

Kawata, Ohtsu, and Irie (2002): *Nano-Optics*

Moerner and Fromm (2003): *Methods of Single-Molecule Fluorescence Spectroscopy and Microscopy*

Paesler and Moyer (1996), *Near-Field Optics: Theory, Instrumentation, and Applications*

Nanophotonics, by Paras N. Prasad
ISBN 0-471-64988-0 © 2004 John Wiley & Sons, Inc.

Saiki and Narita (2002): *Recent Advances in Near-Field Scanning Optical Microscopy*

Vanden Bout, Kerimo, Higgins, and Barbara (1997): *Near-Field Optical Studies of Thin-Film Mesostructured Organic Materials*

3.1 NEAR-FIELD OPTICS

Near-field optics deals with illumination (and subsequent optical interaction) by light emerging from a subwavelength aperture or scattered by a subwavelength metallic tip or nanoparticle, of an object in the immediate vicinity (or within a fraction of the wavelength of light) of the aperture or scattering source. The light in the near-field contains a large fraction of nonpropagating, evanescent field, which decays exponentially in the far field (far from the aperture or scattering metallic nanostructure). The case where light is passed through a subwavelength aperture as shown in Figure 3.1, or through a tapered fiber (another type of aperture) as shown in Figure 3.2, is also labeled as aperture near-field optics or simply near-field optics. Most near-field studies or near-field microscopy utilize near-field optics that generally involves a tapered fiber. An example of apertureless near-field optics is also provided in Figure 3.2, where a sharp metallic tip is used to scatter the radiation. The enhanced electromagnetic field around the metallic tip is strongly confined. This field enhancement near the surface of a metallic nanostructure is discussed in detail in Chapter 5.

In aperture-controlled near-field optics, light is squeezed through an aperature such as an aluminum-coated tapered fiber to confine the light from leaking out. Light then emanates through a tip opening, which is generally anywhere from 50 to 100 nm in diameter, and is incident on a sample within nanometers of the tip. Thus, the sample senses the near-field distribution of the light field. The interesting aspect is that light is confined to a dimension much smaller than its wavelength. Even if

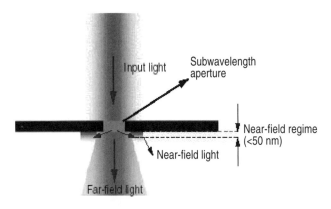

Figure 3.1. Principle of aperture controlled near-field optics.

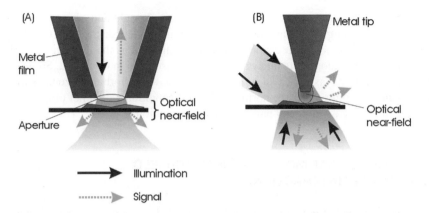

Figure 3.2. Near-field optics. The aperture-controlled near-field optics using a metal-coated tapered fiber is shown on the left. The schematic on the right-hand side shows apertureless near-field optics utilizing scattering from a metal tip.

one uses IR light of 800 nm in wavelength, it can be squeezed down to 50 or 100 nm. Thus, the near-field approach allows one to break the diffraction barrier that limits the focusing in the far field to the dimension of the incident optical wavelength.

The field that comes out of the nanoscopic aperture (fiber tip) has some very interesting, unique properties. The field distribution of light emanating from a fiber tip is shown in Figure 3.3. There is a region of propagation in which the wavevector, **k**, of light is real. This light is just the normal far-field-type light which has an

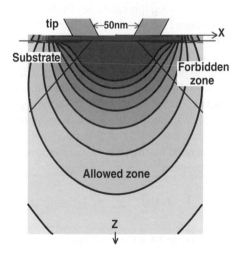

Figure 3.3. Field distribution of light emanating from a fiber.

oscillating character. Toward the edges, on the extreme right and extreme left, the wavevector, **k**, of light is imaginary. The term e^{ikr} in the field distribution then has a form $e^{-|k|r}$ which predicts an exponentially decaying field as the distance from the tip is increased, analogous to an evanescent field discussed in Chapter 2. Some call it "forbidden light," as indicated in Figure 3.3. The term "forbidden zone" simply implies that the light under the normal (far-field) condition would not propagate in this region, hence it would not have any field distribution.

3.2 THEORETICAL MODELING OF NEAR-FIELD NANOSCOPIC INTERACTIONS

Although Maxwell's equations provide a general description of electromagnetic phenomena, their analytical solutions are limited to relatively simple cases and rigorous treatment of nanoscale optical interactions presents numerous challenges. The various ways of approaching the theory of near-field optics can be classified according to the following considerations (Courjon, 2003):

- The physical model of the light beam
- The space chosen to carry out the modeling (i.e., the direct space or Fourier space modeling)
- Global or nonglobal way of treating the problem (e.g., performing separate calculation for the field in the sample and then computing the capacity of the tip to collect the field)

Among several methods used for electromagnetic field calculation, one can distinguish techniques derived from the rigorous theory of gratings like the differential method (Courjon, 2003) and the Reciprocal-Space Perturbative Method (RSPM), as well as techniques that operate in direct space like the Finite-Difference Time-Domain Method (FDTD) and the Direct-Space Integral Equation Method (DSIEM).

In general, analytical solutions can provide a good theoretical understanding of simple problems, while a purely numerical approach (like that of the FTDT method) can be applied to complex structures. A compromise between a purely analytical and a purely numerical approach is the multiple multipole (MMP) model (Girard and Dereux, 1996). With the MMP model, the system being simulated is divided into homogeneous domains having well-defined dielectric properties. Within individual domains, enumerated by the index i, the electromagnetic field $f^{(i)}(\mathbf{r}, \omega_0)$ is expanded as a linear combination of basis functions

$$f^{(i)}(\mathbf{r}, \omega_0) \approx \sum_j A_j^{(i)} f_j(\mathbf{r}, \omega_0) \tag{3.1}$$

where the basis functions $f_j(\mathbf{r}, \omega_0)$ are the analytical solutions for the field within a homogeneous domain. These basic functions satisfy the eigenwave equation for the eigenvalue q_j (analogous to the equation in Table 2.1):

$$-\nabla \times \nabla \times f_j(\mathbf{r}, \omega_0) + q_j^2 f_j(\mathbf{r}, \omega_0) = 0 \tag{3.2}$$

MMP can use many different sets of basis fields, but fields of multipole character are considered the most useful. The parameters $A_j^{(i)}$ are obtained by numerical matching of the boundary conditions on the interfaces between the domains.

As an example of the use of this technique for investigations of nonlinear optical processes in the near field, we show here investigations of second harmonic generation in a noncentrosymmetric nanocrystal exposed to fundamental light from a near-field scanning tip (Jiang et al., 2000).

One notes that a consequence of nonlinear optical interaction in the near-field is that the phase-matching conditions do not need to be fulfilled because the domains are much smaller than the coherence length. Starting from Maxwell's equations, the electric fields of the fundamental and the second harmonic (SH) wave can be shown to satisfy the nonlinear coupled vector wave equations

$$\nabla \times \nabla \times \mathbf{E}(\mathbf{r}, \omega_0) - \frac{\omega_0^2}{c^2} \varepsilon(\mathbf{r}, \omega_0) \mathbf{E}(\mathbf{r}, \omega_0) = 4\pi \frac{\omega_0^2}{c^2} \mathbf{P}^2(\mathbf{r}, \omega_0) \tag{3.3}$$

$$\nabla \times \nabla \times \mathbf{E}(\mathbf{r}, 2\omega_0) - \frac{4\omega_0^2}{c^2} \varepsilon(\mathbf{r}, 2\omega_0) \mathbf{E}(\mathbf{r}, 2\omega_0) = 4\pi \frac{4\omega_0^2}{c^2} \mathbf{P}^2(\mathbf{r}, 2\omega_0) \tag{3.4}$$

where $\varepsilon(\mathbf{r}, \omega_0)$ and $\varepsilon(\mathbf{r}, 2\omega_0)$ are linear dielectric functions for the fundamental and the SH waves, respectively.

The propagation constant k_z along the z direction is

$$k_z = (\mathbf{k}^2 - \mathbf{k}_{\parallel}^2)^{1/2} = k_0(1 - n_1^2 \sin^2 \theta)^{1/2} \tag{3.5}$$

where $k_0 = 2\pi/\lambda$, λ is the wavelength of illumination light in free space; n_1 is the refractive index of the tip, and θ is the incident angle. If $1 - n_1^2 \sin^2 \theta > 0$ (i.e., k_z is real), the waves will propagate with constant amplitude between the probe and the sample, which corresponds to the "allowed light" in the sample. In the areas where k_z is imaginary, the waves will decay exponentially within distances comparable to the wavelength, thus such waves have evanescent character and produce the "forbidden light" in the sample. From the electrical field distribution of the fundamental wave calculated with the MMP method, we can obtain the electrical field distribution of the SH wave and the different contributions of "allowed light" and "forbidden light."

Figure 3.4 shows the three-dimensional perspective view of the optical near-field intensity of the fundamental and the SH wave, respectively. The field intensity of SH wave is orders of magnitude weaker than that of the fundamental wave (FW), and it is highly localized within the area of the probe tip center—that is, about 50 nm × 50 nm. The fundamental wave is more delocalized compared to the SH wave.

Figure 3.5 shows the sectional plot of Figure 3.4 along the x-axis direction, and Figure 3.6 shows the integration of $|E|^2$ for the SHG over the total solid angle. It is

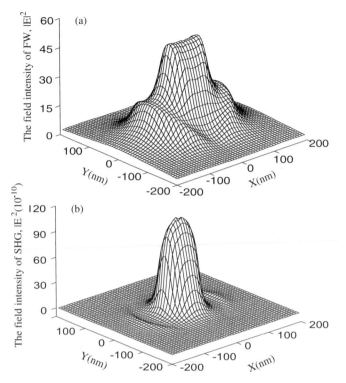

Figure 3.4. Three-dimensional perspective view of the optical near-field intensity of (a) the fundamental wave and (b) the second harmonic generation. The calculation is performed for p-polarization and for the tip-sample distance of 10 nm.

clear that the field intensity close to the probe center comes almost entirely from the allowed light, while a field enhancement appearing at the edge of the tip is due to the field components from the forbidden light. The field intensity decreases very rapidly with the tip-sample distance, and its typical decay length is approximately equal to the tip size—that is, about 50 nm. Furthermore, the field intensity of the forbidden light, which decays exponentially, exhibits a much larger variation with the probe-sample distance than does the field intensity of the allowed light. Figure 3.6 also indicates that when the probe is very close to the sample surface—that is, $d < 50$ nm—the intensity from the forbidden light dominates. However, when the probe–sample distance is larger than 50 nm, the intensity from the allowed light becomes the main contribution to the total field intensity. Because the allowed light only contains the low spatial frequencies of the sample surface, the detection of the forbidden light is essential to investigate details for both linear and nonlinear optical interactions.

3.2 THEORETICAL MODELING OF NEAR-FIELD NANOSCOPIC INTERACTIONS **47**

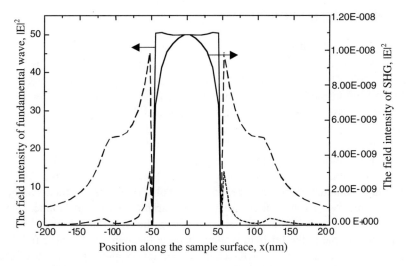

Figure 3.5. The electric field intensity of the fundamental wave and the second harmonic generation along the sample surface for tip-sample distance of 10 nm. The solid curve denotes the field intensity of the allowed light; the dashed curve denotes the field intensity of the forbidden light.

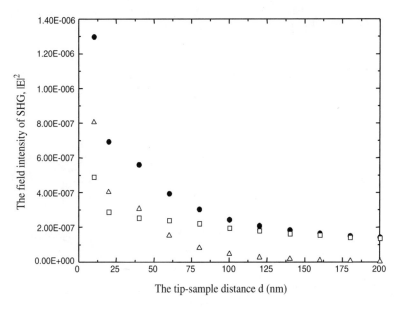

Figure 3.6. The effect of the tip-sample distance d on the near-field intensity of the SHG from the total field (●), allowed light (□), and forbidden light (△).

3.3 NEAR-FIELD MICROSCOPY

Near-field optical microscopy utilizes near-field interactions that allow one to achieve a resolution of <100 nm, significantly better than that permitted by the diffraction limit. The resolution of conventional (far-field) optical imaging techniques is limited by diffraction of light. The concept of using the near field for imaging was first discussed in 1928 by Synge, who suggested that by combining a subwavelength aperture to illuminate an object, together with a detector very close to the sample (≪ one wavelength, or in the "near field"), high resolution could be obtained by a non-diffraction-limited process (Figure 3.1) (Synge, 1928). The implementation of this principle in practice (Ash and Nicholls, 1972; Pohl et al., 1984; Betzig and Trautman, 1992; Heinzelmann and Pohl, 1994) created the technique of near-field microscopy. Now there are different variations of this technique. One can illuminate the sample in the near field, but collect the signal in the far field or illuminate the sample in the far field while collecting the signal in the near field or do both in the near field. In most methods, the important component is the use of a subwavelength aperture that can be achieved by using a tapered optical fiber with a tip radius of <100 nm.

The most commonly used near-field probe consists of an optical fiber that is tapered and coated on the outside with a reflective aluminum coating. The tip of the fiber is typically about 50 nm. Light propagating through this fiber, either for excitation or for collection of emission, produces a resolution determined by the size of the fiber tip and the distance from the sample. The image is collected point to point by scanning either the fiber tip or the sample stage. Hence the technique is called *near-field scanning microscopy* (NSOM) or *scanning near-field microscopy* (SNOM). Different modes of near-field microscopy are shown in Figure 3.7.

In illumination-mode NSOM, the excitation light is transmitted through the probe and illuminates the sample in the near field. A typical setup used for near-

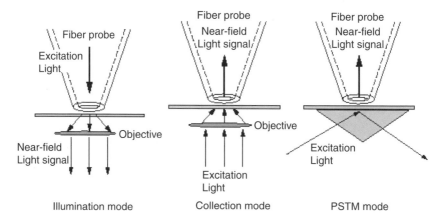

Figure 3.7. Different modes of near-field microscopy.

field imaging is shown in Figure 3.8. In collection-mode NSOM, the probe collects the optical response (transmitted or emitted light) in the near field. Another mode used in near-field imaging is photon scanning tunneling microscopy (PSTM) in which the sample is illuminated in a total internal reflection geometry using an evanescent wave (described here as due to photon tunneling as discussed in Chapter 2); the emitted light is collected by a near-field optical probe. The photon scanning tunneling geometry, or simply exciting from the bottom in the far field, is also much more convenient if one is using ultrashort femtosecond laser pulses. In the case of excitation through the fiber tip, there are complications due to broadening of the short pulses as they propagate through a length of the fiber. Therefore, a pair of gratings is often used to correct for pulse broadening in a tapered fiber. The second reason to choose a photon scanning tunneling microscopy geometry (PSTM) for excitation and the near-field for collection is that this geometry also avoids damage of the fiber tip caused by high peak power of the laser pulse. When passing a very short pulse through a 50-nm tip, intensity may be sufficiently high to damage the tip. In contrast, excitation provided in the PSTM geometry, or in the far field from the bottom, minimizes optical damage.

The resolution in NSOM and PSTM is determined by two factors: the probe aperture (opening) size and the probe–sample distance. Because most samples exhibit some topography, it is important to keep the optical probe at a constant dis-

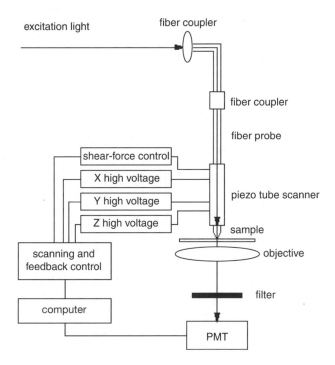

Figure 3.8. Typical instrumentation used for near-field imaging setup.

tance from a reference point on the sample surface so that any change in the optical signal is attributed to a topographic feature, and not to variation in the probe–sample distance. Two different types of arrangement used to maintain a constant probe-sample distance are shown in Figure 3.9. One is a shear-force feedback technique that can be used for distance regulation in cases of both conductive and nonconductive samples. In a shear-force feedback, the optical probe is attached to a tuning fork and oscillates laterally at its resonance frequency, with an amplitude of a few nanometers. As the probe approaches the sample surface, the probe–sample interaction dampens the amplitude and shifts the phase of the resonance. The change in the amplitude normally occurs over a range of 0–10 nm from the sample surface and is monotonic with the distance, which can be used in a feedback loop for distance regulation. The shear-force feedback can also be used to simultaneously obtain the topographic (AFM) image of the sample, to provide a monitoring reference for NSOM and PSTM. In an alternate arrangement, light reflection from the surface is used while dithering the fiber.

In regard to a tapered fiber geometry, some of the designs use a straight fiber geometry with the tip at the narrow end. This is shown in the top illustration of Figure 3.10. Some use a cantilever geometry, shown on the bottom of Figure 3.10, where the fiber tip is bent. This cantilever arrangement allows one to use the same probe for atomic force microscopy (AFM). The quality of the tapered fiber tip probe determines both the spatial resolution and the sensitivity of measurements. Hence tip fabrication is of major importance. The two methods used for tip fabrication are as follows: (i) *The heating-and-pulling method.* Here an optical fiber is locally heated by a CO_2 laser and pulled uniformly on both sides of the heated region. (ii) *The chemical etching method.* This method utilizes a hydrofluoric acid (HF) solution to etch the glass fiber. The desired taper angle is achieved by adjusting the

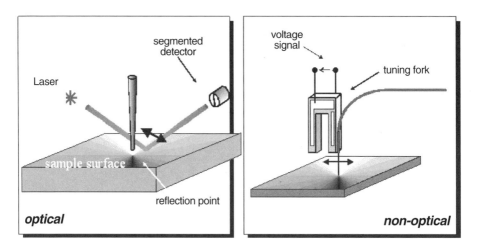

Figure 3.9. The two types of arrangements used to maintain a constant probe–sample distance.

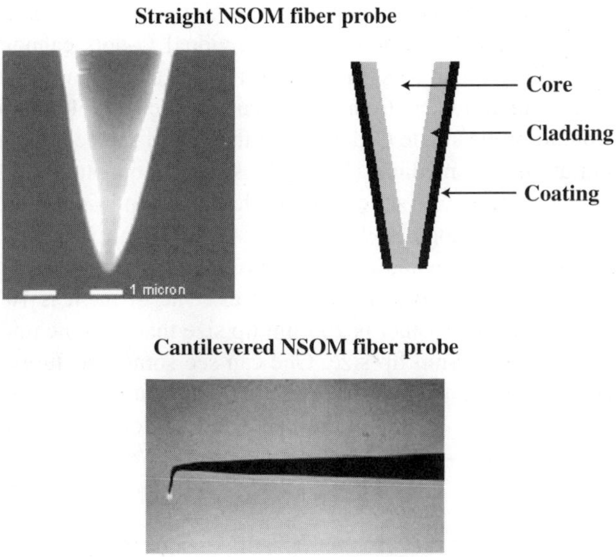

Figure 3.10. Different types of optical fiber geometries used for NSOM.

composition of a buffered HF solution. It has been shown that the optical transmission efficiency of a double-tapered structure with a large cone angle is two-orders of magnitude greater than that of a single-tapered probe with a small cone angle (Saiki et al., 1998). Such a double-tapered structure can be obtained by using a multi-step HF etching process (Saiki and Narita, 2002).

Another approach that utilizes a metallic tip or a metallic nanoparticle to localize the field of light is called *apertureless near-field microscopy* and is described in Section 3.5.

3.4 EXAMPLES OF NEAR-FIELD STUDIES

3.4.1 Study of Quantum Dots

Here are some examples of studies using near-field microscopy. One involves quantum dots (often abbreviated as Q-dots), which we will discuss later in Chapter 4 on nanoscale materials. For the present we can simply assume that these are nanoparticles of dimensions in the 1- to 10-nm range that exhibit fairly narrow optical transitions that are particle size-dependent. Therefore, particles of a certain size show optical transition (absorption and emission) at a given frequency; but for the increasing size of the particles, the optical resonance shifts to a lower frequency (a longer wavelength). If one looks at a sample in a far field, one samples a large number of quantum dots (an ensemble) of different sizes and in different environments, each with its own optical resonance. This produces an inhomogeneous (statistical) broadening of optical transition. One will find a convolution of the distribution of various Q-dot sizes as individual Q-dots cannot be resolved. Consequently, a fairly broad spectrum is observed as shown in Figure 3.11. Near-field microscopy can allow one to probe domains, which are 50 nm in size, whereby it is possible to probe single Q-dots when they are dispersed homogeneously at dilute concentration in a medium (film). Thus, single quantum dot spectroscopy can be achieved. Figure 3.11 shows near-field spectra of single Q-dots obtained using a near-field microscope (website of D. Awschalom). One can image an individual quantum dot region and take a spectrum using near-field excitation. Here the results are shown for two different tip sizes. One of them is 100-nm tip size that is shown at the top, the other is 200-nm tip size that is in the middle. The bottom spectrum is with 300-nm tip size. One can see some structures related to the subsets of Q-dots that are excited when the tip is 100 nm. Then going to 200 and 300 nm, this resolution is lost as more than a few Q-dots are simultaneously excited.

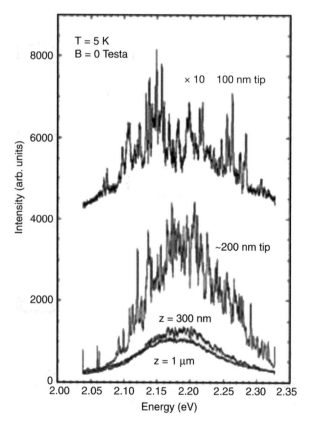

Figure 3.11. Near-field microscopy/spectroscopy of quantum dots. (Reproduced from D. Awschalom: www.iquest.ucsb.edu/sites/awsch/research/nonmag.html, reproduced with permission.)

Saiki et al. (1998) and Matsuda et al. (2001) conducted room temperature photoluminescence study on a single quantum dot from InGaAs quantum dots grown on a GaAs substrate. Their result is shown in Figure 3.12. Because of the spectral resolution obtained by sampling only a single quantum dot (no inhomogeneous broadening), they were able to observe, at an appropriate excitation density, emission not only from the lowest level (subband) of the conduction band but also from higher levels. (See Chapter 4 for a description of these bands.) They were able to study the homogeneous line width, determined by the dephasing time of excitation (see Chapter 6 for a description of dephasing time), as a function of the interlevel spacing energy. They found that the line width was larger for a smaller-size quantum dot for which the interlevel spacing is larger. (This is predicted by a simple particle in a box model as the length of the box becomes smaller, see Chapters 2 and 4.)

3.4.2 Single-Molecule Spectroscopy

An exciting direction is to push the limit of spatial probing to single molecule detection. A single atom or a molecule represents the ultimate goal of nanoscopic resolution. This means that we are not just looking at individual Q-dots, which are assemblies of hundreds or thousands of atoms, but we can look at individual molecules and atoms.

Single-molecule detection using spectroscopic methods is a highly active field (Moerner, 2002). Single-molecule spectroscopy is a powerful technique to probe individual nanoscale behavior of atoms and molecules in a complex local environment of a condensed phase. The ability to detect a single molecule and study its structure and functions provides the opportunity to elucidate single-molecule prop-

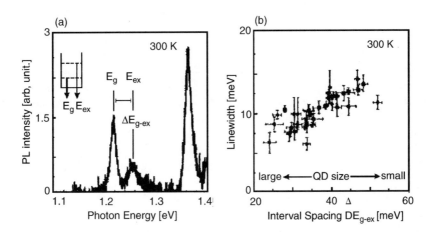

Figure 3.12. Photoluminescence spectrum of single QD at room temperature (a), and dependence of the homogeneous linewidth of ground-state emission on interval spacing, which is closely related to size of Qd's (b). From Saiki and Narita (2002), reproduced with permission.

erties that are not available in an averaging measurement on an ensemble containing a large number of molecules. Single-molecule study thus allows investigation of hidden heterogeneity and provides information on dynamics of photophysical and photochemical changes in a single molecule. Furthermore, a single molecule can be used as an ultimate local reporter of a "nanoenvironment." This is particularly important in the case of biomolecules where heterogeneity can be derived from various individual copies of a protein or oligonucleotide in different folded states, configurations, or stages of an enzymatic cycle (Moerner, 2002).

Single-molecule spectroscopy has two requirements:

- There is only one molecule present in the volume probed by the light source. This condition is met by using appropriate dilution together with microscopic techniques to probe a small volume. Near-field microscopy, providing the ultimate resolution possible by an optical microscopy, allows one to optically probe the smaller nanoscopic domain and thus more readily meet the condition of having a single molecule in the spatial domain of interrogation. Near-field microscopy thus has emerged as a tool for single-molecule spectroscopy.
- The signal-to-noise (SNR) ratio for the single-molecule signal is sufficiently greater than unity for a reasonable averaging time to provide adequate sensitivity. For this purpose, large absorption cross sections, high photostability, operation below saturation of absorption, and (in the case of fluorescence detection) a high fluorescence quantum yield are needed. In addition, the use of a small focal volume also provides an enhancement of SNR. For this reason also, near-field microscopy is desirable.

Since the original work of Moerner et al. in the 1980s (Moerner and Kador, 1989; Moerner, 1994; Moerner et al., 1994), this field has seen an explosion of activities and reports. For single-molecule spectroscopy, various spectroscopic techniques such as absorption, fluorescence, and Raman have been used. Fluorescence has been the most widely utilized spectroscopic method for single-molecule detection. Single-molecule fluorescence detection has been successfully extended to biological systems (Ha et al., 1996, 1999; Dickson et al., 1997). Excellent reviews of the applications of single-molecule detection to bioscience are by Ishii and Yanagida (2000) and Ishijima and Yanagida (2001). Single-molecule detection has been used to study molecular motor functions, DNA transcription, enzymatic reactions, protein dynamics, and cell signaling. The single molecule detection permits one to understand the structure–function relation for individual biomolecules, as opposed to an ensemble average property that is obtained by classical measurements involving a large number of molecules.

Near-field excitation can provide enhancement of the fluorescence to increase SNR for single molecule spectroscopy. Fluorescence probes used for single-molecule detection are fluorescence lifetime, two-photon excitation, polarization anistropy, and fluorescence resonant energy transfer (FRET). Single-molecule fluorescence lifetime measurements have been greatly aided by the use of a technique called *time-correlated single-photon detection* (Lee et al., 2001).

Figure 3.13 shows fluorescence NSOM images of single molecules from Barbara's group (Higgins and Barbara, private communication). The variation of fluorescence intensity from one molecular emitter to another may be due to possible variations of the molecule on the substrate.

Ambrose et al. (1994) reported an abrupt irreversible photobleaching in Rhodamine-6G dye after repeated excitations of the same molecule. This abrupt bleaching is not observed for an ensemble of molecules which exhibit a gradual decrease of fluorescence. Betzig and Chichester (1993) reported a similar irreversible photobleaching on lipophilic carbocyanine.

A measurement of Stark shift in a single molecule can be used as a local sensor of the electric field distribution in nanoscopic domains. For this purpose, Moerner et al. (1994) used a 60-nm-diameter near-field optical probe consisting of an aluminum-coated optical fiber tip. A static electric potential applied to the Al-coated tip produced the Stark shifts of the absorption line of a single molecule.

3.4.3 Study of Nonlinear Optical Processes

Nonlinear optical processes provide information about nanoscopic-level organization and interactions. The nonlinear optical processes, discussed in Chapter 2, depend on higher powers of the field strength and involve more than one photon interacting at one time in a medium. For sake of clarity, Figure 3.14 shows some examples of nonlinear optical processes previously covered in Chapter 2. In the case of second harmonic generation (SHG), two photons of the same frequency, ω, interact with the medium and generate an output at 2ω. This interaction occurs without the absorption of light. Thus, this is a coherent process where a medium simply interacts with light and converts the light, at its fundamental at frequency ω (wavelength λ), to second harmonic at doubled frequency 2ω (hence wavelength, $\lambda/2$). The SHG process is limited by symmetry requirements to occur only in a noncentrosymmetric medium or at an interface which by nature is asymmetric (two different media on opposite sides).

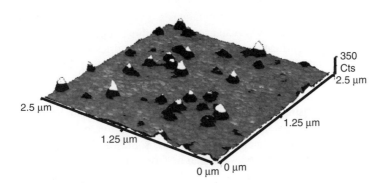

Figure 3.13. Fluorescence NSOM images of single molecules. From Professor D. Higgins and Professor P. Barbara, unpublished results.

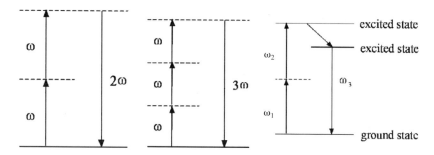

Figure 3.14. Examples of nonlinear optical processes.

The second represented process is third harmonic generation (THG). It is derived from third-order nonlinear optical interactions. In this case, the medium interacts with light of fundamental frequency ω, utilizing three photons together to generate an output of a single photon with frequency 3ω. If ω is in the near infrared at wavelength 1.06 μm, the output is at wavelength ~355 nm in the UV. This process can occur in any media because no symmetry restriction on the medium is imposed.

Another process is two-photon excitation (TPE), where the material absorbs two photons simultaneously to reach an excited state. In the case represented in the figure, the two photons have different frequencies, ω_1 and ω_2, to produce nondegenerate two-photon absorption, or they can have the same frequency ω to produce two-photon absorption at 2ω. In the case represented, $\omega_1 + \omega_2$ generates a real excited state, which can then relax to another lower state and emit a photon of frequency, ω_3. ω_3 is of higher value than the initial photons of frequencies ω_1 or ω_2. The TPE thus produces an up-converted emission process. It is different from SHG, because in SHG the optical output is fixed at 2ω and is a very sharp line at 2ω, related to the line width only of the fundamental at frequency ω. In the TPE, ω_3 is a higher frequency compared to either ω_1 or ω_2, but it is not $2\omega_1$, $2\omega_2$, or $\omega_1 + \omega_2$. In most cases, $\omega_3 < (\omega_1 + \omega_2)$. TPE fluorescence is an incoherent process. This process is also derived from the third-order nonlinear optical interactions and has no symmetry restriction.

Some of these nonlinear optical techniques are very sensitive to the orientation of molecules on the surface. One can use these nonlinear optical processes, particularly SHG, to probe surface structures and surface modifications, because SHG is very sensitive to interfacial characteristics. Another application is that one can fabricate structures on the surface. Two-photon excited fluorescence can conveniently be used to image nanodomains. Two-photon excitation can be used to fabricate nanostructures by using photolithographic techniques, as shall be described in Chapter 11. TPE provides the advantage that this excitation, using two photons, is quadratically dependent on intensity, so it will be significantly more localized near the focal point. Therefore, nanodevices can be fabricated with great precision using nonlinear optical excitations.

A number of studies of SHG using near-field optics have been reported (Bozhevolnyi and Geisler, 1998; Kajikawa et al., 1998; Shen et al., 2000; Shen et al., 2001a). For the sake of convenience, the work reported here is that performed at our Institute (Shen et al., 2000; Shen et al., 2001a).

Figure 3.15 provides an example of SHG in an organic crystal called NPP (structure represented in Figure 3.16) using near-field microscopy. Many different nanocrystals of NPP are formed on the surface of the substrate. The NPP crystallites are ~100 nm lengthwise and well-oriented (Shen et al., 2001a). NPP has the appropriate symmetry and structure to generate the second harmonic, so we can use it to

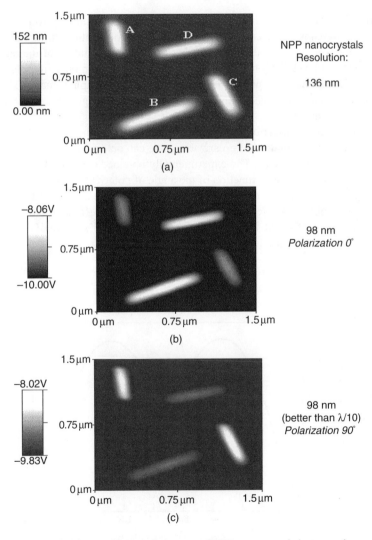

Figure 3.15. Near-field second harmonic images of NPP nanocrystals in two orthogonal polarizations.

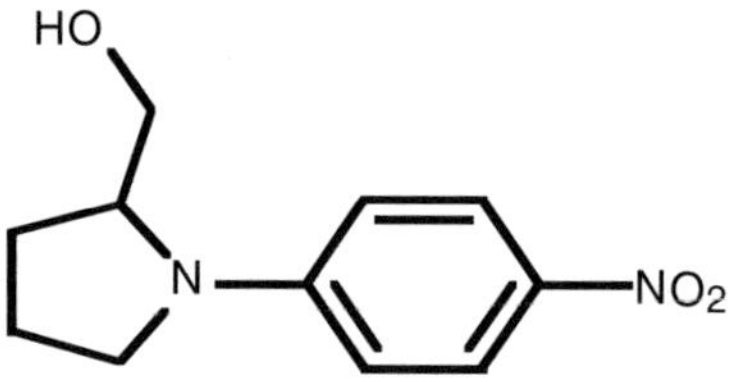

Figure 3.16. Structure of NPP.

map out the local second harmonic domain distribution and discover how these crystalline domains are oriented on the surface. The polarization of the light is varied from an arbitrary zero degree to further probe the orientation of the nanocrystals. At 90° the crystallites that are bright are not so active in the zero polarization. The top illustration in Figure 3.15 is a shear force topographic image. The topographic image correlates well with the second harmonic image, thus confirming the absence of any artifacts.

By analyzing polarization dependence obtained from the study of SHG as a function of angle of rotation of polarization, one can obtain information on the orientation of the nanocrystals. The experimentally obtained curve is shown in Figure 3.17, which plots SHG as a function of the angle of polarization of light by rotating the polarizer to different angles (Shen et al., 2001a). The second-order nonlinear optical susceptibility is often described for SHG by a d coefficient that is a tensor. The effective d coefficient, d_{eff}, is a measure of the material's figure of merit for the SHG. The d_{eff} and its angular distribution is described by Eq. (3.6):

$$d_{\text{eff}} = \{d_{21}^2 \sin^2 2(\theta + \alpha) + [d_{21} \sin^2(\theta + \alpha) + d_{22} \cos^2(\theta + \alpha)]^2\}^{1/2} \quad (3.6)$$

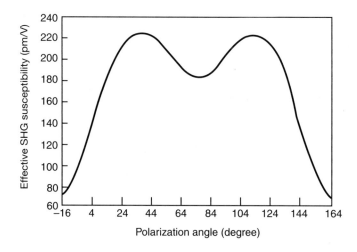

Figure 3.17. Polarization dependence of second harmonic generation in the NPP nanocrystal.

A fit of these equations yields d_{eff} = 224 pm/V and provides two tensor components d_{21} and d_{22} of the d coefficient. The other information one gets, by the fit of this anisotropy, is that the crystallographic b axis and the CT axis of this crystal are parallel to the plane of the substrate. Therefore, this study provides a very clear indication of how these nanocrystal domains are oriented on the substrate plane.

This study provided details of structural information on the nanoscopic order of the crystal which are listed in Table 3.1. The result shows that there is a uniform SHG intensity distribution over the entire nanocrystal for each of them, as the nanocrystals do not exhibit bright spots and dark spots. This means that there is a nanoscopic order in these domains. Second, the polarization dependence conforms to this nanoscopic order as demonstrated by the applicability of Eq. (3.6). The fit of d_{eff} shows the crystal orientation on the surface. Different locations in the same crystal yield the same d_{eff}. It means the crystals are mono-domain with a well-defined orientation.

Other examples of nonlinear processes are third harmonic generation (THG) and two-photon excitation (TPE) (Shen et al., 2001b). As described above, THG involves generation of a beam at a wavelength ~ 355 nm (frequency 3ω) for the incident beam, called the fundamental beam, of wavelength 1064 nm (frequency ω). TPE, on the other hand, may involve simultaneous absorption of two photons, each of frequency ω, and generates an up-converted fluorescence of higher frequency ω_3 ≫ ω (see Chapter 2). It might appear that THG is a three-photon process and that TPE is a two-photon process. But the theory of nonlinear optical interactions shows that both these processes are derived from third-order nonlinear optical interactions. THG depends on the square of the absolute value of the third-order nonlinear optical susceptibility $\chi^{(3)}$ which is a measure of the strength of third-order nonlinear response of a medium. This susceptibility is a higher-order (rank four) tensor (i.e., it relates four vectors). $\chi^{(3)}$ is a complex quantity near a two-photon absorption. Two-photon absorption relates to the imaginary part of the third-order nonlinear optical susceptibility. THG does not involve the absorption of light by the medium. The medium simply interacts and transforms the fundamental light into its third harmonic.

Figure 3.18 shows the study of both THG and two-photon fluorescence, observed simultaneously in an organic nanocrystal called DEANST (structure shown in Figure 3.19). The topographic image on the top is shear force generated (as ob-

Table 3.1. Nanoscopic Information from Second Harmonic Imaging of Nanocrystals

- Uniform SH intensity distribution → Nanoscopic order
- Polarization dependence → Nanoscopic order
- Applicability of equation for d_{eff} → Crystallographic b axis (P_{21}) and the molecular CT axis parallel to the plane of the substrate
- Locations in the same nanocrystal or different nanocrystals give same d_{eff} → Same symmetry and order

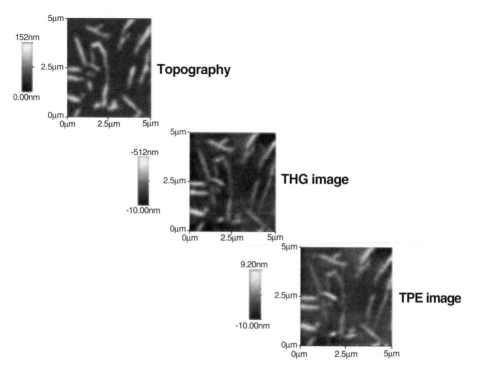

Figure 3.18. Near-field third harmonic and two-photon microscopy and spectroscopy in DEANST crystals.

tained by atomic force microscopy, AFM). The middle image is obtained by monitoring the third harmonic signal. In this case the fundamental light is at 1.064 μm. Hence, the third-harmonic output is generated at ~355 nm in the UV. This image correlates fairly well with the topographic image shown on the top. The bottom image is a TPE fluorescence image. Here the fluorescence emission is in the red region with a maximum at ~600 nm. All three of the images correlate well.

Figure 3.20 shows the spectral distribution of the near-field signal. A very sharp line at ~355 nm is due to THG. The line width of the spectroscopic feature is related to the line width of the fundamental at 1.064 μm. Then, at around 600 nm there

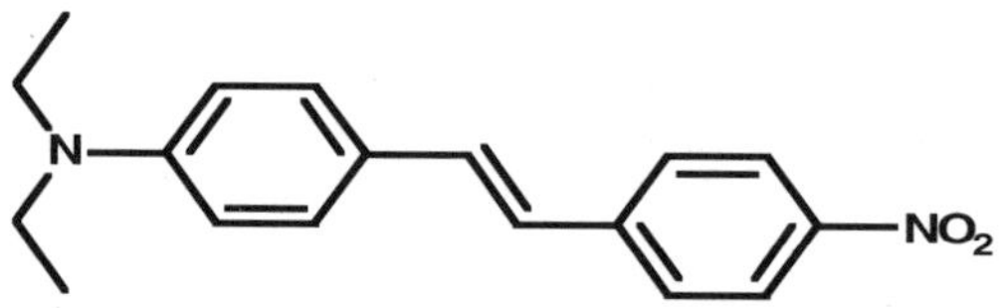

Figure 3.19. Structure of DEANST.

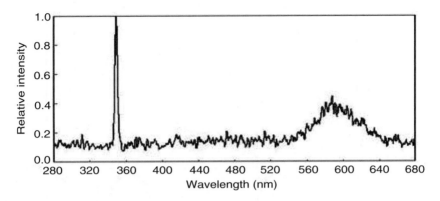

Figure 3.20. Spectral distribution of THG and TPE near-field signal, in the DEANST nanocrystals.

is a broad emission peak. This broad curve is due to TPE fluorescence. Hence, the DEANST crystal generates both third harmonic and TPE fluorescence. Thus, both the absolute value (for THG) and the imaginary part (for TPE) of the third-order nonlinear optical interactions are manifested. Both are strong, when a fundamental light of 1.064 μm is used. An intensity dependence study shows that TPE appears first, in agreement with the fact that TPE should depend on the square of the input intensity while THG depends on the cube. Detailed spectral study also demonstrates that no SHG is generated, as expected because DEANST crystals are centrosymmetric (see Chapter 2). It also establishes that THG is generated directly by a $\chi^{(3)}$ process, rather than two coupled (cascaded) second-order processes that first produce 2ω and then sums it with ω.

In order to get information on the orientation of the crystal, one can rotate the polarization of light incident on the crystal and can map out the anisotropy. Figure 3.21 shows the results of such a study for both THG (*top*) and TPE fluorescence (*bottom*) (Shen et al., 2001b). There is a one-to-one correlation between the two curves. Because these angular variations relate to the anisotropy of the third-order nonlinear optical susceptibility, this correlation is not surprising. The observed anisotropy is determined by the relative contributions of various tensor components of $\chi^{(3)}$. The fitted result shows that the ratio of the two in-plane diagonal tensor components $\chi^{(3)}_{yyyy}$ to $\chi^{(3)}_{xxyy}$ is ~ 4 (Shen et al., 2001b). The angular distribution is fitted by using an effective susceptibility $\chi^{(3)}_{\text{eff}}$ given by

$$\chi^{(3)}_{\text{eff}} = \chi^{(3)}_{yyyy} \cos^4(\theta + \alpha) + 6\chi^{(3)}_{xxyy} \cos^2(\theta + \alpha)\sin^2(\theta + \alpha) + \chi^{(3)}_{xxxx} \sin^4(\theta + \alpha) \quad (3.7)$$

where α is the reference angle, and $\chi^{(3)}_{xxyy}$ is the in-plane off-diagonal $\chi^{(3)}$ tensor component. Furthermore, at a two-photon resonance, the dominant contribution to $\chi^{(3)}$ may be from the imaginary components. Thus even THG may be determined by the imaginary part of $\chi^{(3)}$.

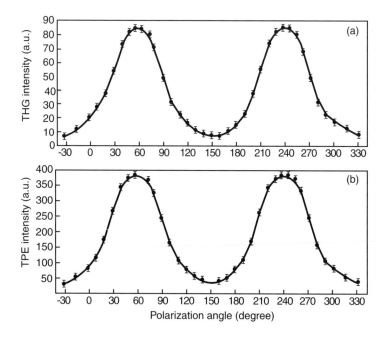

Figure 3.21. In-plane THG and TPE anisotropy for the DEANST nanocrystals.

3.5 APERTURELESS NEAR-FIELD SPECTROSCOPY AND MICROSCOPY

As mentioned in Section 3.3, an emerging approach is the apertureless near-field spectroscopy and microscopy (Novotny et al., 1998; Sanchez et al., 1999; Bouhelier et al., 2003). The use of an aperture such as a tapered fiber opening poses a number of experimental limitations. Some of these are:

- Low light throughput due to the small fiber aperture and the finite skin depth (light penetration) into the aluminum metal coating around the tapered fiber.
- Absorption of light in the metal coating; this can produce significant heating that can create a problem in imaging, particularly of biological samples.
- Pulse broadening in the fiber, when using short pulses for nonlinear optical studies. Also, the fiber tip may be damaged by the high peak intensity as already discussed in Section 3.3.

The apertureless approach overcomes these limitations, at the same time providing a significantly improved resolution. It has been demonstrated by Novotny, Xie, and co-workers (Sanchez et al., 1999; Hartschuh et al., 2003; Bouhelier et al., 2003) that optical images and spectra of nanodomains ≤25 nm can be obtained using the apertureless near-field approach involving a metal tip of end diameter ≤10 nm.

The two approaches used for apertureless NSOM are:

1. Scattering type, which involves nanoscopic localization and field enhancement of the electromagnetic radiation by scattering of the light from a metallic nanostructure. An example is provided by Figure 3.2 where the light is scattered by a sharp metallic tip. Scattering and field localization can also be produced by a metallic nanoparticle within nanometers of distance from the sample surface. The localization and enhancement of electromagnetic field by plasmon coupling to a metallic nanoparticle is discussed in Chapter 5 under "Plasmonics." This principle of obtaining nanoscopic resolution using scattering from a metallic nanoparticle also forms the basis of "plasmonic printing," discussed in Chapter 11 on "Nanolithography".
2. Field-enhancing apertureless NSOM, where a metallic tip is used to enhance the field of an incident light in the near field. In this case, the light is incident on the tip as a normal propagating mode (far-field). The strongly enhanced electric field at the metal tip produces nanoscopic localization of optical excitation. This approach offers simplicity and versatility of using light by just focusing on the metallic tip through a high-numerical-aperture lens. Hence it is described here in detail, with examples of some recent studies utilizing this approach.

A schematics of "field-enhancing" apertureless NSOM, as used by Novotny, Xie, and co-workers (Novotny et al., 1998; Sanchez et al., 1999), is presented in Figure 3.22. This method, as pointed above, combines both the far-field and the near-field techniques. The principle utilized is that if the incident radiation has a polarization component (E field) along the tip axis, a strong surface charge density is induced at the tip end, producing a large field enhancement (Novotny et al., 1997). For this purpose, a normal Gaussian beam cannot be used, because it does not provide any polarization component along the tip axis, in the geometry of an on-axis optical illumination as shown in Figure 3.22. Therefore, laser beams of higher order such as Hermite–Gaussian are needed, which provide a longitudinal (along the tip axis) electric field in the focal region. The highly confined fields close to the tip then interact with the neighboring nanodomains. The optical response, whether fluorescence, second harmonic generation, or Raman scattering (at a shifted wavelength of λ_2), is also collected by the same high N.A. (numerical aperature) lens used for illumination.

Novotny et al. (1997) showed that a gold tip with an end diameter of 10 nm can produce an intensity enhancement factor of over 3000 over the incident intensity. This enhancement is effective even though the incident wavelength is far away from the surface plasmon resonance of the metal (for surface plasmon resonance, see Chapters 2 and 5). As described above, the field enhancement is produced by the high surface charge density at the tip due to the incident light component polarized along the tip. Sanchez et al. (1999) showed that with an asymmetrical metal tip (bent shape), even a focused Gaussian beam (TEM$_{00}$) can be used to produce field enhancement at the tip. They used this geometry with femtosecond pulses from a

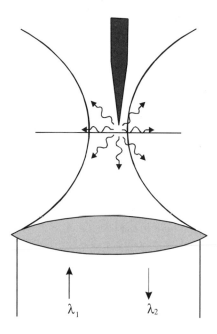

Figure 3.22. Metallic tip enhancing the local field by interacting with the focused beam at λ_1. The optical response at another wavelength λ_2 is collected by the same objective lens.

Ti-sapphire laser to produce efficient two-photon excited emission. This two-photon, near-field microscopy was used to image fragments of photosynthetic membranes as well as J-aggregates (see Chapter 2 for J-aggregates) with spatial resolutions of ~20 nm. Figure 3.23 shows the near-field, two-photon excited fluorescence image (Figure 3.23b) together with the topographic image (Figure 3.23a) of J-aggregates of the PIC dye in a polyvinyl sulfate (PVS), coated on a glass substrate (Sanchez et al., 1999). The lower portions of the figure show the simultaneous cross sections taken across the aggregate strands. The emission image cross section has a full width at half-maximum (FWHM) of ~30 nm, indicating a superior spatial resolution obtained by this method.

More recently, Hartschuh et al. (2003) reported high-resolution, near-field Raman microscopy of single-walled carbon nanotubes using a sharp silver tip (10–15 nm in diameter) as a probe. They reported a spatial resolution of ~25 nm and obtained the Raman image with the excitation beam of λ = 633 nm, using the Raman bands at 1596 cm^{-1} (the so-called G band corresponding to the tangential stretching mode, which is sharp) and 2615 cm^{-1} (the so-called G′ band corresponding to an overtone of the disorder-induced mode at 1310 cm^{-1}, which is broad).

Bouhelier et al. (2003) used the local field enhancement by a gold tip to study second harmonic generation. They showed that the second harmonic generation at an infinite tip can be represented by a dipole oscillating at the second harmonic fre-

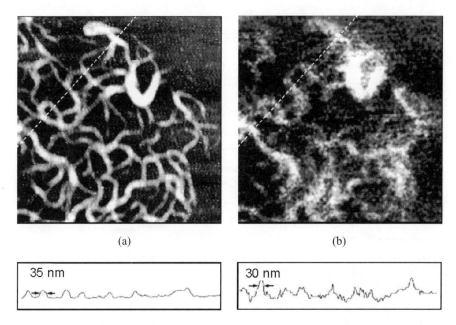

Figure 3.23. Simultaneous topographic image (a) and near-field two-photon excited fluorescence image (b) of J-aggregates of PIC dye in a PVS film on a glass substrate. The topographic cross section along the dashed line has a particular feature of 35-nm FWHM (indicated by arrows) and a corresponding 30-nm FWHM in the emission cross section. From Sanchez et al. (1999), reproduced with permission.

quency. The second harmonic generation is dominated by excitation fields polarized along the tip axis. Thus the tip can act as a probe for longitudinal (along the tip axis) fields.

3.6 NANOSCALE ENHANCEMENT OF OPTICAL INTERACTIONS

There are a number of ways to enhance optical interactions. One of them is surface plasmon enhancement. Surface plasmons are oscillatory charge waves in a metallic thin film (see Chapter 2) or metallic nanosized domain (see Chapter 5) that can be excited by coupling light into the metal film at a specific angle (Raether, 1988). The surface plasmon enhancement can be induced at the metal–dielectric interface or within nanometer distances from this interface. This enhancement can be utilized in a near-field geometry to enhance weak optical interactions such as nonlinear optical interactions. Figure 3.24 shows the well-known Kretschmann geometry (see Chapter 2) utilizing a prism, which is located right below the tip of the fiber (Shen et al., 2002). The prism is coated with a silver film, on the top of which is placed the nanostructured material. The light is coming from the right, denoted as the excitation light. It bounces off a mirror and then is incident on the base of the prism from

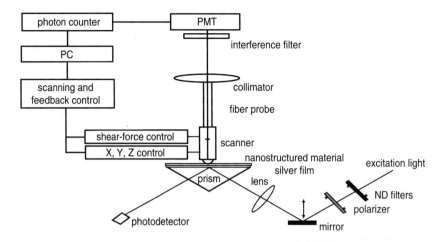

Figure 3.24. The experimental arrangement for studying surface-plasmon enhancement effect using a photon scanning tunneling microscope.

where it is reflected, because it is incident at an angle of total internal reflection. The reflected light is detected by a photodetector. If one scans the mirror vertically, as the arrow shows, one is changing the angle of incidence of light to the prism base and its interaction with the metal film. At a given angle, called the surface plasmon resonance (SPR) coupling angle, the light couples into the silver film and propagates as a surface plasmon wave. This wave is also evanescent (see Chapter 2). On the other side of the silver film, it has a very high intensity near the surface, and it decays exponentially, as the distance is increased away from the surface of the metal film. Because of this confinement of light as a surface plasmon wave, the optical interaction at the surface plasmon angle, where light is coupled, becomes very strong. Hence one can obtain near-field nanoscopic image at much lower light intensity, because of the surface plasmon enhancement. If one wants to do photofabrication using a photoprocess, one can also fabricate nanostructures using a much lower intensity, and in a much more spatially confined manner, because of the confinement of the surface plasmon interaction.

The results shown in Figure 3.25 illustrate surface plasmon enhancement in a near-field geometry (Shen et al., 2002). There are two curves. The curve on the left is the surface plasmon local field enhancement where light emanating above the silver film is collected by the near-field probe (optical fiber tip) when it is coupled to the film as the surface plasmon wave. At the precise angle of surface plasmon coupling (42.5° in the present case), the strong local field enhancement, exhibited by a peak in the intensity, is observed as the angle of incidence is varied. The enhancement of the local field intensity is ~ 120 times. The curve on the right-hand side shows the complementary feature in the reflected light that, at the angle of surface plasmon coupling, the reflected light exhibits a dip (minimum). The dotted curves represent calculations using Fresnel equations for prism–metal film–air configurations.

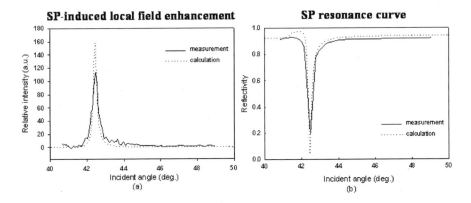

Figure 3.25. Surface plasmon enhancement in the near-field geometry.

Figure 3.26 shows the surface plasmon resonance enhanced two-photon excited emission when an organic film containing nanocrystals of a two-photon dye PRL-701 (structure represented in Figure 3.27) is coated on the surface of the silver film (Shen et al., 2002). The resonance angle has shifted a little because of the presence of the organic film that absorbs light by two-photon excitation (TPE) and exhibits up-converted emission. There is a strong enhancement of the TPE at the angle at which the light is propagating as a surface plasmon wave. By coupling light as a surface plasmon wave, the optical response, in this case the two-photon excited emission, is enhanced by more than two orders of magnitude. The surface plasmon

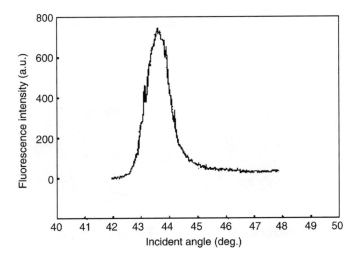

Figure 3.26. Surface plasmon resonance enhanced two-photon excited emission.

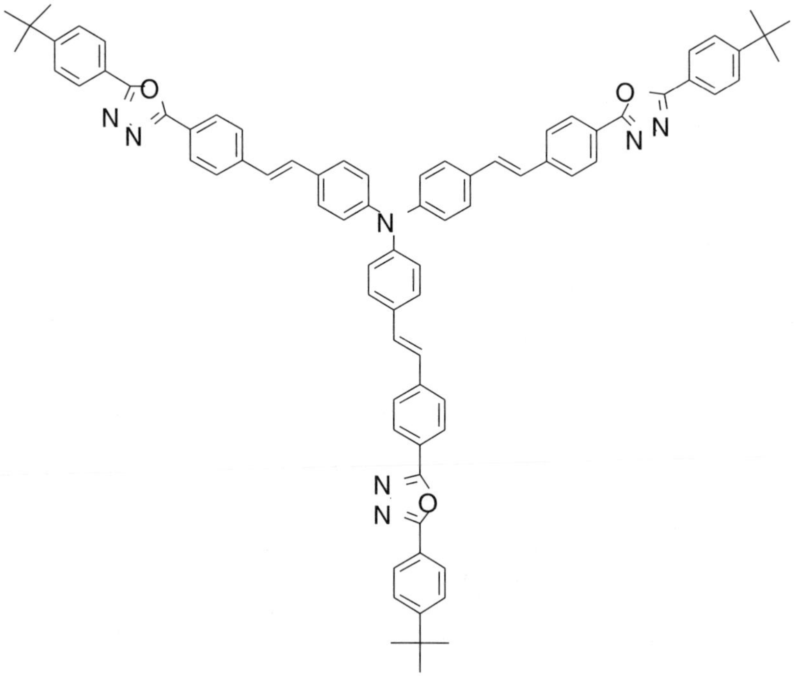

Figure 3.27. Structure of the dye PRL-701.

enhancement can be used for both linear and nonlinear optical processes. Therefore, sensitivity of either imaging or photofabrication can be significantly increased using the surface plasmon enhancement.

Figure 3.28 provides an example of surface-plasmon enhanced two-photon fluorescence images using the experimental arrangement of Figure 3.24 (Shen et al., 2002). It shows the topographic image on the left and shows the two-photon fluo-

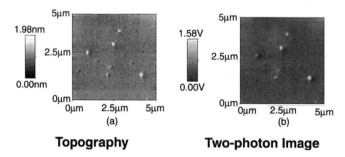

Figure 3.28. Topography and two-photon fluorescence images under surface plasmon resonance condition.

rescence image on the right, with the latter enhanced by using the surface plasmon coupling angle. In this case the two-photon excitation is provided at an angle at which light is coupled as the surface plasmon wave. The up-converted fluorescence used for imaging is obtained in the near-field collection mode. The bright domains are the ones generating two-photon fluorescence. Again, a very good correlation is found between the topographic image and the two-photon fluorescence of the nanodomains formed in the organic film on the prism surface.

3.7 TIME- AND SPACE-RESOLVED STUDIES OF NANOSCALE DYNAMICS

Investigation of dynamics of ultrafast processes in nanostructures require high resolution in both time and space. For this purpose, one can conduct a time-resolved pump-probe experiment using a near-field geometry. One can excite one nanodomain and collect a time-resolved optical response, from the same or another nanodomain. In this way, optical communication between two domains can be monitored. One can utilize this type of study to probe nanoscale excitation dynamics or to monitor optical cross-talks between two domains. An example of time-resolved study, using a near-field geometry, is provided here (Shen et al., 2003). In this study, nanocrystals of a two-photon active dye, PRL-701 (Figure 3.27), were deposited on a substructure by rapid evaporation of the solvent, tetrahydrofuran (THF). These nanocrystals behave as a saturable absorber. In other words, at high intensity the absorption saturates (the population in the ground and the excited states becomes nearly equal). At this point, the exciting beam passes through the sample without any absorption. The medium in this condition is also referred to as physically photobleached. This case is to be distinguished from a chemical bleaching, where increased light transmission occurs due to a chemical change in the absorbing molecule.

An experimental arrangement for this type of study is shown in Figure 3.29. A mode-locked Ti-sapphire laser pumped by a diode pumped laser provides ~100-fs pulses of 800-nm light which are split into two parts. One part at 800 nm is used to produce two-photon excitation in the sample. The other part goes through a nonlinear crystal, called BBO, which frequency doubles it to 400 nm. The two beams are combined together and are incident on the sample from below in the far-field mode. The optical response of the light is the transmission of the 400-nm light from the sample which is collected in the near-field geometry using the tapered fiber.

The PRL-701 nanocrystal has a strong linear absorption at ~400 nm and exhibits a strong two-photon absorption at 800 nm. The 800-nm light source is used to saturate the optical transition in the sample by two-photon absorption. The 400-nm light as a weaker probe light passes through the sample at various time delays to probe the saturation recovery. If the bleaching due to saturation is present, the 400-nm light does not get absorbed by one-photon excitation and will be transmitted. Thus by varying delay of the 400-nm pulse with respect to the 800-nm pulse, one can get the dynamics of saturation recovery and hence the excited-state lifetime. One can then

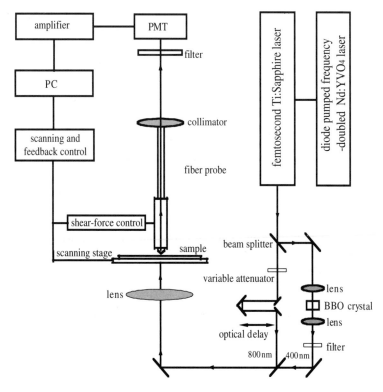

Figure 3.29. Schematic of femtosecond time-resolved collection-mode near-field microscopy and spectroscopy for studying local transient absorption dynamics in organic nanostructures.

change the illumination point from one spot to another to investigate the dynamics of saturation recovery from one nanodomain to another. One can excite in one nanodomain and probe another nanodomain to see if photobleaching spreads out.

The saturation behavior is illustrated by the near-field transmission images shown in Figure 3.30. The top left is the topographic image. The middle is the transmission image at time delay between pump and probe being equal to 360 fs. The one below it is a transmission image at 200-fs time delay; and the last one, at the bottom of the transmission image, is at zero time delay. The image of nanodomains, where the optical saturation takes place, is contrasted against the surrounding medium by the transmission of the probe beam at 400 nm. Saturation leading to maximum photobleaching occurs at zero time delay. Under saturation conditions, there is very little contrast between the sample nanodomains and the surrounding transparent medium. Hence the image of the nanodomains (absorber) is very faint. As the time delay between the pump beam and the probe beam increases, these nanodomains begin to recover from saturation and start absorbing the probe beam. As a result, they appear darker in regard to the surrounding medium. Then for a delay of 360 fs, one can see

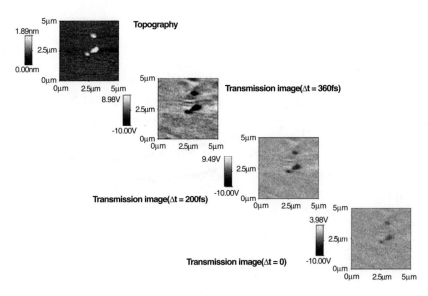

Figure 3.30. Topographic image is displayed with transmission images at various time delays between the pump (800-nm two-photon excitation) and the probe (400-nm one-photon excitation) pulses.

very dark regions because of strong absorption of the probe pulses at 400 nm. From the transient transmission imaging as a function of time delay, one can obtain the excited-state lifetime (represented by the saturation recovery time).

The transient transmittances of probe pulses for three different probe–sample distances, d, are shown in Figure 3.31. The transient transmittance is measured by locating the near-field probe at a local site of the nanostructures and then detecting the transmitted probe intensity as a function of time delay between the pump and probe pulses. The population depletion recovery provides the excited-state relaxation time, which is strongly dependent on the probe sample distance. The distance dependence of the excited-state relaxation time, τ, is shown in Figure 3.32. The solid curve in the figure is a best fit to the following empirical function:

$$\tau = \tau_0 \frac{Ad^4}{B + Cd + Dd^4} \tag{3.8}$$

which can be derived for dipole–metal surface interactions (Drexahage, 1974; Wokaun et al., 1983; Gersten and Nitzan, 1981) and where $A = D = 3.88 \times 10^{-15}/\text{nm}^4$, $B = 7.11 \times 10^{-10}$, $C = -2.35 \times 10^{-11}/\text{nm}$, and $\tau_0 = 1.75$ ns is the excited state decay time at an infinite distance from the metal surface. τ_0 is determined in bulk PRL-701 sample using a streak camera in an independent measurement. Both Figures 3.31 and 3.32 indicate that the relaxation of excited-state population is reduced dramatically with decreasing probe–sample distance. The reduction of relax-

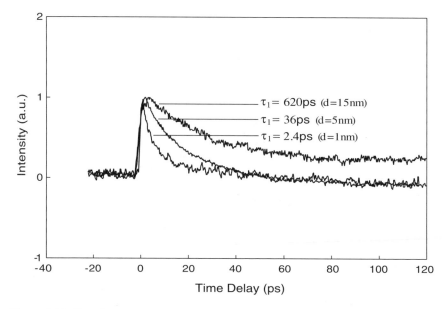

Figure 3.31. Transient transmittance of single PRL-701 nanostructure as a function of time-delay between the pump and the probe pulses for three different distances between the tip and the sample.

ation is due to the interactions between the excited molecules and the metal coating of the near-field probe. When excited molecules are located close to a metal layer, they can be treated as damped-harmonic dipoles, which decay either by spontaneous emission of photons or by nonradiative energy transfer to the metal layer. The probabilities for both those processes depend on the distance between the excited molecules and the metal surface (Gersten and Nitzan, 1981). Equation (3.8) indicates that the decay rate is much faster when excited dipoles are located in the near-field of the metal surface. Therefore, smaller probe-sample distance significantly accelerates the relaxation rate. Equation (3.8) also implies that, with increasing d to infinity, τ becomes independent of d and is equal to τ_0.

Another example of use of near-field optics to investigate nanoscopic dynamics is the study of carrier generation, recombination, and drift in a conjugated polymer, MEH-PPV (McNeill and Barbara, 2002). Near-field optical excitation using a metal-coated optical fiber served as a ~ 100-nm-diameter electrode that acted to either collect or repel majority charge carrier, depending on the sign of the applied voltage. Carrier-induced photoluminescence quenching was used to determine the local carrier density. A modulation of the photoluminescence up to 30% was observed, which was assigned to exciton quenching by positively charged polarons (a composite particle containing an electronic hole excitation and a surrounding lattice deformation). These charged polarons (hole polaron P^+ and negative polaron P^-) are generated by dissociation of singlet excitons (excitons are defined in Chapter 2,

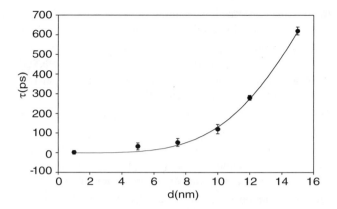

Figure 3.32. Observed distance dependence of excited-state relaxation (●) is displayed along with the fitted curve using the dipole–metal surface interaction model.

Section 2.1.5; singlet refers to the paired spin) as a result of field-assisted tunneling of a carrier to a nearby site. The transient response was found to be biexponential (time constants 26 ± 2 and 520 ± 40 μs) which suggested the presence of at least two distinct types of hole polarons with widely different mobilities.

An example of a space resolved study of excitation dynamics is the nanoscopic fluorescence resonance energy transfer (FRET), discussed in Chapter 2, using NSOM (Shubeita et al., 1999). These authors reported FRET occurring between the donor dye molecules deposited onto substrate and the acceptor dye molecules deposited onto the NSOM tip. FRET occurred only when the tip was in contact with the substrate and in a region of few tens of square nanometers.

3.8 COMMERCIALLY AVAILABLE SOURCES FOR NEAR-FIELD MICROSCOPE

Near-Field Imaging:
Thermomicroscopes: http://www.tmmicro.com/
Nanonics Imaging Ltd.:   http://www.nanonics.co.il
WITec Wissenschaftliche: http://www.witec.de
Triple-O Microscopy GmbH: http://www.triple-o.de

3.9 HIGHLIGHTS OF THE CHAPTER

- Near-field interactions probe optical interactions of a sample, distant from the source or an aperture, by a fraction of the wavelength of light. It forms the basis for near-field scanning optical microscopy (NSOM).

- NSOM provides an imaging resolution of ≤100 nm, which is significantly better than the diffraction limit imposed in a classical optical microscopy.
- The two approaches used for implementing NSOM are: (a) passing light through a subwavelength aperture, also known as aperture NSOM, and (b) localization and field enhancement around a metallic nanoparticle or a metallic tip, which is called apertureless NSOM.
- A number of theoretical methods are available to model the near-field distribution of the electromagnetic field.
- One theoretical method, presented here, is a multiple multipole (MMP) model that considers, for example, an aperture through which light emanates as a collection of individual domains of radiating multipoles as the source.
- The field distribution in the near-field has two components: (i) the central core for which the wavevector of light is a real quantity, and this portion of propagating light is also called "allowed light," and (ii) the outer periphery where the wavevector of light is imaginary; thus this field component, being evanescent, is sometimes called forbidden light.
- An aperature with a tapered fiber, coated with a metallic aluminum layer and having an opening of <100 nm, is often used for NSOM imaging.
- When the tapered fiber opening is used to illuminate the sample, it is called illumination mode NSOM. If the tapered fiber opening is used to collect the optical signal generated from the sample, it is called collection mode NSOM.
- Another variation for NSOM microscopy utilizes illumination of the sample in a total internal reflection geometry of a prism, using the evanescent wave generated at its base. The optical signal generated is collected by the opening of a tapered fiber. Hence this geometry, which in a sense utilizes tunneling of photons from the prism to the fiber tip, is also called photon scanning tunneling microscopy (PSTM).
- Near-field excitation and microscopy have proved valuable for the study of the spectroscopic properties of single molecules and nanometer-size single quantum dots.
- NSOM can be used to study nonlinear optical processes (such as second harmonic generation, third harmonic generation, and two-photon excited emission) in nanocrystals and nanoscopic domains.
- Apertureless NSOM utilizing a 5 to 15-nm-diameter metallic tip provides the convenient advantage of utilizing a far-field light illumination. Here, the field enhancement around the metallic tip, which is localized within a nanoscopic domain determined by the tip diameter, provides resolutions of ≤25 nm.
- Nanoscale enhancement of optical interactions in a near-field geometry can be obtained by using surface-plasmon resonance, when the sample is mounted on a metal film that has been deposited on a prism base.
- NSOM also provides the opportunity to conduct time-resolved and space-resolved studies of dynamics of excited state in nanodomains. An example is

the recovery of photobleaching in a nanoscopic domain, created by a strong pump beam and studied by absorption of a time-delayed weak probe beam.

REFERENCES

Ambrose, W. P., Goodwin, P. M., Martin, J. C., and Keller, R. A., Single Molecule Detection and Photochemistry on a Surface Using Near-Field Optical Excitation, *Phys. Rev. Lett.* **72,** 160–163 (1994).

Ash, E. A., and Nicholls, G., Super-Resolution Aperture Scanning Microscope, *Nature* **237,** 510–513 (1972).

Betzig, E., and Chichester, R. J., Single Molecules Observed by Near-Field Optical Microscopy, *Science* **262,** 1422–1425 (1993).

Betzig, E., and Trautman, J. K., Near-Field Optics: Microscopy, Spectroscopy, and Surface Modification Beyond the Diffraction Limit, *Science* **257,** 189–195 (1992).

Bouhelier, A., Beversluis, M., Hartschuh, A., and Novotny, L., Near-Field Second Harmonic Generation Induced by Local Field Enhancement, *Phys. Rev. Lett.* **90,** 13903-1–13903-4 (2003).

Bozhevolnyi, S. I., and Geisler, T., Near-Field Nonlinear Optical Spectroscopy of Langmuir Blodgett Films, *J. Opt. Soc. Am. A* **15,** 2156–2162 (1998).

Courjon, D., *Near-Field Microscopy and Near-Field Optics,* Imperial College Press, London, 2003.

Dickson, R. M., Cubitt, A. B., Tsien, R. Y., and Moerner, W. E., On/Off Blinking and Switching Behavior of Single Molecules of Green Fluorescent Protein, *Nature* **388,** 355–358 (1997).

Drexahage, K. H., *Progress in Optics,* North-Holland, New York, 1974.

Fillard, J. P., *Near Field Optics and Nanoscopy,* World Scientific, Singapore, 1997.

Gersten, J., and Nitzan, A., Spectroscopic properties of molecules interacting with small dielectric particles, *J. Chem. Phys.* **75,** 1139–1152 (1981).

Girard, C., and Dereux, A., Near-Field Optics Theories, *Rep. Prog. Phys.* **59,** 657–699 (1996).

Ha, T. J., Enderle, T., Ogletree, D. F., Chemla, D. S., Selvin, P. R., and Weiss, S., Probing the Interaction Between Two Single Molecules: Fluorescence Resonance Energy Transfer Between a Single Donor and a Single Acceptor, *Proc. Natl. Acad. Sci. USA* **93,** 6264–6268 (1996).

Ha, T. J., Ting, A. Y., Liang, Y., Caldwell, W. B., Deniz, A. A., Chemla, D. S., Schultz, P. G., and Weiss S., Single-Molecule Fluorescence Spectroscopy of Enzyme Conformational Dynamics and Cleavage Mechanism, *Proc. Natl. Acad. Sci. USA* **96,** 893–898 (1999).

Hartschuh, A., Sanchez, E. J., Xie, X. S., and Novotny, L., High-Resolution Near-Field Raman Microscopy of Single-Walled Carbon Nanotubes, *Phys. Rev. Lett.* **90,** 95503-1–95503-4 (2003).

Heinzelmann, H., and Pohl, D. W., Scanning Near-Field Optical Microscopy, *Appl. Phys. A* **59,** 89–101 (1994).

Ishii, Y., and Yanagida, T., Single Molecule Detection in Life Science, *Single Mol.* **1,** 5–16 (2000).

Ishijima, A., and Yanagida, T., Single Molecule Nanobioscience, *Trends Biomed. Sci.* **26**, 438–444 (2001).

Jiang, Y., Jakubczyk, D., Shen, Y., and Prasad, P. N., Nanoscale Nonlinear Optical Processes: Theoretical Modeling of Second-Harmonic Generation for Both Forbidden and Allowed Light, *Opt. Lett.* **25**, 640–642 (2000).

Kajikawa, K., Seki, K., and Ouchi, Y., *Tech. Dig. 5th Int. Conf. Near Field Opt. Rel. Tech.*, Shirahama, Japan, 2351–2352 (1998).

Kawata, S., Ohtsu, M., and Irie, M., eds., *Nano-Optics*, Springer, Berlin, 2002.

Lee, M., Tang, J., and Hoshstrasser, R. M., Fluorescence Lifetime Distribution of Single Molecules Undergoing Förster Energy Transfer, *Chem. Phys. Lett.* **344**, 501–508 (2001).

Matsuda, K., Ikeda, K., Saiki, T., Tsuchiya, H., Saito, H., and Nishi, K., Homogeneous Linewidth Broadening in a $In_{0.5}Ga_{0.5}As/GaAs$ Single Quantum Dot at Room Temperature Investigated Using a Highly Sensitive Near-Field Scanning Optical Microscope, *Phys. Rev. B* **63**, 121304 121307 (2001).

McNeill, J. D., and Barbara, P. F., NSOM Investigation of Carrier Generation, Recombination, and Drift in a Conjugated Polymer, *J. Phys. Chem. B* **106**, 4632–4639, (2002).

Moerner, W. E., Examining Nanoenvironments in Solids on the Scale of a Single, Isolated Impurity Molecule, *Science* **265**, 46–53 (1994).

Moerner, W. E., A Dozen Years of Single-Molecule Spectroscopy in Physics, Chemistry, and Biophysics, *J. Phys. Chem. B* **106**, 910–927 (2002).

Moerner, W. E., and Fromm, D. P., Methods of Single-Molecule Fluorescence Spectroscopy and Microscopy, *Rev. Sci. Int.* **74**, 3597–3619 (2003).

Moerner, W. E., and Kador, L., Optical Detection and Spectroscopy of Single Molecules in a Solid, *Phys. Rev. Lett.* **62**, 2535–2538 (1989).

Moerner, W. E., Plakhotnik, T., Imgartinger, T., Wild, U. P., Pohl, D., and Heckt, B., Near-Field Optical Spectroscopy of Individual Molecules in Solids, *Phys. Rev. Lett.* **73**, 2764–2767 (1994).

Novotny, L., Brian, R. X., and Xie, X. S., Theory of Nanometric Optical Tweezers, *Phys. Rev. Lett.* **79**, 645–648 (1997).

Novotny, L., Sanchez, E. J., and Xie, X. S., Near-Field Optical Imaging Using Metal Tips Illuminated by Higher-Order Hermite-Gaussian Beams, *Ultramicroscopy* **71**, 21–29 (1998).

Paesler, M. A., and Moyer, P. J., *Near-Field Optics: Theory, Instrumentation, and Applications*, John Wiley & Sons, New York, 1996.

Pohl, D. W., Denk, W., and Lanz, M., Optical Stethoscopy: Image Recording with Resolution $\lambda/20$, *Appl. Phys. Lett.* **44**, 651–653 (1984).

Raether, H., *Surface Plasmons*, Springer, New York, 1988.

Saiki, T., and Narita, Y., Recent Advances in Near-Field Scanning Optical Microscopy, *JSAP Int.* **5**, 22–29 (2002).

Saiki, T., Nishi, K., and Ohtsu, M., Low Temperature Near-Field Photoluminescence Spectroscopy of InGaAs Single Quantum Dots, *Jpn. J. Appl. Phys.* **37**, 1638–1642 (1998).

Sanchez, E. J., Novotny, L., and Xie, X. S., Near-Field Fluorescence Microscopy Based on Two-Photon Excitation with Metal Tips, *Phys. Rev. Lett.* **82**, 4014–4017 (1999).

Shen, Y., Lin, T.-C., Dai, J., Markowicz, P., and Prasad, P. N., Near-Field Optical Imaging of Transient Absorption Dynamics in Organic Nanostructures, *J. Phys. Chem. B* **107**, 13551–13553 (2003).

REFERENCES

Shen, Y., Swiatkiewicz, J., Winiarz, J., Markowicz, P., and Prasad, P. N., Second-Harmonic and Sum-Frequency Imaging of Organic Nanocrystals with Photon Scanning Tunneling Microscope, *Appl. Phys. Lett.* **77,** 2946–2948 (2000).

Shen, Y., Markowicz, P., Winiarz, J., Swiatkiewicz, J., and Prasad, P. N., Nanoscopic Study of Second Harmonic Generation in Organic Crystals with Collection-Mode Near-Field Scanning Optical Microscopy, *Opt. Lett.* **26,** 725–727 (2001a).

Shen, Y., Swiatkiewicz, J., Markowicz, P., and Prasad, P. N., Near-Field Microscopy and Spectroscopy of Third-Harmonic Generation and Two-Photon Excitation in Nonlinear Organic Crystals, *Appl. Phys. Lett.* **79,** 2681–2683 (2001b).

Shen, Y., Swiatkiewicz, J., Lin, T.-C., Markowicz, P., and Prasad, P. N., Near-Field Probing Surface Plasmon Enhancement Effect on Two-Photon Emission, *J. Phys. Chem. B* **106,** 4040–4042 (2002).

Shubeita, G. T., Sekafskii, S. K., Chergui, M., Dietler, G., and Letokhov, V. S., Investigation of Nanolocal Fluorescence Resonance Energy Transfer for Scanning Probe Microscopy, *Appl. Phys. Lett.* **74,** 3453–3455 (1999).

Synge, E. H., A Suggested Method for Extending Microscopic Resolution into the Ultra-microscopic Region, *Philos. Mag.* **6,** 356–362 (1928).

Vanden Bout, D. A., Kerimo, J., Higgins, D. A., and Barbara, P. F., Near-Field Optical Studies of Thin-Film Mesostructured Organic Materials, *Acc. Chem. Res.* **30,** 204–212 (1997).

Wokaun, A., Lutz, H. P., King, A. P., Wild, U. P., and Ernst, R. R., Energy transfer in surface enhanced luminescence, *J. Chem. Phys.* **79,** 509–514 (1983).

CHAPTER 4

Quantum-Confined Materials

Nanomaterials constitute a major area of nanophotonics. As we shall see from this chapter and a number of others to follow, nanoscale optical materials involve a highly diverse range of nanostructure designs and cover a broad range of optical applications. By manipulation of their nanostructure, optical properties can be judiciously controlled to enhance a specific photonic function and/or introduce a new photonic manifestation as well as to allow integration of many functions to achieve multifunctionality.

The contents of this chapter deal with the effect of dimensional confinement on the optical properties of materials whose bulk phases exhibit a relatively free motion of electrons. A more precise description of these materials is in terms of delocalized electrons that have their wavefunctions spread over a large distance. Examples are (a) semiconductors with delocalized electrons and holes with an imposed periodic potential and (b) conjugated organic molecules containing delocalized π electrons. These materials, when confined with a reduction in their sizes, exhibit strong effects on their optical properties. Sections 4.1 to 4.6 deal with nanostructures of inorganic semiconductors. Although every effort has been made to minimize theoretical details and introduce the concepts rather qualitatively, these sections still utilize a fair amount of semiconductor physics. Readers having little familiarity with this topic may find it difficult to fully appreciate the contents of Sections 4.1 to 4.6. However, in order to proceed to further chapters, it is not necessary for a reader to fully understand all the details provided in these sections. These sections may even be skipped by those not having familiarity with or interest in inorganic semiconductors.

Section 4.1 deals with inorganic semiconductors, nanoscale confined in one, two, or three dimensions to produce artificial structures (not naturally occurring) called quantum wells, quantum wires, and quantum dots. A donut-shaped arrangement of quantum dots, called a *quantum ring,* is also discussed.

The quantum confinement effect, conceptually described in Chapter 2 by the model of an electron in a box, produces many optical manifestations that are useful for various technological applications. These manifestations are described for inorganic semiconductors in Section 4.2. Section 4.3 discusses the effect caused by differences between the dielectric constants of the quantum-confined structures and the surrounding media.

Section 4.4 describes an ordered periodic (repeat) arrangement, a superlattice, of the quantum-confined structures such as a stack of quantum-confined wells, each

Nanophotonics, by Paras N. Prasad
ISBN 0-471-64988-0 © 2004 John Wiley & Sons, Inc.

being confined in one dimension. The periodic structures called *multiple quantum wells* represent the case where the separation between two quantum wells is smaller than the electron mean free path, a distance traveled by an electron without being scattered from its path. The quantum wells in these superlattices are electronically coupled; their electronic states interact to give rise to new optical transitions and manifestations. Section 4.4 covers superlattices of quantum wells, quantum wires, and quantum dots, along with their electronic and optical properties.

Section 4.5 describes core-shell-type quantum dots, which are nanostructures comprised of a quantum dot with a shell of a wider bandgap semiconductor around it. This overcoating improves the luminescence efficiency of the quantum dot. Quantum dot–quantum well type structures are also discussed.

A major commercial application of the quantum-confined semiconductors is in producing energy-efficient and highly compact lasers. These lasers also provide the advantage of a broad spectral coverage by controlling the type and composition of the semiconductors, the size and the dimensionality of confinement, and, in some cases, the temperature. This subject is covered in Section 4.6.

Section 4.7 discusses the confinement effect in organic structures. An analogy between the quantum-confined inorganic semiconductors and organic conjugated structures involving π-electron delocalization is drawn in this section. Some examples are provided to illustrate the confinement effect and its utility.

Some suggested general references on this subject are:

Weisbuch and Vinter (1991): *Quantum Semiconductor Structures*
Singh (1993): *Physics of Semiconductors and Their Heterostructures*
Gaponenko (1999): *Optical Properties of Semiconductor Nanocrystals*
Borovitskaya and Shur, eds. (2002): *Quantum Dots*

4.1 INORGANIC SEMICONDUCTORS

Inorganic semiconductors constitute the most widely studied examples of quantum-confined structures that are already in the marketplace. Quantum-confined semiconductors provide an added dimension to the highly active area of "bandgap engineering," which deals with the manipulation of the semiconductor bandgap. A brief description of the bandgap properties of a semiconductor has already been provided in Section 2.1.4 of Chapter 2.

To date, quantum-confined structures of many different types of semiconductors have been produced. Semiconductors are often classified by the periodic table groups to which they belong. Table 4.1 lists some semiconductors, their bandgap energies in the bulk phase, the corresponding wavelengths, the excitation binding energies, and the exciton Bohr radii.

These properties are relevant in the description of quantum-confined structures. The bandgap can also be tuned by varying the composition of a ternary semiconductor such as $Al_xGa_{1-x}As$, which for $x = 0.3$ has a bandgap of 1.89 eV compared with 1.52 eV for pure GaAs. The different types of confinement used for inorganic semiconductors are described here in a number of subsections.

4.1 INORGANIC SEMICONDUCTORS

Table 4.1. Semiconductor Material Parameters

Material	Periodic Table Classification	Bandgap Energy (eV)	Bandgap Wavelength (μm)	Exciton Bohr Radius (nm)	Exciton Binding Energy (meV)
CuCl	I–VII	3.395	0.36	0.7	190
CdS	II–VI	2.583	0.48	2.8	29
CdSe	II–VI	1.89	0.67	4.9	16
GaN	III–V	3.42	0.36	2.8	
GaP	III–V	2.26	0.55	10–6.5	13–20
InP	III–V	1.35	0.92	11.3	5.1
GaAs	III–V	1.42	0.87	12.5	5
AlAs	III–V	2.16	0.57	4.2	17
Si	IV	1.11	1.15	4.3	15
Ge	IV	0.66	1.88	25	3.6
$Si_{1-x}Ge_x$	IV	$1.15 - 0.874x + 0.376x^2$	$1.08 - 1.42x + 3.3x^2$	$0.85 - 0.54x + 0.6x^2$	$14.5 - 22x + 20x^2$
PbS	IV–VI	0.41	3	18	4.7
AlN	III–V	6.026	0.2	1.96	80

Sources of Data

GaN: H. Morkoç, *Nitride semiconductors and Devices,* Springer-Verlag, New York, 1999.
PbS: I. Kang and F. W. Wise, Electronic Structure and Optical Properties of PbS and PbSe Quantum Dots, *J. Opt. Soc. Am. B* **14**(7), 1632–1646, (1997).
PbS: H. Kanazawa and S. Adachi, Optical Properties of PbS, *J. Appl. Phys.* **83**(11), 5997–6001 (1998).
SiGe: D. J. Robbins, L. T. Canham, S. J. Barnett, and A. D. Pitt, Near-Band-Gap Photoluminescence from Pseudomorphic $Si_{1-x}Ge_x$ Layers on Silicon, *J. Appl. Phys.* **71**(3), 1407–1414 (1992).
InP: P. Y. Yu and M. Cardona, *Fundamentals of Semiconductors,* Springer-Verlag, New York, 1996.
Ge: D. L. Smith, D. S. Pan, and T. C. McGill, Impact Ionization of Excitons in Ge and Si, *Phys. Rev. B* **12**(10), 4360–4366, (1975).
AlN: K. B. Nam, J. Li, M. L. Nakarmi, J. Y. Lin, and H. X. Jiang, Deep Utraviolet Picosecond Time-Resolved Photoluminescence Studies of AlN Epilayers, *Appl. Phys. Lett.* **82**(11), 1694–1696 (2003).
AlAs: S. Adachi, GaAs, AlAs, and $Al_xGa_{1-x}As$: Material Parameters for Use in Research and Device Applications, *J. Appl. Phys.* **58**(3), R1–R29 (1985).
GaP: D. Auvergne, P.Merle, and H. Mathieu, Phonon-Assisted Transitions in Gallium Phosphide Modulation Spectra, *Phys. Rev. B* **12**(4), 1371–1376 (1975).

4.1.1 Quantum Wells

Quantum wells are structures in which a thin layer of a smaller bandgap semiconductor is sandwiched between two layers of a wider bandgap semiconductor. The heterojunction between the smaller and the wider bandgap semiconductors forms a potential well confining the electrons and the holes in the smaller bandgap material region. This is the case of a type I quantum well. In a type II quantum well, the electrons and the holes are confined in different layers. Thus the motions of the electrons and the holes are restricted in one dimension (along the thickness direction). The model of an electron in a one-dimensional box discussed in Chapter 2 can be applied in this case, except that the potential barriers instead of being infinite are finite, determined by the difference in the bandgaps of the two semiconductors. This system represents a two-dimensional electron gas (2DEG), when electrons are present in the conduction band.

Figure 4.1 shows the schematic of a representative quantum well, which here is the extensively studied case of AlGaAs/GaAs. In this case the narrower bandgap semiconductor, GaAs of the thickness l, is confined between two layers of the wider bandgap ternary semiconductor $Al_xGa_{1-x}As$, where the composition x can be varied to control the potential barrier height. These III–V semiconductors are more popular quantum well systems because a large difference in the bandgap yields a large confinement effect. Also, the two constituents are nearly lattice matched over a wide range of x which minimizes lattice strain between the two layers.

As the left-hand side of Figure 4.1 shows, a finite potential barrier modifies the behavior of the energy eigenvalues and the wavefunctions compared to that for an infinite potential barrier. The following features of the quantum well are to be noted.

- For energies $E < V$, the energy levels of the electron are quantized for the direction z of the confinement; hence they are given by the model of particle in a one-dimensional box. The electronic energies in the other two dimensions (x and y) are not discrete and are given by the effective mass approximation discussed in Chapter 2. Therefore, for $E < V$, the energy of an electron in the conduction band is given as

$$E_{n,k_x,k_y} = E_C + \frac{n^2h^2}{8m_e^*l^2} + \frac{\hbar^2(k_x^2 + k_y^2)}{2m_e^*} \tag{4.1}$$

where $n = 1, 2, 3$ are the quantaum numbers. The second term on the right-hand side represents the quantized energy; the third term gives the kinetic energy of the electron in the x–y plane in which it is relatively free to move. The

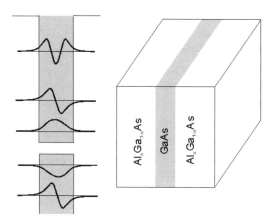

Figure 4.1. Right: Example of a type I quantum well. **Left:** electron (*top*) and hole (*bottom*) wavefunctions.

symbols used are as follows: m_e^* is the effective mass of electron, and E_C is the energy corresponding to the bottom of the conduction band.

Equation (4.1) shows that for each quantum number n, the values of wavevector components k_x and k_y form a two-dimensional band structure. However, the wavevector k_z along the confinement direction z takes on only discrete values, $k_z = n\pi/l$. Each of the bands for a specific value of n is called a sub-band. Thus n becomes a sub-band index. Figure 4.2 shows a two-dimensional plot of these sub-bands.

- For $E > V$, the energy levels of the electron are not quantized even along the z direction. Figure 4.1 shows that for the AlGaAs/GaAs quantum well, the quantized levels $n = 1$–3 exist, beyond which the electronic energy level is a continuum. The total number of discrete levels is determined by the width l of the well and the barrier height V.
- The holes behave in analogous way, except their quantized energy is inverted and the effective mass of a hole is different. Figure 4.1 also shows that for the holes, two quantized states with quantum numbers $n = 1$ and 2 exist for this particular quantum well (determined by the composition of AlGaAs and the width of the well). In the case of the GaAs system, two types of holes exist, determined by the curvature (second derivative) of the band structure. The one with a smaller effective mass is called a *light hole* (lh), and the other with a heavier effective mass is called a *heavy hole* (hh). Thus the $n = 1$ and $n = 2$ quantum states actually are each split in two, one corresponding to lh and the other to hh.
- Because of the finite value of the potential barrier ($V \neq \infty$), the wavefunctions, as shown for levels $n = 1, 2$, and 3 in the case of electrons and levels $n = 1$ and 2 in the case of holes, do not go to zero at the boundaries. They extend into the region of the wider bandgap semiconductor, decaying exponentially

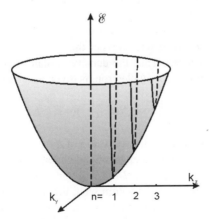

Figure 4.2. Plot of energy sub-bands for electrons in the conduction band of a quantum well.

into this region. This electron leakage behavior has already been discussed in Section 2.1.3 of Chapter 2.

- The lowest-energy band-to-band optical transition (called the interband transition) is no longer at E_g, the energy gap of the smaller bandgap semiconductor, GaAs in this case. It is at a higher energy corresponding to the difference between the lowest energy state ($n = 1$) of the electrons in the conduction band and the corresponding state of the holes in the valence band. The effective bandgap for a quantum well is defined as

$$E_g^{\text{eff}} = (E_C - E_V) + \frac{h^2}{8l^2}\left(\frac{1}{m_e^*} + \frac{1}{m_h^*}\right) \quad (4.2)$$

In addition, there is an excitonic transition below the band-to-band transition. These transitions are modifications of the corresponding transitions found for a bulk semiconductor. In addition to the interband transitions, new transitions between the different sub-bands (corresponding to different n values) within the conduction band can occur. These new transitions, called intraband or inter-sub-band transitions, find important technologic applications such as in quantum cascade lasers. The optical transitions in quantum-confined structures are further discussed in the next section.

- Another major modification, introduced by quantum confinement, is in the density of states. The density of states $D(E)$, defined by the number of energy states between energy E and $E + dE$, is determined by the derivative $dn(E)/dE$. For a bulk semiconductor, the density of states $D(E)$ is given by $E^{1/2}$. For electrons in a bulk semiconductor, $D(E)$ is zero at the bottom of the conduction band and increases as the energy of the electron in the conduction band increases. A similar behavior is exhibited by the hole, for which the energy dispersion (valence band) is inverted. Hence, as the energy is moved below the valence band maximum, the hole density of states increases as $E^{1/2}$. This behavior is shown in Figure 4.3, which also compares the density of states for electrons (holes) in a quantum well. The density of states is a step function because of the discreteness of the energy levels along the z direction (confinement direction). Thus the density of states per unit volume for each sub-band, for example for an electron, is given as a rise in steps of

$$D(E) = m_e^*/\pi^2 \quad \text{for } E > E_1 \quad (4.3)$$

The steps in $D(E)$ occur at each allowed value of E_n given by Equation (4.1), for k_x and $k_y = 0$, then stay constant for each sub-band characterized by a specific n (or k_z). For the first sub-band with $E_n = E_1$, $D(E)$ is given by Eq. (4.3). This step-like behavior of $D(E)$ implies that for a quantum well, the density of states in the vicinity of the bandgap is relatively large compared to the case of a bulk semiconductor for which $D(E)$ vanishes. As is discussed below, a major manifestation of this modification of the density of states is in the strength of optical transition. A major factor

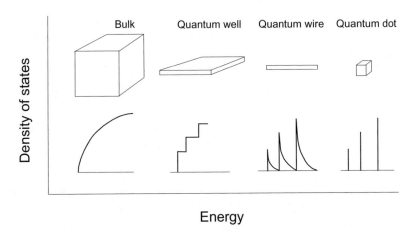

Figure 4.3. Density of states for electrons in bulk conduction band together with those in various confined geometries.

in the expression for the strength of optical transition (often defined as the oscillator strength) is the density of states. Hence, the oscillator strength in the vicinity of the bandgap is considerably enhanced for a quantum well compared to a bulk semiconductor. This enhanced oscillator strength is particularly important in obtaining laser action in quantum wells, as discussed in Section 4.4.

4.1.2 Quantum Wires

Quantum wires represent two-dimensional confinement of electrons and holes. Such confinement permits free-electron behavior in only one direction, along the length of the wire (say the y direction). For this reason, the system of quantum wires describes a one-dimensional electron gas (IDEG) when electrons are present in the conduction band. Both III–V (e.g., InP) and II–VI (e.g., CdSe) quantum wires have been prepared. Although quantum wires can be cylindrical with a circular cross section, as well as rectangular or square in the lateral x–z plane, a representative model system presented here consists of a rectangular wire with dimensions l_x and l_z.

The energy of a one-dimensional electron gas thus consists of a sum of two parts: (i) that due to a continuous band value given by the effective mass approximation and (ii) those corresponding to the quantized values of an electron in a two-dimensional box of dimensions l_x and l_z, as described in Chapter 2. Therefore, the energy is labeled by two quantum numbers n_1 and n_2 corresponding to confinements along the x and the z directions, with k_y corresponding to the wavevector along the y direction. Hence,

$$E_{n_1,n_2,k_y} = E_C + \frac{n_1^2 h^2}{8 m_e^* l_x^2} + \frac{n_2^2 h^2}{8 m_e^* l_z^2} + \frac{\hbar^2 k_y^2}{2 m_e^*} \qquad (4.4)$$

where both n_1 and n_2 can assume values 1, 2, 3. . . . The lowest sub-band then corresponds to $n_1 = n_2 = 1$. The energy corresponding to the bottom of each sub-band ($k_y = 0$) is given as

$$E_{n_1,n_2} = E_C + \frac{h^2}{8m_e^*}\left(\frac{n_1^2}{l_x^2} + \frac{n_2^2}{l_z^2}\right) \quad (4.5)$$

If l_x is significantly larger than l_z, the n_1 levels form a staircase of small steps within the widely separated sub-bands corresponding to various values n_2—that is, quantization along the z direction. If $l_x = l_z$, which describes a square-cross-section wire, the energy steps corresponding to the same values of n_1 and n_2 are indistinguishable. In this case, many energy levels corresponding to different possible sets of n_1 and n_2 (e.g., $n_1 = 1$, $n_2 = 2$, and $n_1 = 2$, $n_2 = 1$) have the same energy value, a case called *degeneracy*.

Quantum rod is another term used to represent two-dimensional confinement of a semiconductor (Li et al., 2001; Htoon et al., 2003). Quantum rods can be described as quantum wires, where the lengths l_x and l_z are not very small compared to the length l_y of the wire. It still may be a factor of 10 smaller than the length of the rod to produce a self-supporting rod structure.

The two-dimensional confinement and thus a one-dimensional electron gas behavior produces, again, a major modification for the density of states that is different from that of a two-dimensional electron gas (quantum well). The density of states, $D(E)$, for a quantum wire can be shown to have an inverse energy dependence $E^{-1/2}$ compared to the $E^{1/2}$ dependence for a three-dimensional electron gas (bulk semiconductor). Thus

$$D(E) \propto (E - E_{n_1,n_2})^{-1/2} \quad (4.6)$$

For each sub-band, characterized by a specific set of values n_1 and n_2, the density of states has a singularity near $k_y = 0$; that is, it has a large value, which then decays as $E^{-1/2}$, as k_y assumes nonzero values for that sub-band. The $D(E)$ behavior for a quantum wire is compared with that for a quantum well and a bulk semiconductor in Figure 4.3 which also exhibits $D(E)$ for a quantum dot to be discussed later.

A major manifestation of a large $D(E)$ at the bottom of each sub-band ($k_y = 0$) is again an increase of the strength of optical transition (increase in the oscillator strength). The singularity (sharp peak) in the density of states produces sharp peaks in the optical spectra, hence improved optical efficiency (e.g., emission) as compared to their two-dimensional (quantum well) and three-dimensional (bulk) counterparts. Additional confinement in the quantum wire case, compared to a quantum well, also leads to an increase in the exciton binding energy.

4.1.3 Quantum Dots

Quantum dots represent the case of three-dimensional confinement, hence the case of an electron confined in a three-dimentional quantum box, typically of dimensions ranging from nanometers to tens of nanometors. These dimensions are smaller than the de Broglie wavelength thermal electrons. A 10-nm cube of GaAs would contain about 40,000 atoms. A quantum dot is often described as an artificial atom because the electron is dimensionally confined just like in an atom (where an electron is confined near the nucleus) and similarly has only discrete energy levels. The electrons in a quantum dot represent a zero-dimensional electron gas (0DEG). The quantum dot represents a widely investigated group of confined structures with many structural variations to offer. The size dependence of the lower excited electronic state of semiconductor nanocrystals has been a subject of extensive investigation for a long time (Brus, 1984). Quantum dots of all types of semiconductors listed in Table 4.1 have been made. Also, as we shall see in Chapter 7, a wide variety of methods have been utilized to produce quantum dots. Recent efforts have also focused on producing quantum dots in different geometric shapes to control the shapes of the potential barrier confining the electrons (and the holes) (Williamson, 2002).

A simple case of a quantum dot is a box of dimensions l_x, l_y, and l_z. The energy levels for an electron in such a case have only discrete values given as

$$E_n = \frac{h^2}{8m_e}\left[\left(\frac{n_x}{l_x}\right)^2 + \left(\frac{n_y}{l_y}\right)^2 + \left(\frac{n_z}{l_z}\right)^2\right] \qquad (4.7)$$

where the quantum numbers l_x, l_y, and l_z, each assuming the integral values 1, 2, 3, characterize quantization along the x, y, and z axes, respectively. Consequently, the density of states for a zero-dimensional electron gas (for a quantum dot) is a series of δ functions (sharp peaks) at each of the allowed confinement state energies.

$$D(E) \propto \sum_{E_n} \delta(E - E_n) \qquad (4.8)$$

In other words, $D(E)$ has discrete (nonzero) values only at the discrete energies given by Eq. (4.8). This behavior for $D(E)$ is also shown in Figure 4.3. The discrete values of $D(E)$ produce sharp absorption and emission spectra for quantum dots, even at room temperature. However, it should be noted that this is idealized, and the singularities are often removed by inhomogeneous and homogeneous broadening of spectroscopic transitions (see Chapter 6).

Another important aspect of a quantum dot is its large surface-to-volume ratio of the atoms, which can vary as much as 20%. An important consequence of this feature is strong manifestation of surface-related phenomena.

Quantum dots are often described in terms of the degree of confinement. The strong confinement regime is defined to represent the case when the size of the quantum dot (e.g., the radius R of a spherical dot) is smaller than the exciton Bohr radius a_B. In this case, the energy separation between the sub-bands (various quantized levels of electrons and holes) is much larger than the exciton binding energy. Hence, the electrons and holes are largely represented by the energy states of their respective sub-bands. As the quantum dot size increases, the energy separation between the various sub-bands becomes comparable to and eventually less than the exciton binding energy. The latter represents the case of a weak confinement regime where the size of the quantum dot is much larger than the exciton Bohr radius. The electron–hole binding energy in this case is nearly the same as in the bulk semiconductor.

4.1.4 Quantum Rings

A quantum ring represents a donut-shaped quantum dot (Warburton et al., 2000). It can be visualized as a quantum wire bent into a loop. This difference in the topology of a quantum ring compared to a regular quantum dot (i.e., having a hole in the middle) produces important consequences when an external magnetic field is applied. The magnetic flux penetrating the interior of the ring strongly influences the nature of the electronic states.

In the case of a quantum ring, a donut-shaped ring of a narrower bandgap semiconductor (e.g., InAs) is surrounded by a wider bandgap semiconductor (GaAs), which produces a potential barrier to confine the electrons and holes in the ring (or donut)-shaped region. There has been considerable interest in recent years on the tunability of the electronic states of quantum rings and their optical properties (Warburton et al., 2000; Pettersson et al., 2000).

4.2 MANIFESTATIONS OF QUANTUM CONFINEMENT

Quantum confinement produces a number of important manifestations in the electronic and optical properties of semiconductors. These manifestations have found technological applications. Since the focus of this book is on photonics, only those manifestations that are important in relation to the optical properties will be highlighted in this section. Discussed here in various subsections are the optical properties, the nonlinear optical behavior, and the quantum-confined Stark effect (electric-field-induced changes) on the optical spectra.

4.2.1 Optical Properties

The optical properties discussed in this subsection are summarized in Table 4.2. First, important manifestations of quantum confinement on the optical properties are described here.

Size Dependence of Optical Properties. As discussed above, quantum confinement produces a blue shift in the bandgap as well as appearance of discrete sub-bands corresponding to quantization along the direction of confinement. As the dimensions of confinement increase, the bandgap decreases; hence the interband transitions shift to longer wavelengths, finally approaching the bulk value for a large width.

4.2 MANIFESTATIONS OF QUANTUM CONFINEMENT

Table 4.2. Optics of Quantum Confined Semiconductors

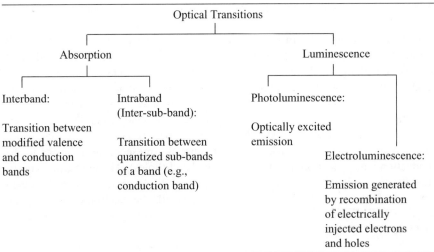

Increase of Oscillator Strength. As has been explained in Section 4.1, quantum confinement produces a major modification in the density of states both for valence and conduction bands. Instead of a continuous, smooth distribution of the density of states, the energy states are squeezed in a narrow energy range. This packing of energy states near the bandgap becomes more pronounced as the dimensions of confinement increase from quantum well, to quantum wire, to quantum dots. For the last, the density of states has nonzero values only at discrete (quantized) energies. The oscillator strength of an optical transition for an interband transition depends on the joint density of states of the levels in the valence band and the levels in the conduction bands, between which the optical transition occurs. Furthermore, it also depends on the overlap of the envelope wavefunctions of electrons and holes. Both these factors produce a large enhancement of oscillator strength upon quantum confinement. This effect is quite pronounced for quantum wires and quantum dots, which are more confined structures (two-dimensionally and three-dimensionally).

New Intraband Transitions. These transitions, corresponding to promotion of electrons from one level to another in the conduction band or hole from one level to another in the valence band, are also known as free carrier absorption in the bulk semiconductor. They depend on the presence of free carriers (electrons in the conduction band or holes in the valence band) as a result of impurity doping (excess electrons or holes) or as a result of charge injection introduced by a bias field (photoinjection). In the bulk, such transitions from one k level to another k level of the conduction band (or the valence band) require a change of quasi-momentum k and thus become allowed only by coupling with lattice phonons, which can provide or take up this momentum change. Thus, these processes are generally weak compared to interband transitions in the same bulk, because interband transitions do not require a change in k.

In quantum-confined structures such as a quantum well, there are sub-bands characterized by the different quantum numbers ($n = 1, 2, \ldots$) corresponding to quantization along the direction of confinement (growth). Thus for the conduction band, an electron can make a jump from one sublevel to another without changing its now two-dimensional quasi-momentum k. These new transitions are in IR and have been utilized to produce inter sub-band detectors and lasers, the most interesting of which are quantum cascade lasers described in Section 4.6. These intraband (or inter-sub-band) transitions still require the presence of a carrier in the conduction band (electrons) or in the valence band (holes). The absorption coefficient for an intraband transition increases rapidly with decreasing width of a quantum well. However, as the well size becomes small, the electronic states are no longer confined within the well, which produces a leveling off of the absorption coefficient.

Increased Exciton Binding. Quantum confinement of electrons and holes also leads to enhanced binding between them and thereby produces increased exciton binding energy compared to the exciton binding energy for the bulk sample, listed in Table 4.1.

For example, a simple theoretical model (variational calculations) predicts that the Coulomb interactions between free electrons and holes in a two-dimensional system (quantum wells) is four times that in a three-dimensional system (bulk). However, the actual binding energy is somewhat smaller than the four times bulk value, because of the wavefunction penetration into the barriers (McCombe and Petrou, 1994). As discussed in Chapter 2, this binding produces excitonic states just below the bandgap, giving rise to sharp excitonic peaks at temperatures where the exciton binding energy is higher than the thermal energy. Thus, excitonic resonances are very pronounced in quantum-confined structures and, in the strong confinement conditions, can be seen even at room temperature.

Increase of Transition Probability in Indirect Gap Semiconductor. As we discussed in Chapter 2, an optical transition for an indirect bandgap semiconductor requires a change of quasi-momentum and thus involves the participation of phonons. Silicon is an example of an indirect gap semiconductor. Consequently, the emission produced by the transition of an electron from the conduction band to the valence band is either extremely weak or nonexistent in the bulk form of an indirect gap semiconductor. However, in the quantum-confined structures, confinement of electrons produces a reduced uncertainty Δx in its position and, consequently, produces a larger uncertainty Δk in its quasi-momentum. Confinement, therefore, relaxes the quasi-momentum Δk selection rule, thus allowing enhanced emission to be observed in porous silicon and silicon nanoparticles. The area of luminescent silicon nanoparticles is currently witnessing a great deal of activity.

4.2.2 Examples

Figure 4.4 provides an example of the quantum confinement effect on optical absorption of a GaAs/AlGaAs quantum well. The bulk GaAs absorption is shown in Figure 4.4a, where the sharp transition at the band edge is due to excitons. The dashed curve represents the distribution of the density of states. Figure 4.4b represents the schematic absorption band for a quantum well. In Figure 4.5a–c, the absorption spectra at 2K, of GaAs/$Al_{0.2}Ga_{0.8}$As quantum wells of different widths are shown (Dingle et al., 1974). The spectrum in Figure 4.5a is that of a thick quantum well (l_z = 400 nm) which is essentially like that of bulk GaAs (Figure 4.4a). For a much thinner quantum well (width 21 nm), the quantization effect can readily be seen in the spectra (Figure 4.5b), where transitions involving quantized energy sub-bands having sub-band indices as high as $n = 4$ can be seen. Hence up to $n = 4$, the electronic energy states are bound states (quantized). For each sub-band, the sharp peak again corresponds to excitonic transitions, again just at the band edge. For an even thinner quantum well of width 14 nm the spectrum shown in Figure 4.5c exhibits a shift of the features toward higher energy (blue shift) and the separation between the transitions corresponding to different sub-bands (different n values) increases, as expected from an l_z^{-2} dependence. In addition, for the thinner sample, the $n = 1$ absorption shows some structures that get further resolved in other reports of even narrower width ($l_z < 14$ nm). This splitting corresponds to absorptions involving heavy- and light-hole bands. These two types of holes with different effective masses have been discussed in Section 2.1.1.

Quantum wires have been prepared both in free-standing form as well as embedded in a dielectric medium of a higher bandgap. They provide a strong, confining cylindrical potential for electrons and holes and offer substantial promises as building blocks for nanoelectronic and nanophotonic devices (Lieber, 2001; Hu et al., 1999). The high aspect ratio offered by the cylindrical symmetry produces a giant polarization anisotropy in the photoluminescence as reported for InP by Lieber's group (Wang et al., 2001).

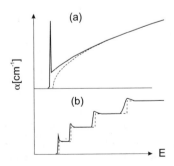

Figure. 4.4. (a) Bulk GaAs band-edge absorption including exciton effects. (b) Schematic band-edge absorption for a quantum well. The dashed curves in each case are the density of states.

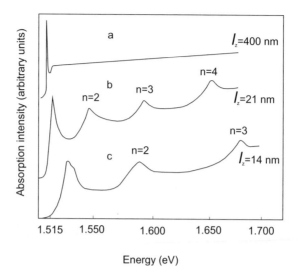

Figure 4.5. Absorption spectra at 2 K of GaAs/Al$_{0.2}$Ga$_{0.8}$As quantum wells of different width: (a) 400 nm, (b) 21 nm, (c) 14 nm. From Dingle et al. (1974), reproduced with permission.

Lieber and co-workers (Wang et al., 2001) synthesized single-crystal InP nanowires via a laser-assisted catalytic growth to obtain monodisperse nanowires of diameters 10, 15, 20, 30, and 50 nm. Figure 4.6 shows their results of the excitation and photoluminescence studies on 15-nm-diameter InP nanowires. These spectra are recorded in two orthogonal polarizations and clearly exhibit a strong anisotropy. Lieber and co-workers used a dielectric contrast model to explain the polarization anisotropy. In this model, a nanowire is treated as an infinite dielectric cylinder in a

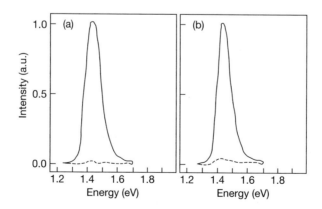

Figure 4.6. Excitation (a) and photoluminescence (b) spectra of 15-nm-diameter InP nanowires for two orthogonal polarizations are represented by the solid and dotted curves in each case. From Wang et al. (2001), reproduced with permission.

vacuum. The field intensity $|E^2|$ due to light in such a dielectric is strongly attenuated inside the nanowire for the perpendicular polarization $E_\perp$, but is unaffected for the parallel polarization $E_\parallel$.

The work by Lieber's group (Wang et al., 2001) also exhibits the quantum confinement effect as a diameter-dependent shift in the photoluminescence energy (wavelength of emission maximum) from the bulk bandgap value of 1.35 eV for InP, of diameters < 20 nm. A giant polarization anisotropy is reported for nanowires with diameters 10–50 nm. Lieber and co-workers (Zhong et al., 2003) have produced p-doped (to create holes) and n-doped (to add electrons) GaN nanowires for electronic and photonic nanodevices.

Quantum dots have been prepared by a wide variety of methods, as discussed in Chapter 7. They have been prepared as encapsulated species that can be dispersed in a liquid medium (for imaging), self-assembled in the form of a superlattice (discussed in Section 4.3), grown on a substrate in various geometric shapes (square, spherical, pyramidal), embedded in a wider bandgap semiconductor, or incorporated in polymers and glasses. The quantum dots provide spectacular examples of tunability of their optical properties by changing the size and composition of the nanocrystals (Brus, 1984). An excellent coverage of the luminescence photophysics in quantum dot have been given by Brus and co-workers (Nirmal and Brus, 1999). Figure 4.7 shows the luminescence properties of CdSe, InP, and InAs nanocrystals of various sizes as reported by Alivisatos and co-workers (Bruchez et al., 1998). These nanocrystals are encapsulated (surface coated) by surfactants as discussed in Chapter 7. For the smallest-size CdSe nanoparticles, a blue color emission is seen. On increasing the size of the nanocrystals of CdSe, the bandgap (gap between quantized levels) decreases, resulting in a red shift (shift to longer wavelength) of the emission peak. Further shift to longer wavelengths beyond 700 nm is achieved by changing the material from CdSe to InP and then to InAs. Thus, a combination of changes in the size and material com-

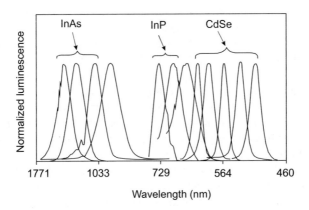

Figure 4.7. The luminescence properties of InAs, InP and CdSe nanocrystals of decreasing sizes from left to right in each case. From Bruchez et al. (1998), reproduced with permission.

position of the quantum dot can allow one to tune the emission wavelength from UV to IR.

Silicon nanoparticles are another class (group IV) of semiconductor nanostructures extensively studied and continuing to attract worldwide attention. The interest in silicon nanostructures was initiated by the reports on luminescence from porous silicon (which consist of nanodomains of silicon) (Canham, 1990; Lockwood, 1998). Brus and co-workers published a series of papers in which they prepared silicon nanoparticles by high-temperature decomposition of disilane and studied their photoluminescence mechanism (Wilson et al., 1993; Littau et al., 1993; Brus, 1994). A variety of methods have been used to prepare and characterize silicon nanoparticles (Heath, 1992; Bley and Kauzlarich, 1996; Carlisle et al., 2000; Ding et al., 2002; English et al., 2002; Baldwin et al., 2002).

In order to achieve efficient visible photoluminescence, it is believed that silicon nanoparticles must be smaller than 5 nm, and their surface must be 'properly passivated' such that there are no nonradiative recombination sites on it. The mechanism of photoluminescence in silicon nanocrystals and the effect of surface passivation on light emission from them remain topics of active research and debate.

In our lab, silicon nanoparticles with bright visible photoluminescence are being prepared by a new combined vapor-phase and solution-phase process, using only inexpensive commodity chemicals (Li et al., 2003; Swihart et al., 2003). CO_2 laser-induced pyrolysis of silane is used to produce Si nanoparticles at high rates (20–200 mg/h). Particles with an average diameter as small as 5 nm can be prepared directly by this method. Etching these particles with mixtures of hydrofluoric acid (HF) and nitric acid (HNO_3) reduces the size and passivates the surface of these particles such that they exhibit bright visible luminescence at room temperature. The wavelength of maximum photoluminescence (PL) intensity can be controlled from above 800 nm to below 500 nm by controlling the etching time and conditions. Particles with blue and green emission are prepared by rapid thermal oxidation of orange-emitting particles. These silicon nanoparticles have been successfully homogeneously dispersed in a number of matrix for photonics applications. Also, stable dispersion of these nanoparticles in a variety of solvents has been accomplished for biophotonics (bioimaging) applications.

An area of considerable activity focuses on wide bandgap semiconductors such as GaN, which provides lasing in the UV-blue region. Such shorter wavelengths are useful for high-density optical recording, because shorter wavelengths can be focused to a smaller beam spot (thus creating smaller-size pixels).

A quantum rod represents an intermediate form between a zero-dimensional quantum dot (0DEG) and a one-dimensional quantum wire (1DEG) and offers, in some way, a combination of properties exhibited by a quantum dot and a quantum wire. Thus, their bandgaps can be tuned by precise control of both the length and the diameter of the rod. Alivisatos and co-workers have produced CdSe quantum rods of various diameters (3.5–6.5 nm) and lengths (7.5–40 nm) (Li et al., 2001). They have reported that the photoluminescence emission maximum shifts to lower energy (longer wavelength) with an increase either in the width or the length. The observed room-temperature quantum efficiency for these rods is typically

5–10%. Also, as in the case of a quantum wire, the emission is highly linearly polarized.

4.2.3 Nonlinear Optical Properties

The two types of nonlinear optical effects exhibiting enhanced manifestations in quantum-confined structures are as follows:

- *Electro-optic Effect.* In this case a change in optical properties is produced by application of electric field. This effect is derived from the Stark effect—that is, change of energy states by application of electric field. It is discussed in detail in Section 4.2.4.
- *Optically Induced Refractive Index Change.* This is an effective third-order nonlinear optical effect (Kerr effect as discussed in Chapter 2).

Both of these effects in quantum-confined structures can be described as dynamic nonlinearities that are derived from a change ($\Delta\alpha$) in the optical absorption due to applied electric or optical field (Abram, 1990; Chemla et al., 1988). The Kramer–Kronig relation relates the change, $\Delta\alpha$, in the absorption to the corresponding change, Δn, in the real part of the refractive index by the following equation:

$$\Delta n(\omega) \frac{c}{\pi} \text{ p.v.} \int_0^\infty \frac{d\omega'\, \Delta\alpha(\omega')}{\omega'^2 - \omega^2} \qquad (4.9)$$

where p.v. stands for the principal value of the integral, and ω is the angular frequency of light.

The change $\Delta\alpha$ and, consequently, Δn can be used for all optical switching, gating, and signal processing. A self-action effect, where a single optical beam changes its own propagation characteristics as a function of intensity (due to the change in the refractive index), can be used in optical signal processing or memory application. Alternatively, a strong optical pump pulse can control a weaker optical signal. The response time of the nonlinearity—that is, the time in which $\Delta\alpha$ and Δn return to the zero—is determined by the decay of the excited-state population. This type of nonlinearity, determined by the number density of excitation (population density), is also called an *incoherent nonlinearity* (Prasad and Williams, 1991) because no phase coherence of excited state is involved.

Some of the main mechanisms producing $\Delta\alpha$ are discussed below.

Phase-Space Filling. This effect arises at high concentrations of electrons (or holes) or excitons (Chemla et al., 1988; Schmitt-Rink et al., 1989). Excitons in semiconductors are neutral, weakly bound electron hole pairs (see Chapter 2). The k levels of excitons (also described as the phase-space) are described by a linear combination of the k levels of the free electrons and holes. If these states (k levels of the free electrons and holes) are already occupied, they cannot be used again to describe another exciton state, under high exciton density, because electrons (and holes) are fermions and must obey the Pauli exclusion principle. In other words, two electrons (or holes) of the same spin cannot occupy the same k level. Thus at high exciton densities the probability of forming additional excitons decreases with increasing laser intensity. Consequently, the excitonic absorption saturates, leading to a large $\Delta\alpha$ with eventually no absorption taking place at the excitonic energy. This state with no absorption is also referred to as a bleached state, and the process, called photophysical bleaching due to saturation, can lead to complete transmission of the exciting light. Such a medium is called a saturable absorber.

Bandgap Renormalization. Under high excitation density, which produces a dense electron hole plasma, there is an effective bandgap shrinkage or narrowing that results from the many body interactions. The main effect is due to a combination of so-called screened electron–hole interaction and exchange. In the absence of these many body effects, the average energy of the recombining carriers is just the "bare" energy gap E_g plus the average kinetic energies of the electrons and the holes. In the presence of many-body effects involving exchange interaction, there is a competition between electron–hole interaction contributing negative energy and electron–electron and hole–hole repulsion adding positive energy. This results in a net negative energy term per particle (implying the average separation between the electrons or the holes is smaller than it would be in the absence of exchange) that increases faster with carrier density than the single-electron (hole) kinetic energy. Thus at high carrier densities one observes a narrowing of the effective band gap. As a result, the combination of these energy terms gives an effective energy gap that decreases with carrier density (excitation density). This effect is called "bandgap renormalization" (Abram, 1990).

Formation of Biexcitons. These are higher-order complexes, formed by binding of excitons, and have been discussed in Chapter 2. The new optical transitions corresponding to biexcitons provide another mechanism to produce $\Delta\alpha$ and consequently Δn (Levy et al., 1988).

In most optical experiments, all sources of nonlinearity contribute simultaneously. The saturation of absorption has been extensively investigated using commercial glasses doped with CdS_xSe_{1-x} nanocrystals. Figure 4.8 shows the change $\Delta\alpha$ for two pump intensities, 3 MW/cm² and 200 kW/cm², in a glass sample containing $CdS_{0.9}Se_{0.1}$ crystallites of 11-nm size (Olbright and Peyghambarian, 1986). The sample is at room temperature. Figure 4.8 also shows the Δn (or nonlinear refractive index n_2), calculated using the Kramer–Kronig transformation described by Eq. (4.9) (the solid curve) and that obtained by an interferometric technique (dotted points).

4.2.4 Quantum-Confined Stark Effect

The effect of an applied electric field on the energy levels and, therefore, on the optical spectra is called the *Stark effect*. Quantum-confined structures also exhibit pronounced changes in their optical spectra when an electric field is applied along the direction of confinement (Weisbuch and Vinter, 1991). This subsection describes

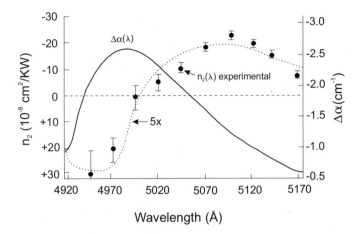

Figure 4.8. Change in absorption coefficient, $\Delta\alpha$, and nonlinear refractive index n_2 in a sample containing 11-nm-size nanocrystals of $CdS_{0.9}Se_{0.1}$. From Olbright and Peyghambarian (1986), reproduced with permission.

this effect through the example of a quantum well. For a bulk semiconductor, the absorption edge exhibits a low-energy tail below the energy gap, when an electric field is applied. This effect is known as the *Franz–Keldysh effect*. It results from the shift of the valence and the conduction bands under the influence of the electric field. In a simple treatment, the electric field can be used as an added interaction potential, and the well-known approximate method of perturbation theory of quantum mechanics can be used to determine the energy shift as well as the mixing of energy levels. The applied electric field can also ionize the exciton by pulling apart the electron–hole pair, causing a broadening of the exciton peak. The excitonic peaks can even disappear under condition of complete ionization of exciton.

In the case of a quantum well, the applied electric field can be either in the plane of the well (longitudinal), where it behaves as a delocalized two-dimensional electron gas, or in the direction of confinement (transverse). The longitudinal field effect is similar to the bulk problem; the exciton dissociation (thus disappearance of excitonic absorption) occurs at a fairly low field, and the absorption edge shifts to lower energy.

The effect of an electric field along the transverse direction is of significant interest. Because of the confinement effect, the excitons are not ionized even at applied fields greater than 100 kV/cm. The major manifestations of the quantum-confined Stark effect produced by the application of an electric field along the confinement direction are as follows:

- The interband separation changes as the electric field pushes the electron and the hole wavefunctions to opposite sides of the quantum-confined region. This behavior is shown in Figure 4.9a for a quantum well.

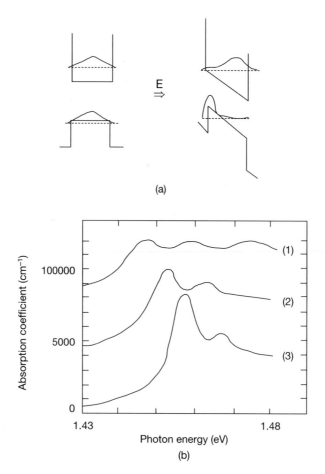

Figure 4.9. (a) Effect of electric field on wavefunctions of electron and hole. (b) Optical absorption spectra of a AlGaAs/GaAs quantum well. From Miller et al. (1985), reproduced with permission.

- Due to the separation of the electron and hole wavefunctions, the binding energy of the exciton decreases. This effect is also shown by the spectral changes at various field strength for a AlGaAs/GaAs quantum well. This leads to broadening of excitonic peaks. However, the shift of the excitonic peak is larger than the broadening. This behavior is also shown in Figure 4.9b (Miller et al., 1985).
- The electric field can also mix different quantized states and lead to redistribution of oscillator strength between optically allowed and optically forbidden excited states. For example, in the absence of an electric field, only $\Delta n = 0$ transitions between the quantized levels of the conduction and the valence bands (such as $n = 1 \rightarrow n = 1$) are allowed for a symmetric quantum well (po-

tential barrier on both sides the same). However, in the presence of an electric field, $\Delta n = \pm 1$ can also become optically allowed.

- A major consequence of all these manifestations is a large change in the optical absorption corresponding to excitonic transitions, as a function of applied electric field in the confinement direction. This effect, also called *electro-absorption,* leads to a corresponding change Δn in the real part of the refractive index and can be used to modulate the propagation of light by the application of an electric field. A major application of the quantum well devices is in electro-optic modulators, which utilize this principle.

Extensive investigations of electro-absorption in quantum dots have also been carried out (Gaponenko, 1999). Cotter et al. (1991) have examined electro-absorption of CdS_xSe_{1-x} nanocrystals and reported a quadratic dependence of induced absorption, which becomes linear at the highest applied field.

Another structure used to enhance the manifestation of electric field is an asymmetric quantum well where the potential barriers on both sides of the quantum well are not the same. This feature can be assumed by using compositional variation (x) of the wider bandgap semiconductor regions $Al_xGa_{1-x}As$ such that x has one value on one side of the quantum well GaAs and has another value on the other side. In an asymmetric quantum well, a linear Stark effect on the energy levels can be realized.

4.3 DIELECTRIC CONFINEMENT EFFECT

Quantum-confined structures also exhibit dielectric confinement effects produced by the difference in the dielectric constants of the confined semiconductor region and the confining potential barrier around it. This effect may be small in the case of a quantum well where the potential barriers, produced by compositional variations (such as $Al_xGa_{1-x}As$ surrounding GaAs), do not introduce a large change in the dielectric constant and are, therefore, often neglected. However, a quantum wire, a quantum rod, or a quantum dot, depending on the method of fabrication and processing, may be embedded in another semiconductor or a dielectric such as glass or polymer. These quantum structures may be encapsulated by organic ligands, dispersed in a solvent or simply surrounded by air. Such media may yield a wide variation in the dielectric constant and lead to important manifestations due to dielectric confinement when the dielectric constant of the surrounding medium is significantly lower than that of the confined semiconductor region (Wang and Herron, 1991; Takagahara, 1993).

The two major manifestations derived from dielectric confinement are as follows:

- Enhancement of Coulomb interaction between quantum-confined states by virtue of the polarization charges which form at the dielectrically mismatched interfaces (Keldysh, 1979). This occurs when the DC (or low-frequency) dielectric constant of the surrounding medium is lower than that of the quantum-confined region. An example is a quantum dot surrounded by organic ligands. Using the variation in the relative dielectric constant provides additional degree of freedom to tailor optical properties of these quantum structures (quantum dots, quantum wires, etc.).

- Enhancement of the local field inside the quantum structure, when illuminated by light. The local field, often estimated by using the Lorentz approximation (Prasad and Williams, 1991), depends on the refractive index of a medium. A lower refractive index of the surrounding medium confines the optical wave near the quantum structure and subsequently enhances the local field, which has important consequences on the optical properties. More pronounced effects can be expected to be manifested in the nonlinear optical properties that involve higher-order interactions with the local field (Prasad and Williams, 1991).

4.4 SUPERLATTICES

A superlattice is formed by a periodic array of quantum structures (quantum wells, quantum wires, and quantum dots). An example of such a superlattice is a multiple quantum well, produced by growth of alternate layers of a wider bandgap (e.g., AlGaAs) and a narrower bandgap (GaAs) semiconductors in the growth (confinement) direction. This type of multiple quantum wells is shown in Figure 4.10a,b by a schematic of their spatial arrangement as well as by a periodic variation of their conduction and valence band edges.

When these quantum wells are widely separated so that the wavefunctions of the electrons and the holes remain confined within individual wells, they can be treated as a set of isolated quantum wells. In this case, the electrons (or the holes) can not tunnel from one well to another. The energies and wavefunctions of electrons (and

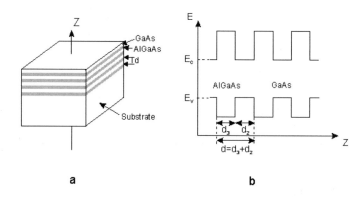

Figure 4.10. Schematics of the arrangement (a) and the energy bands (b) of multiple quantum wells.

holes) in each well remain unchanged even in the multiple quantum well arrangement. However, such noninteracting multiple quantum wells (or simply labeled multiple quantum wells) are often utilized to enhance an optical signal (absorption or emission) obtainable from a single well. An example is lasing, to be discussed in the next section, where the stimulated emission is amplified by traversing through multiple quantum wells, each well acting as an independent medium.

To understand the interaction among the quantum wells, one can use a perturbation theory approach similar to treating identical interacting particles with degenerate energy states. As an example, let us take two quantum wells separated by a large distance. At this large separation, each well has a set of quantized levels E_n labeled by quantum numbers $n = 1, 2, \ldots$ along the confinement direction (growth direction). As the two wells are brought close together so that the interaction between them becomes possible, the same energy states E_n of the two wells are no longer degenerate. Two new states E_n^+ and E_n^- result from the symmetric (positive) overlap and antisymmetric (negative) overlap of the wavefunctions of the well. The $E_n^+ = E_n + \Delta_n$ and $E_n^- = E_n - \Delta_n$ are split by twice the interaction parameter Δ_n for level n.

The magnitude of the splitting, $2\Delta_n$, is dependent on the level E_n. It is larger for higher energy levels because the higher the value of n (the higher the energy value E_n), the more the wavefunction extends in the energy barrier region allowing more interaction between the wells.

The case of two wells now can be generalized into the case of N wells. Their interactions lift the energy degeneracy to produce splitting into N levels, which are closely spaced to form a band, the so-called miniband. In an infinite multiple quantum well limit, the width of such a miniband is $4\Delta_n$, where Δ_n is the interaction between two neighboring wells for the level n. This result is shown in Figure 4.11 for the two levels E_1 and E_2 for the case of a superlattice consisting of alternate layers of GaAs (well) and $Al_{0.11}Ga_{0.89}As$ (barrier), each of width 9 nm. For this system, the miniband energies are $E_1 = 26.6$ meV and $E_2 = 87$ meV, with the respective bandwidth of $\Delta E_1 = 2.3$ meV and $\Delta E_2 = 20.2$ meV (Barnham and Vvedensky, 2001). As explained above, the higher-energy miniband (E_2) has a greater bandwidth (ΔE_2) than the lower energy miniband (ΔE_1).

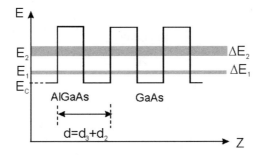

Figure 4.11. Schematics of formation of minibands in a superlattice consisting of alternate layers of GaAs (well) and AlGaAs (barrier).

The formation of minibands by interaction among quantum wells also leads to modification of the density of states. The modified density of states is shown in Figure 4.12. The modification is along the staircase where the density of states jumps from one value to another in the case of a single well. The miniband contribution spreads the density of states around the staircase as shown in Figure 4.12. The minibands are in the energy region $(E_a - E_b)$ and $(E_c - E_d)$ for levels E_1 and E_2, and the modification in the density of states in this region can be seen in Figure 4.12.

As we shall see in Section 4.6, an important consequence of the interacting well superlattice is electron tunneling from one well to another. Electron tunneling among appropriately designed multiple quantum wells has led to the development of quantum cascade lasers, which are also discussed in Section 4.6.

Another area of intense activity is superlattices of quantum dots, which are also called *nanocrystal superlattices* or *quantum dot solids* (Gaponenko, 1999). Various methods have been used to produce quantum dot solids. Some of them utilize *in situ* growth to produce an ordered arrangement. Others utilize self-assembling of spherical nanocrystals to a close-packed arrangement (Murray et al., 2000). Yet another approach is to use a template with specific sites at which quantum dots can be anchored through preferred interactions or chemical bonding (Mirkin et al., 1996; Alivisatos et al., 1996). The self-assembling of spherical nanoparticles, surface capped by organic ligands, offers the flexibility to manipulate interaction between the quantum dots (Murray et al., 2000). These nanoparticles are prepared by colloidal chemistry as described in Chapter 7. The close-packing of these crystals is, therefore, also labeled self assembling of colloidal crystals. The nanoparticles thus prepared have their surfaces capped (covered) by organic ligands that can bind with the surface through various electron coordinating groups. By manipulation of the size (length) of the ligand, the interparticle distance can be controlled, leading to variation in the nature as well as in the magnitude of interaction. Furthermore, by using a volatile organic ligand (such as pyridine, which in the bulk form is a liquid), one can close-pack the nanocrystals and drive the ligand off thermally (and/or by dynamic vacuum) to create a direct contact (closest packing) between the quantum dots.

Bawendi and co-workers have extensively investigated the close-packing of colloidal quantum dots (Murray et al., 2000). Figure 4.13 shows the high-resolution

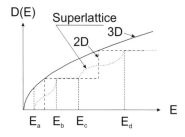

Figure 4.12. Modified density of states, $D(E)$ in a multiple quantum well superlattice due to miniband formation.

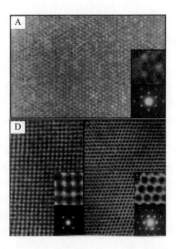

Figure 4.13. (A) TEM image of three-dimensional superlattice of 48-Å CdSe NCs. (B) At high magnification, the internal lattice structure of the NC building blocks is resolved. (C) Small-angle electron diffraction pattern demonstrates the lateral perfection of the superlattice domain. (D) TEM image showing the 101 projection for a superlattice of 48-Å CdSe NCs. (E) High magnification shows lattice imaging of the individual NCs. (F) Small-angle electron diffraction demonstrates the perfection of another characteristic fcc orientation. (G) TEM image of an fcc superlattice of 64 CdSe NCs viewed along the $\{111\}_{SL}$ axis. (H) Higher magnification image showing the individual NCs in the superlattice. (I) Small-angle electron diffraction. From Murray et al. (2000), reproduced with permission.

transmission electron microscopy image of a three-dimensional superlattice of CdSe quantum dots of 4.8-nm diameter. The image under high magnification and the small-angle electron diffraction pattern are shown in the insets. The quantum dots assemble in a face-centered cubic lattice. The image corresponds to the (101) crystallographic plane of the superlattice. Bawendi and co-workers have shown that for large interparticle distances (such as in the case where organic ligands are present between the quantum dots), dipole–dipole interaction between them produce Forster energy transfer, discussed in Chapter 2. In this case, if there is a distribution of sizes of quantum dots, this type of energy transfer upon photoexcitation produces emission from the larger-size particles having lower excitation energy (bandgap decreases as the inverse of the square of the diameter). Hence, a red shift of the emission peak results. This type of interaction dominates for interparticle separation in the range of 0.5–10 nm. At separations of <0.5 nm, the wavefunctions of electrons in the neighboring quantum dots overlap, allowing exchange interactions to occur. This leads to delocalization of electronic excitations over many quantum dots, similar to the delocalization of electronic wavefunctions over many unit cells in the bulk semiconductor. Hence, the optical absorption spectrum of the close-packed quantum dots of CdSe shows a red shift (to a longer wavelength) as the interparticle distance is reduced from 0.7 nm, eventually approaching the absorption spectrum of the bulk CdSe (Murray et al., 2000).

4.5 CORE-SHELL QUANTUM DOTS AND QUANTUM DOT-QUANTUM WELLS

This section describes semiconductor nanostructures that consist of a quantum dot core with one or more overlayers of a wider bandgap semiconductor. The core-shell quantum dots consist of a quantum dot with one overcoating (shell) of a wider bandgap semiconductor (Wilson et al., 1993; Hines and GuyotSionnest, 1996). The quantum dot–quantum well (QDQW) structure involves an onion-like nanostructure composed of a quantum dot core surrounded by two or more shells of alternating lower and higher bandgap materials (Schooss et al., 1994). These types of hierarchical nanostructures introduce a new dimension to bandgap engineering (modifications of the bandgap profile, charge-carrier properties, and luminescence feature). In the case of a core-shell structure, changing the shell can be used to manipulate the nature of carrier confinement in the core, thus affecting the optical properties. Furthermore, an overcoating with a wider bandgap semiconductor passivates the surface nonradiative recombination sites, thereby improving the luminescence efficiency of the quantum dot. In the case of QDQW structures, additional tuning of the energy levels and carrier wavefunctions can be obtained by varying the nature and the width of the quantum well surrounding the quantum dot.

Numerous reports of core-shell quantum dots exist. Some examples are ZnS (shell) on CdSe (quantum dots) (Dabbousi et al., 1997; Ebenstein et al., 2002), CdS on CdSe (Tian et al., 1996; Peng et al., 1997), ZnS on InP (Haubold et al., 2001), and $ZnCdSe_2$ on InP (Micic et al., 2000). The use of a wider bandgap semiconductor shell permits the light emitted from the core quantum dot to be transmitted through it without any absorption in the shell. These core-shell structures exhibit many interesting modifications of their luminescence properties. As an example, the results reported by Dabbousi et al. for the CdSe–ZnS core-shell quantum dots are summarized below:

- A small red shift in the absorption spectra of the core-shell quantum dot compared to that for the bare quantum dot is observed. This red shift is explained to arise from a partial leakage of the electronic wave function into the shell semiconductor. When a ZnS shell surrounds the CdSe quantum dot, the electron wavefunction is spread into the shell but the hole wavefunction remains localized in the quantum dot core. This effect produces a lowering of the bandgap and consequently a red shift. The red shift becomes smaller when the difference between the bandgaps of the core and the shell semiconductor increases. Hence, a CdSe–CdS core-shell structure exhibits a large red shift compared to the bare CdSe quantum dot of the same size (CdSe core). Also, the red shift is more pronounced for a smaller-size quantum dot where the spread of the electronic wavefunction to the shell structure is increased.
- For the same size CdSe quantum dot (~4 nm), as the coverage of the ZnS overlayer increases, the photoluminescence spectra show an increased red shift of emission peak compared to the absorption peak, with an increase in

broadening. This effect may arise from an inhomogeneous distribution of the size and preferential absorption into larger dots. The photoluminescence quantum yield first increases with the ZnS coverage reaching to 50% at approximately ~1.3 monolayer coverage. At higher coverage, it begins to decrease. The increase in the quantum yield is explained to result from passivation of surface vacancies and nonradiative recombination sites. The decrease at higher coverage was suggested to arise from defects in the ZnS shell producing new nonradiative recombination sites.

The reported core-shell structure fabrication methods have utilized procedures for formation of the core (quantum dots) which produce polydispersed samples of core nanoparticles. These core particles then must be size-selectively precipitated before overcoating, making the process inefficient and time-consuming.

More recently, we developed a rapid wet chemical approach to produce core-shell structures using novel chemical precursors which allows us to make III–V quantum dots (a more difficult task than the preparation of II–VI quantum dots) rapidly, (less than two hours) (Lucey and Prasad, 2003). Then without the use of surfactants or coordinating ligands, the quantum dots are overcoated by the shells rapidly using inexpensive and commercially available II–VI precursors that are not air-sensitive. This process was used to produce CdS on InP, CdSe on InP, ZnS on InP, and ZnSe on InP. The optical properties of these core-shell structures were strongly dependent on the nature of the shell around InP. A ZnS shell around the InP quantum dot produced the most efficient and narrower emission among all these core-shell structures, as shown in Figure 4.14 (Furis et al., 2003). This synthe-

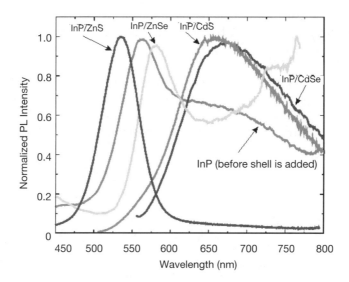

Figure 4.14. Photoluminescence spectra of InP quantum dots and the various core-shell structures involving InP core of some diameter (~3 nm).

sis method also provides the advantage of producing large quantities of core-shell materials in a relatively short time.

A widely studied QDQW system is CdS/HgS/CdS. A number of experimental as well as theoretical reports exist on the optical properties of these structures and their tunability as a function of the layers thickness. Some of them cited here are by Schooss et al. (1994), Mews et al. (1996), Braun et al. (2001), Borchert et al. (2003), and Bryant and Jaskolski (2003). All report a red shift of the emission and/or absorption feature of the QDQW structure after adding the shells, in comparison to the features associated with the core before adding the shells. This illustrates the tunability of the emission wavelength. A second type of QDQW structure, studied to a somewhat lesser extent, is ZnS/CdS/ZnS (Little et al., 2001; Bryant and Jaskolski, 2003). These studies also find that the wavelength is red-shifted as soon as the first shell of lower bandgap material is added.

4.6 QUANTUM-CONFINED STRUCTURES AS LASING MEDIA

The most widely used application of quantum-confined structures is as an active medium for highly efficient and compact solid-state lasers. Quantum well lasers have become commonplace in the consumer market, their uses range from being in compact-disc players to being in laser printers. They serve as a useful laser source or laser pump source in telecommunications. The quantum well lasers cover a wide spectral range, from 0.8 to 1.5 μm. More recently, using a wide bandgap semiconductor, GaN, lasing output in the blue region (~400 nm) has also been produced, which holds promise for high-density optical data-storage applications. Table 4.3 lists the quantum well semiconductors, together with the barrier materials and the corresponding lasing wavelength region. Recent entry of quantum cascade lasers, utilizing multiple quantum wells, covers the mid-infrared range for many environmental monitoring and sensing applications.

This section provides a brief overview of the principles of lasers and utilization of quantum structures as lasing media. A comparison of the merits of lasers utilizing

Table 4.3. Quantum-Confined Semiconductors and the Lasing Wavelength Region

Quantum-Confined Active Layer	Barrier Layer	Substrate	Lasing Wavelength (nm)
InGaN	GaN	GaN	400–450
InGaP	InAlGaP	GaAs	630–650
GaAs	AlGaAs	GaAs	800–900
InGaAs	GaAs	GaAs	900–1000
InAsP	InGaAsP	InP	1060–1400
InGaAsP	InGaAsP	InP	1300–1550
InGaAs	InGaAsP	InP	1550
InGaAs	InP	InP	1550

quantum wells, quantum wires, and quantum dots is presented. Specific examples presented also include vertical cavity lasers, VCSELS, and quantum cascade lasers.

The principle of lasers and laser technology is discussed in many books (Siegman, 1986). Here only a simplified description is presented for those who are not familiar with this topic. A laser, an acronym for *light amplification by stimulated emission of radiation*, utilizes stimulated emission that is triggered by an incident photon of the same energy. This occurs when a medium has more population of electrons in the excited quantum level than in the ground level. This artificially created situation, called *population inversion*, is produced by either electrical stimulation (electroluminescence) or optical stimulation and is different from the spontaneous emission, whereby the electron returns to the ground state in the natural course (within the lifetime of the excited state), even in the absence of any photon to stimulate it. These two processes are schematically represented in Figure 4.15.

In a laser, the stimulated emission is amplified by passing the emitted photons to stimulate emission at other locations. A simple diagrammatic description of a laser design is shown in Figure 4.16 (Prasad, 2003). The various components of a laser are as follows:

- An active medium, also called a gain medium (in the present case is a quantum well), where a suitable pumping mechanism creates population inversion. The spontaneously emitted photons at some site in the medium stimulates emission at other sites as it travels through it.
- An energy pump source, which in the case of quantum structure lasers is an electrical power supply.
- Two reflectors, also called rear mirror and output coupler, to reflect the light in phase (determined by the length of the cavity) so that the light will be fur-

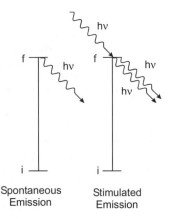

Figure 4.15. Schematics of spontaneous emission and stimulated emission from excited level f to lower level i. In stimulated emission, the incoming photon stimulating the emission and the photon produced by stimulated emission have their waves in phase.

ther amplified by the active medium in each round-trip (multipass amplification). The output is partially transmitted through a partially transmissive output coupler from where the output exits as a laser beam (e.g., $R = 80\%$ as shown in Figure 4.16).

Both the processes of stimulated emission and cavity feedback impart coherence to the laser beam. Only the waves reflected in phase and in the direction of the incident waves contribute to multipass amplification and thus build up their intensity. Emission coming at an angle (like from the sides) does not reflect to amplify. This process provides directionality and concentration of beam in a narrow width, making the laser highly directional with low divergence (compared to fluorescence) and great coherence. For an incident beam and reflected beam to be in phase in a cavity, the following cavity resonance conditions must be met (Svelto, 1998), as shown in Eq. (4.10):

$$m\frac{\lambda}{2} = l \tag{4.10}$$

where λ = the wavelength of emission, l = the length of the cavity, and m is an integral number. The range of wavelengths over which sufficient stimulated emission and lasing action can be achieved defines what is called the *gain curve*.

Within the gain curve of an active medium, Eq. (4.10) can be satisfied for many wavelengths with different integral numbers m. These are called the *longitudinal cavity modes*. Wavelength selection can be introduced by replacing the rear mirror in Figure 4.16 with a spectral reflector (grating or prism) that permits only a certain narrow wavelength of light (monochromatic beam) to be multipass-amplified. However, these optical elements do not provide the resolution necessary to select a single longitudinal mode of the cavity (narrowest bandwidth possible). One intro-

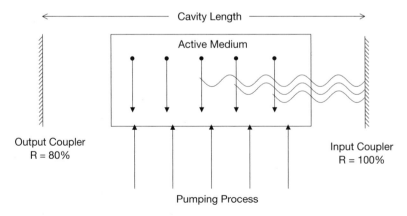

Figure 4.16. The schematics of a laser cavity. R represents percentage reflection.

duces other optical elements such as a Fabry–Perot etalon or Liot filter in the cavity to isolate a single longitudinal mode. In semiconductor lasers, distributed feedback using a Bragg grating is commonly used for wavelength selection, together with a waveguide arrangement to confine light.

In the case of a semiconductor quantum well laser, the active medium is a narrow bandgap semiconductor (e.g., GaAs) that is sandwiched between a p-doped (excess holes) AlGaAs and an n-doped (excess electrons) AlGaAs barrier layer. However, in most semiconductor lasers, the doped layers of AlGaAs are separated from GaAs by a thin undoped layer of same composition AlGaAs. The cavity is formed by the cleaved crystal surfaces that act as reflectors. The general principle for the semiconductor lasers can be illustrated by the schematic shown in Figure 4.17 for a double heterostructure semiconductor laser. The ohmic contacts on the top and bottom inject electrons and holes into the active region, which in this design is significantly thicker (>100 nm) than a quantum well. The electrons and holes combine in the active region (GaAs in the present case) to emit a photon. The threshold current density (current per cross-sectional area) is defined as the current density required for the onset of stimulated emission and hence lasing. At this current density, the optical gain produced in the medium by stimulated emission just balances the optical loss due to various factors (e.g., scattering, absorption loss, etc.). A thick active layer in the double heterostructure also acts as a waveguide to confine the optical waves. The output emerges from the edges, hence this configuration is also referred to produce edge emitting laser action.

The case of a quantum well laser, where a single quantum well of dimensions <10 nm (depends on the material) is used as the active layer, is referred to as SQW (single quantum well) laser. This thickness is too small to confine the optical wave, which then leaks into the confining barrier region. To simultaneously confine the carriers in the quantum well and the optical wave around it, one uses a layer for optical confinement in which the quantum well is embedded.

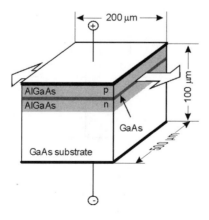

Figure 4.17. Scheme of double heterostructure semiconductor (DHS) laser.

The quantum well laser offers the advantage of a considerably lower threshold current density for laser action because of a reduction in the active-layer thickness. For example, SQW laser current threshold is ~0.5 mA compared to ~20 mA for a double heterostructure laser with the active medium of thickness 100 nm. Furthermore, the quantum well lasers provide a narrower gain spectrum and a smaller line width (<10 MHz) of the laser modes. The quantum well lasers also provide much higher modulation frequencies at which the lasing output can be modulated electrically. This feature is of great importance for high-speed telecommunications. The modulation speed is determined by the differential gain (gain as a function of current) that is larger in a quantum well compared to a 3D bulk because of the staircase type density of states behavior for the quantum well. However, the power output in a quantum well saturates at low value. A SQW laser thus typically produces ~100 mW of power. Arrays of quantum-well lasers that produce over 50 W are commercially available.

Another approach for generating high laser power output is the use of multiple quantum wells (MQW) as an amplifying medium. A schematics of a MQW laser is shown in Figure 4.18. However, at low current densities, a SQW laser produces a superior performance, while the MQW are useful at high current densities.

An approach to produce lower threshold current density for a quantum well laser is to introduce lattice strain. The lasers thus produced are called *strained-layer lasers*. In this case the active quantum well layer, such as InGaAs, is grown on a confining layer, AlGaAs, which has a different lattice constant. This lattice mismatch creates a strain that can be accommodated by the thin active layer without creating any dislocations. The quantum well layer of InGaAs experiences a biaxial compression in the plane of the layer. The net effect of this strain is to reduce the threshold current density through changes in band profile. By altering the band structure, the strain also shifts the lasing wavelength.

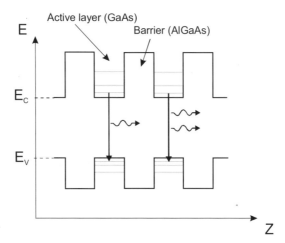

Figure 4.18. The energy diagram for multiple quantum wells (MQW) laser.

4.6 QUANTUM-CONFINED STRUCTURES AS LASING MEDIA

Lower-dimensional systems such as quantum wires and quantum dots offer the prospect of even lower threshold current. These lasers thus can also be modulated at higher frequencies, transmitting information for telecommunications at a higher rate. Furthermore, increased exciton binding energy for the 1DEG quantum wires and 0DEG quantum dots also provides the prospect of lasing utilizing photons produced from excitons, which are already formed by association of free electrons and holes. Quantum wire lasers have been produced by using quantum wires in different shapes, such as on a V-groove substrate. However, a real challenge is to grow an array of wires of nearly equal cross section to minimize the inhomogeneous broadening of emission that arises from a large variation in the cross-section.

Because of discrete features in the density of states, the quantum dot lasers have a very narrow gain curve and thus can be operated at even lower drive current and lower threshold current density compared to quantum wells (Arakawa, 2002). Figure 4.19 illustrates these features by comparing a quantum well laser with a quantum dot laser. It shows that for any output power, the drive current for a quantum dot laser is lower than that for a quantum well laser. Another important factor is temperature dependence. The quantum dot lasers, because of the widely separated discrete quantum states, show much more tolerance for temperature variation, compared to quantum well lasers. Since the first report of self-assembled InAs/InGaAs quantum dot lasers showing reduced temperature dependence of the threshold current (Bimberg et al., 1997), many other quantum dot lasers have been demonstrated. A more recent advancement is the report of InGaN quantum dots epitaxially grown on GaN to produce lasing action at room temperature (Krestnikov et al., 2000a,b).

In order for quantum dot lasers to compete with quantum well lasers, two major issues have to be addressed. Since the active volume provided by a single quantum dot is extremely small, a large array of quantum dots have to be used. A major challenge here is to produce an array of quantum dots with a very narrow size distribution to reduce inhomogeneous broadening and that are without defects that degrade

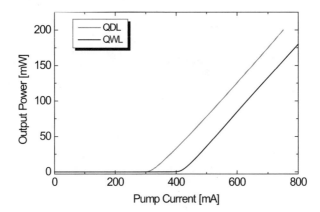

Figure 4.19. Comparison of efficiency between a quantum well laser and a quantum dot laser.

the optical emission by providing alternate nonradiative defect channels. Another issue is the phonon bottleneck created by confinement which limits the number of states that are coupled efficiently by phonons (due to energy conservation) and thus limits the relaxation of the excited carriers into lasing states. This bottleneck causes a degradation of stimulated emission (Benisty et al., 1991). However, other mechanisms such as a process called Auger interaction can be brought into play to suppress this bottleneck effect.

The laser cavity design discussed so far is that of an edge-emitting laser, also known as an in-plane laser, where the laser output emerges from the edge. However, many applications utilizing optical interconnection of systems require a high degree of parallel information throughput where there is a demand for surface emitting laser (SEL). In SEL the laser output is emitted vertically through the surface. Many schematics have been utilized to produce surface emitting lasers. A particularly popular geometry is that of a vertical cavity SEL, abbreviated as VCSEL. This geometry is shown in Figure 4.20. It utilizes an active medium such as multiple quantum wells sandwiched between two distributed Bragg reflectors (DBR), each comprising of a series of material layers of alternating high and low refractive indices. Thus for an InGaAs laser, the DBR typically consists of alternating layers of GaAs with refractive index ~3.5 and AlAs with refractive index 2.9, each layer being a quarter of a wavelength thick. These DBRs act as the two mirrors of a vertical cavity. Thus, both the active layer (InGaAs) and the DBR structures (GaAs, AlAs) can be produced in a continuous growth process.

An advantage offered by a VCSEL is that the lateral dimensions of the laser can be controlled, which offers the advantage that the laser dimensions can be tailored to match the fiber core for fiber coupling. An issue to deal with in VCSEL is the heating effect occurring in a complex multilayer structure, as the current is injected through a high series resistance of the DBRs.

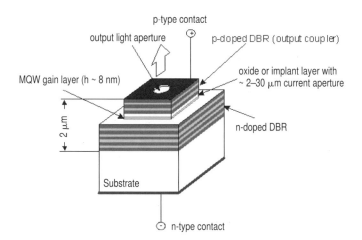

Figure 4.20. The schematics of vertical cavity surface emitting laser (VCSEL).

4.6 QUANTUM-CONFINED STRUCTURES AS LASING MEDIA

Another major development in lasers utilizing quantum structures is that of the "quantum cascade lasers," sometimes abbreviated as QC lasers (Faist et al., 1994; Capasso et al., 2002). The QC lasers utilize fundamentally different principles, compared to the quantum-confined lasers discussed above. Some important differences are as follows:

- In contrast to the lasers discussed above, which involve the recombination of an electron in the conduction band and a hole in the valence band, the QC lasers use only electrons in the conduction band. Hence they are also called unipolar lasers.
- Unlike the quantum-confined lasers discussed above, which involve an interband transition between the conduction band and the valence band, the QC lasers involve intraband (inter-sub-band) transition of electrons between the various sub-bands corresponding to different quantized levels of the conduction band. These sub-bands have been discussed in Section 4.1.
- In the conventional laser design, one electron at the most can emit one photon (quantum yield one). The QC lasers operate like a waterfall, where the electrons cascade down in a series of energy steps, emitting a photon at each step. Thus an electron can produce 25–75 photons.

Figure 4.21 illustrates the schematics of the basic design principle of an earlier version of the QC lasers that produce optical output at 4.65 μm. These lasers are based on AlInAs/GaInAs. It consists of electron injectors comprised of a quantum well superlattice in which each quantized level along the confinement is spread into a miniband by the interaction between wells, which have ultrathin (1–3 nm) barrier layers. The active region is where the electron makes a transition from a higher subband to a lower sub-band, producing lasing action. The electrons are injected from left to right by the application of an electric field of 70 kV/cm as shown in the slope diagram. Under this field, electrons are injected from the ground state g of the miniband of the injector to the upper level 3 of the active region. The thinnest well in the active region next to the injector facilitates electron tunneling from the injector into the upper level in the active region. The laser transition, represented by the wiggly arrow, occurs between levels 3 and 2, because there are more electron populations in level 3 than in level 2. The composition and the thickness of the wells in the active region are judiciously manipulated so that level 2 electron relaxes quickly to level 1.

The cascading process can continue along the direction of growth to produce more photons. In order to prevent accumulation of electrons in level 1, the exit barrier of the active region is, again, made thin, which allows rapid tunneling of electrons into a miniband of the adjacent injector. After relaxing into the ground state g of the injector, the electrons are re-injected into the next active region. Each successive active region is at a lower energy than the one before; thus the active regions act as steps in a staircase. Therefore, the active regions and the injectors are engineered to allow the electrons to move efficiently from the top of the staircase to the

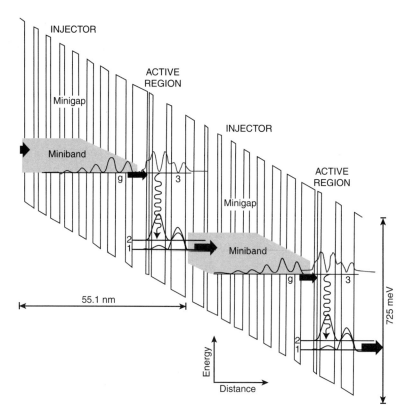

Figure 4.21. The schematics of a quantum cascade laser. From Capasso (2002), reproduced with permission.

bottom in succession, producing multiple photons in this process. This laser design utilized 25 units of injectors and active regions to produce peak power in excess of 100 mW at 4.65 μm at room temperature.

By now a number of reports of room temperature pulse operation of QC lasers covering a broad spectral range from 4 μm to 24 μm in the mid-IR have emerged. This broad spectral coverage is produced by bandgap engineerings, which includes choice of semiconductor, thickness of the wells, and use of quantum well superlattices in the active region. More than 1 W of peak power at room temperature have been produced. Even QC laser operation at 70 μm in the far-IR has been reported. Also, simultaneous lasing action at more than one wavelength has been demonstrated (Tredicucci et al., 1998; Capasso et al., 2002). CW operation at room temperature has also been reported (Beck et al., 2002).

The QC lasers have been successfully used in trace gas analysis, with a sensitivity of parts per billion in volume. The two wavelength regions of 3–5 μm and 8–13 μm are particularly important for chemical sensing of toxic gases, pollutants, and industrial combustion products, because the atmosphere is relatively transparent in this region. Thus, QC lasers have been utilized for the detection of methane, nitrous oxide (N_2O), carbon monoxide, NO, and sulfur oxides.

4.7 ORGANIC QUANTUM-CONFINED STRUCTURES

Conjugated Structures with Delocalized π Electrons. In organic structures the delocalization of electrons, similar to the free electrons in a semiconductor band, is provided by π-bonding in a class of organic compounds called conjugated molecules and polymers. These conjugated molecules involve alternate single and multiple bonds. According to the molecular orbital theory of bonding in molecules, a single covalent bond between two atoms is formed by the axial overlap of the atomic orbitals (wavefunctions) of the two atoms and is called the σ bond. The additional bond(s) involved in a double or triple bond between two atoms (say carbon) is (are) formed by the lateral overlap of the directional *p*-type atomic orbitals situated on the bonding atoms. These bonds are called the π bonds, and the electrons involved are called the π electrons. For a more detailed, basic level description of the bonding and conjugated structures, the readers are referred to the book by this author on Biophotonics (Prasad, 2003). Here it would suffice to say that the π electrons are loosely bound electrons and are spread over the entire conjugated structure, hence behaving similar to free electrons.

Figure 4.22 shows the structures of a π-bonded series of organic structures. Butadiene and hexatriene are examples of conjugated structures involving two and three alternating double bonds. They can be considered as larger-size oligomers (many repeat units chemically bond) of the basic monomeric unit ethene, also shown in Figure 4.22. Higher-length oligomers are also known. As their lengths reach nanometer size, these oligomers are also referred to as nanomers. Thus the nanomers are the organic analogues of quantum wires.

The length of the carbon chain defines the conjugation length, providing a structural framework over which the π electrons can be spread (delocalized). The picture of a particle in a one-dimensional box can describe the conjugation (delocalization) effect. The length of one-dimensional conjugation, as defined by the chain of carbon atoms involved in alternate single and multiple bonds, determines the length of the box. As conjugation increases, the length of the one-dimensional box increases, leading to the following properties:

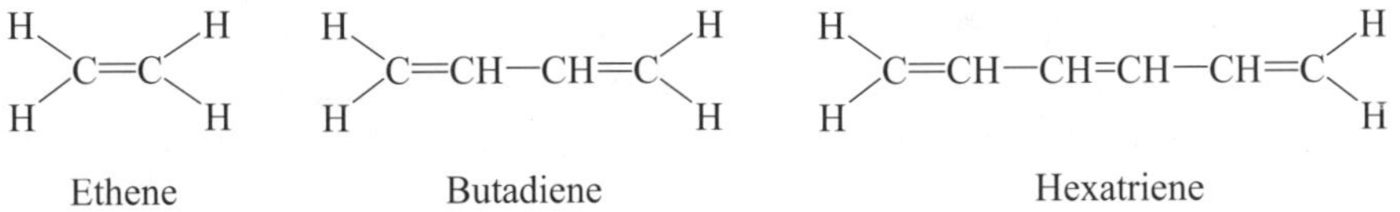

Ethene   Butadiene   Hexatriene

Figure 4.22. Linear conjugated structures.

- The increase of conjugation leads to lowering of the π electron energy (delocalization energy).
- The energy gap between two successive π orbitals decreases as the conjugation increases. This effect gives rise a red shift of the color (wavelength) of the conjugated structure with the increase of conjugation length (i.e., shift of the absorption band corresponding to an electronic transition between the two levels, from UV to a longer wavelength in the visible region).

The π-bonding in the case of a conjugated structure is often described by the Hückel theory (Prasad, 2003). In the case of ethene, the two *p*-type atomic orbitals (abbreviated AO) from the two carbons overlap to form one bonding π and one anti-bonding π* molecular orbital. The pair of bonding electrons is in the lower molecular energy orbital π. Similar descriptions apply to butadiene and hexatriene.

In general, the mixing of N $2p$ orbitals on N carbon atoms in a conjugated structure produces N π orbitals. The gap between the π orbitals decreases as N increases (i.e., as the delocalization length or the length of the one-dimensional box increases). In the limit of very large N, the spacing between successive π levels is very small; the π levels form a closely spaced energy band. Another feature observed from the Hückel theory calculation is that the energy of the π electrons in butadiene is lower than the energy predicted by two isolated ethene type π bonds. This additional energy is called the *delocalization energy*, resulting from the spread of π electrons over all four carbon atoms.

The band formed by closely spaced bonding π levels is occupied by electrons in the lowest energy configuration. This π-band is, therefore, organic analogue of the valence band in a semiconductor. The band formed by the closely spaced antibonding π* levels is empty (energetically unfavorable) in the ground-state configuration and is the organic analogue of the conduction band. The interband transition (valence band to conduction band) in the organic structure is called as a π→π* transition. The transition, analogous to the bandgap in a semiconductor, involves promotion of a π electron from the highest occupied molecular orbital (abbreviated as HOMO) of the π band to the lowest unoccupied molecular orbital (abbreviated as LUMO) of the π* band.

Organic Nanomers as Quantum Wires. As an oligomer (nanomer) allows electron delocalization over its entire length, it will behave just like a nanowire of organics. Therefore, as the length of the wire (oligomer) increases, the bandgap will decrease, which leads to absorbance and emission of different colors for different-length nanomers. This behavior is illustrated by polymerization of phenylene vinylene, which in the absence of any geometric restriction, produces a polymer called poly-paraphenylene vinylene (PPV). It is a conjugated polymer that allows for π-electron delocalization. In a reverse micelle cavity nanoreactor, described in Chapter 7, different nanomers were produced in a series of different-size length reverse micelle cavities by using the reaction shown in Figure 4.23 (Lal et al., 1998). A quantity W_0 (defined as the ratio of amount of water and the

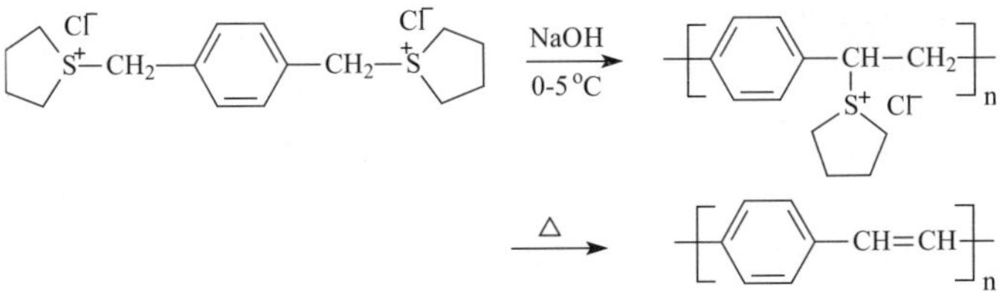

4.7 ORGANIC QUANTUM-CONFINED STRUCTURES

Figure 4.23. Formation of PPV from the sulfonium precursor monomer.

surfactant forming reverse micelle) is an indication of the relative size of these cavities. The smallest-size cavity ($W_0 = 6$) for which the nanomer exhibits the absorption in the ultraviolet thus shows no color. The cavity corresponding to $W_0 = 19$ produces a nanomer that exhibits a yellowish green color absorption because the size is much larger. Figure 4.24 shows a plot of the cavity (particle) size determined by light scattering (left vertical axis) and the emission peak (right vertical axis) as a function of the magnitude of W_0. A pronounced shift of the emission peak wavelength toward red is seen as the size of the cavity (hence the size of the nanomer formed) increases.

A very active area of current research is organic and polymeric light-emitting diodes for high brightness and flexible displays (Akcelrud, 2003). Some of these

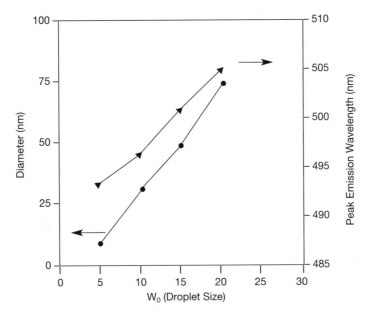

Figure 4.24. Size measurement of polymeric nanoparticles within reverse micellar system having various droplet sizes and their corresponding peak emission wavelengths. From Lal et al. (1998), reproduced with permission.

devices are already in the marketplace. The use of these selectively engineered nanomers as active media provides one with a convenient method to achieve different colors. Karasz and co-workers have used a block copolymer approach to achieve color tunability (see Chapter 8). A block copolymer consists of blocks of repeat units of different kinds (such as -A-A-A- and -B-B-B- in a di-block copolymer). By adjusting the block length (oligomer length) of the emitting block conjugated segment (say -A-A-A-), the color of emission can be varied as described in detail in Chapter 8.

Organic Nanotubes and Nanorods. Other forms of nano objects of conjugated structures have also been prepared. Examples are nanotubes and nanorods. Carbon nanotubes and its various functionalized form have been extensively investigated (Harris, 2000). However, the main focus on carbon nanotubes have been on its electronic and mechanical properties.

Nanotubes of conjugated structures can be considered as folded quantum wells where delocalization of π electrons is along the walls of the tube. Jin and co-workers have produced nanotubes and nanorods of the conjugated polymer PPV, already described earlier in this section under "Organic Nanomers as Quantum Wires" (Kim and Jin, 2001). They used a different synthetic approach in which the polymer was produced by chemical vapor deposition (CVD) polymerization of a different precursor polymer. The polymerization was conducted on the inner surface or inside the nanopores of a porous alumina or polycarbonate membrane filter. They used the same polymerization process on the surface of silicon wafers to produce PPV nanofilms. The filter membranes were removed by dissolution to separate the PPV nanotubes and naorods. The thickness of the PPV nanotubes could be controlled by the reaction condition such as evaporation temperature of the monomer, flow rate of the carrier gas and the reaction time. A typical wall thickness obtained using alumina with pore sizes of 200 nm was ~28 nm. The luminescence properties of these nanotubes were different from that of a bulk PPV film in that the shorter-wavelength (520-nm) emission was more pronounced in the nanotubes and nanorods.

Organic Quantum Dots. Nakanishi and co-workers have prepared and characterized nanoparticles of a class of conjugated polymers called polydiacetylenes which may be considered organic analogues of quantum dots (Katagi et al., 1997). Polydiacetylenes are conjugated structures that involve a sequence of single, double, single, and triple bonds. They are formed in solid state by UV-induced, γ-ray-induced or thermal polymerization of the corresponding monomer as shown in Figure 4.25. Nakanishi and co-workers used a re-precipitation method to produce nanocrystals of the monomer, where a nondissolving solvent was poured into a solution of the monomer to precipitate the nanocrystals. Control of the concentration and temperature was used to produce nanocrystals of various sizes. The monomer nanocrystals were polymerized by radiation polymerization to produce the corresponding polymer nanocrystals which can be redispersed in common organic solvents.

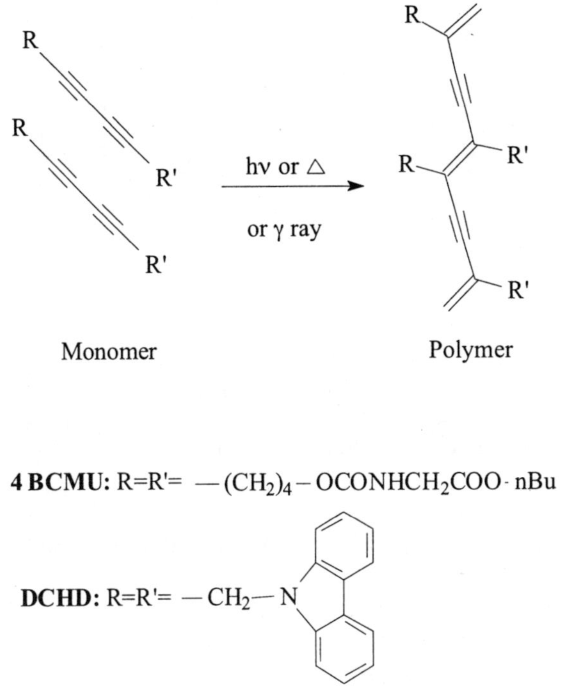

4 BCMU: R=R'= —(CH$_2$)$_4$—OCONHCH$_2$COO-nBu

DCHD: R=R'= —CH$_2$—N(carbazole)

Figure 4.25. Photoinduced, thermal, or radiation-induced polymerization of diacetylenes.

The structure of one polydiacetylene extensively characterized by them, often abbreviated as poly-DCHD, is shown in Figure 4.25. The absorption spectra of polymer nanoparticles of different particle sizes are shown in Figure 4.26. They found that for particle sizes less than 200 nm, the absorption maximum shifts to shorter wavelength (wider bandgap) upon reduction in size. This behavior is, therefore, quite analogous to that of inorganic semiconductor quantum dots. However, the length scale for size-dependent optical properties for the polydiacetylene nanocrystal (<200 nm) is significantly larger than that for inorganic semiconductor quantum dots.

Organic Quantum Wells. Organic analogues of semiconductor quantum wells have also been discussed in the literature (Donovan et al., 1993). Again, a conjugated structure, now in a two-dimensional topology, forms the active well layer, separated by the barrier regions that are made of aliphatic (σ-bonded) groups. The conjugated structures can be vacuum-deposited, as in the case of a planar polynuclear macrocycle structure such as porphyrins. An alternate method is that provided by the Langmuir–Blodgett technique, described in Chapter 8. In this case, a monolayer of the film formed on a water surface is compressed and transferred. The active layer can be built by successive transfer of one monolayer at a time to produce a multi-

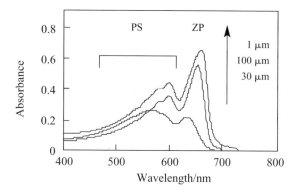

Figure 4.26. Absorption spectra of polydiacetylenes nanocrystals with different crystal sizes (30 nm, 100 nm, 1 mm) dispersed in water. ZP, zero phonon band; PS, phonon side band. Courtesy of Professor H. Nakanishi, Tohoku University.

layer stack of sufficient well thickness. Then multilayers of the barrier film can be deposited in an alternate layer deposition scheme.

The Langmuir–Blodgett film approach has been extensively used for deposition of monolayers and multilayers of polydiacetylenes (derived from BCMU, also shown in Figure 4.25). It has been shown that the color (bandgap) of these polymer layers changes when going from a monolayer to a bilayer and a multilayer (well depth change) (Prasad, 1988). However, a major focus of the work reported on these Langmuir–Blodgett film quantum wells has been on the process of electron transport and tunneling through the barriers (Donovan et al., 1993).

4.8 HIGHLIGHTS OF THE CHAPTER

- Quantum confinement of an inorganic semiconductor provides a means to manipulate its bandgap, and consequently the optical properties.
- Quantum confinement to nanoscale dimensions can be realized in one, two, or three dimensions producing quantum wells, quantum wires, and quantum dots, respectively.
- Quantum wells are structures in which a thin layer of a smaller bandgap semiconductor is sandwiched between two layers of a wider bandgap semiconductor.
- Quantum wells possess quantized (discrete) electronic energy levels existing in the direction of confinement, while the plane of the well, the electronic energy, exhibits a continuous set of closely spaced energy levels to form a two-dimensional band. The combined result is many sub-bands, each starting at a specific quantized energy value.
- The effective bandgap, described by the energy separation between the bottom of the conduction band and the top of the valence band, is increased in a

4.8 HIGHLIGHTS OF THE CHAPTER

quantum well. This effective bandgap reduces as the width of the well increases, finally reaching the limiting value for the bulk semiconductor.
- An excitonic transition occurs below the band-to-band transition.
- The density of states $D(E)$, defined by the number of energy states between E and $E + dE$, is as a series of steps at each quantized energy value for the direction of quantization. In contrast, $D(E)$ for a bulk semiconductor is given as $E^{1/2}$.
- Quantum wires represent confinement of electrons and holes in two dimensions, allowing the free-electron behavior only along the length of the wire. The quantization along the cross section of the wire is obtained by using the model of confinement of an electron in a two-dimensional box.
- The density of states $D(E)$ for a quantum wire exhibits an $E^{-1/2}$ dependence.
- Quantum dots are nanoparticles of semiconductors, representing the case of an electron trapped in a three-dimensional box. They possess only discrete energy levels where the density of states has nonzero values.
- A quantum ring represents a donut-shaped quantum dot, or can be considered as a quantum wire bent into a loop. It exhibits strong response to a magnetic field on its electronic states.
- Optical properties of the quantum-confined structures are dependent on the length of confinement. As the dimension of confinement increases, the bandgap decreases.
- The strength of an optical transition measured by its oscillator strength increases upon quantum confinement. This effect is quite pronounced for quantum wires and quantum dots.
- New transitions occur between sub-bands of the same band (e.g., conduction band). These are characterized by different quantum numbers corresponding to quantization along the direction of confinement.
- These sub-band transitions in quantum wells have been exploited to produce detectors and lasers, the most interesting being quantum cascade lasers.
- Quantum confinement of electrons and holes also causes enhanced binding between them to form excitons.
- In the case of an indirect semiconductor, the bottom of the conduction band and the top of the valence band do not have the same wavevector **k**. Thus emission is not very probable in the bulk form, but it is significantly enhanced upon quantum confinement.
- Quantum-confined materials also exhibit enhanced nonlinear optical effects, compared to the corresponding bulk materials. A primary effect is the intensity dependent changes in absorption, derived from a number of processes such as phase-space filling and bandgap renormalization. These processes are manifested at high excitation densities (high concentrations of electrons and holes).
- New optical transitions are manifested on binding of excitons to produce biexcitons.

- Pronounced changes in optical spectra are observed by applying an electric field along the confinement direction. This is called quantum-confined Stark effect and has been utilized in electro-optic modulators.
- Quantum-confined structures also exhibit a strong dependence of their electronic and optical properties on the dielectric constant of the media surrounding them.
- Superlattices are periodic arrays of quantum structures (e.g., quantum wells, quantum wires, and quantum dots). Interactions between these arrays further split the quantized levels into minibands.
- Core-shell quantum dots are comprised of a quantum dot with an overcoating of a wider bandgap semiconductor. The shell improves the efficiency of luminescence from the core quantum dot.
- Quantum-confined structures are highly efficient lasing media. Quantum well lasers hold a major share of the solid-state laser market. The quantum wire and quantum dot lasers promise to provide significant improvements.
- Quantum cascade lasers utilize transitions between the sub-bands within the same band (e.g., conduction band) of quantum-confined structures. They operate like a waterfall, where the electrons cascade down in a series of steps, emitting a photon at each step. Quantum cascade lasers can be designed to cover a broad spectral range from 4 μm to 24 μm.
- Quantum confinement exists in a conjugated organic structure, consisting of alternate single σ and multiple π bonds. Here, the confinement is effected by restricting the delocalization length (spread) of the labile π electrons in nano-size oligomers comprised of small numbers of repeat units.
- Organic conjugated structures can be produced in various nano-object shapes such as nanowires, nanocrystals, nanorods, nanotubes, and nano-size thickness wells.

REFERENCES

Abram, I. I., The Nonlinear Optics of Semiconductor Quantum Wells: Physics and Devices, in *Nonlinear Optics in Solids,* O. Keller, ed., Springer-Verlag, Berlin, 1990, pp. 190–212.

Akcelrud, L., Electroluminescent Polymers, *Prog. Polym. Sci.* **28**, 875–962 (2003).

Alivisatos, A. P., Johnsson, K. P., Peng, X. G., Wilson, T. E., Loweth, C. J., Bruchez, M. P., and Schultz, P. G., Organization of "Nanocrystal Molecules" Using DNA, *Nature* **382**, 609–611 (1996).

Arakawa, Y., Connecting the Dots, *SPIE's OE Mag.* **January,** 18–20 (2002).

Baldwin, R. K., Pettigrew, K. A., Ratai, E., Augustine, M. P., and Kauzlarich, S. M., Solution Reduction Synthesis of Surface Stabilized Silicon Nanoparticles, *Chem. Commun.* **17**, 1822–1823 (2002).

Barnham, K., and Vvedensky, D., eds., *Low-Dimensional Semiconductor Structures,* Cambridge University Press, Cambridge, 2001, pp. 79–122.

Beck, M., Hofstetter, D., Aellen T., Faist J., Oesterle, U., Ilegems, M., Gini, E., and Melchior, H., Continuous Wave Operation of a Mid-infrared Semiconductor Laser at Room Temperature, *Science* **295**, 301–305 (2002).

Benisty H., Sotomayor-Torres, C. M., and Weisbuch, C., Intrinsic Mechanism for the poor Luminescence Properties of Quantum-Box System, *Phys. Rev. B* **44**, 10945–10948 (1991).

Bimberg, D., Kirstaedter, N., Ledentsov, N. N., Alferov, Z. I., Kopev, P. S., and Ustinov, V. M., InGaAs-GaAs Quantum-Dot Lasers, *IEEE Select Topics in Quantum Electronics* **3**, 196–205 (1997).

Bley, R. A., and Kauzlarich, S. M., A Low-Temperature Solution Phase Route for the Synthesis of Silicon Nanoclusters, *J. Am. Chem. Soc.* **118**, 12461–12462 (1996).

Borchert, H., Dorfs, D., McGinley, C., Adam, S., Moeller, J., Weller, H., and Eychmuller, A., Photoemission of Onion-Like Quantum Dot, Quantum Well and Double Quantum Well Nanocrystals of CdS and HgS, *J. Phys. Chem. B* **107**, 7486–7491 (2003).

Borovitskaya, E., and Shur, M. S., eds., *Quantum Dots,* World Scientific, Singapore, 2002.

Braun, M., Burda, C., and El-Sayed, M. A., Variation of the Thickness and Number of Wells in the CdS/HgS/CdS Quantum Dot Quantum Well System, *J. Phys. Chem. A* **105**, 5548–5551 (2001).

Bruchez, M. Jr., Morronne, M., Gin, P., Weiss, S., and Alivisatos, A. P., Semiconductor Nanocrystals as Fluorescent Biological Labels, *Science* **281**, 2013–2016 (1998).

Brus, L., Luminescence of Silicon Materials: Chains, Sheets, Nanowires, Microcrystals and Porous Silicon, *J. Phys. Chem.* **98**, 3575–3581 (1994).

Brus, L. E., Electron–Electron and Electron–Hole Interactions in Small Semiconductor Crystallites: The Size Dependence of the Lowest Excited Electronic State, *J. Chem. Phys.* **80**, 4403–4409 (1984).

Bryant, G. W., and Jaskolski, W., Tight-Binding Theory of Quantum-Dot Quantum Wells: Single-Particle Effects and Near-Band-Edge Structure, *Phys. Rev. B* **67**, 205320-1–205320-17 (2003).

Canham, L. T., Silicon Quantum Wire Array Fabrication by Electrochemical and Chemical Dissolution of Wafers, *Appl. Phys. Lett.* **57**, 1046–1048 (1990).

Capasso, F., Gmachi, C., Sivco, D. L., and Cho, A. Y., Quantum Cascade Lasers, *Physics Today* **May,** 34–40 (2002).

Carlisle, J. A., Dongol, M., Germanenko, I. N., Pithawalla, Y. B., and El-Shall, M. S., Evidence for Changes in the Electronic and Photoluminescence Properties of Surface-Oxidized Silicon Nanocrystals Induced by Shrinking the Size of the Silicon Core, *Chem. Phys. Lett.* **326**, 335–340 (2000).

Chemla, D. S., Miller, D. A. B., and Schmitt-Rink, S., Nonlinear Optical Properties of Semiconductor Quantum Wells, in *Optical Nonlinearities and Instabilities in Semiconductors,* H. Haug, ed., Academic Press, San Diego, 1988, pp. 83–120.

Cotter, D., Girdlestone, H. P., and Moulding, K., Size-Dependent Electroabsorptive Properties of Semiconductor Microcrystallites in Glass, *Appl. Phys. Lett.* **58**, 1455–1457 (1991).

Dabbousi, B. O., Rodriguez, V. J., Mikulec, F. V., Heine, J. R., Mattoussi H., Ober, R., Jensen, K. F., and Bawendi, M. G., (CdSe)ZnS Core-Shell Quantum Dots: Synthesis and Characterization of a Size Series of Highly Luminescent Nanocrystallites, *J. Phys. Chem. B* **101**, 9463–9475 (1997).

Ding, Z. F., Quinn, B. M., Haram, S. K., Pell, L. E., Korgel, B. A., and Bard, A. J., Electrochemistry and Electrogenerated Chemiluminescence from Silicon Nanocrystal Quantum Dots, *Science* **296,** 1293–1297 (2002).

Dingle, R., Wiegmann, W., and Henry, C. H., Quantum States of Confined Carriers in Very Thin $Al_xGa_{1-x}As$–GaAs–$Al_xGa_{1-x}As$ Heterostructures, *Phys. Rev. Lett.* **33,** 827–830 (1974).

Donovan, K. J., Scott, K., Sudiwala, R. V., Wilson, E. G., Bonnett, R., Wilkins, R. F., Paradiso, R., Clark, T. R., Batzel, D. A., and Kenney, M. E., Determination of Anisotropic Electron Transport Properties of Two Langmuir–Blodgett Organic Quantum Wells, *Thin Solid Films* **232,** 110–114 (1993).

Ebenstein, Y., Makari, T., and Banin, U., Fluorescence Quantum Yield of CdSe/ZnS Nanocrystals Investigated by Correlated Atomic-Force and Single-Particle Fluorescence Microscopy, *Appl. Phys. Lett.* **80,** 4033–4035 (2002).

English, D. S., Pell, L. E., Yu, Z. H., Barbara, P. F., and Korgel, B. A., Size Tunable Visible Luminescence from Individual Organic Monolayer Stabilized Silicon Nanocrystal Quantum Dots, *Nano Lett.* **2,** 681–685 (2002).

Faist, J., Capasso, F., Sivco, D. L., Sirtori, C., Hutchinson, A. L., and Cho, A. Y., Quantum Cascade Laser, *Science* **264,** 553–555 (1994).

Furis, M., Sahoo, Y., Lucey, D., Cartwright, A. N., and Prasad, P. N., unpublished results, 2003.

Gaponenko, S. V., *Optical Properties of Semiconductor Nanocrystals,* Cambridge University Press, Cambridge, 1999.

Harris, P. J. F., *Carbon Nanotubes and Related Structures; New Materials for the Twenty-First Century,* Cambridge University Press, Cambridge, 2000.

Haubold, S., Haase, M., Kornowski, A., and Weller, H., Strongly Luminescent InP/ZnS Core-Shell Nanoparticles, *Chem. Phys. Chem.* **2,** 331–334 (2001).

Heath, J. R., A Liquid-Solution Phase Synthesis of Crystalline Silicon, *Science* **258,** 1131–1133 (1992).

Hines, M. A., and GuyotSionnest, P., Synthesis and Characterization of Strongly Luminescing ZnS-Capped CdSe Nanocrystals, *J. Phys. Chem.* **100,** 468–471 (1996).

Htoon, H., Hollingworth, J. A., Malko, A. V., Dickerson, R., and Klimov, V. I., Light Amplification in Semiconductor Nanocrystals: Quantum Rods Versus Quantum Dots, *Appl. Phys. Lett.* **82,** 4776–4778 (2003).

Hu, J. T., Odom, T. W., and Lieber, C. M., Chemistry and Physics in One Dimension: Synthesis and Properties of Nanowires and Nanotubes, *Acc. Chem. Res.* **32,** 435–445 (1999).

Katagi, H., Kasai, H., Okada, S., Oikawa, H., Matsuda, H., and Nakanishi, H., Preparation and Characterization of Poly-diacetylene Microcrystals, *J. Macromol. Sci. Pure Appl. Chem.* **A34,** 2013–2024 (1997).

Keldysh, L. V., Coulomb Interaction in Thin Semiconductor and Semimetal Films, *JETP Lett.* **29,** 658–661 (1979).

Kim, K., and Jin, J. I., Preparation of PPV Nanotubes and Nanorods and Carbonized Products Derived Therefrom, *Nano Lett.* **1,** 631–636 (2001).

Krestnikov, I. L., Sakharov, A. V., Lundin W. V., Musikhin, Y. G., Kartashova, A. P., Usikov, A. S., Tsatsulnikov, A. F., Ledentsov, N. N., Alferov, Z. I., Soshnikov, I. P., Hahn, E., Neubauer, B., Rosenauer, A., Litvinov, D., Gerthsen, D., Plaut, A. C., Hoffmann, A. A., and Bimberg, D., Lasing in the Vertical Direction in InGaN/GaN/AlGaN Structures with InGaN Quantum Dots, *Semiconductors* **34**, 481–487 (2000a).

Krestnikov, I. L., Sakharov, A. V., Lundin W. V., Usikov, A. S., Tsatsulnikov, A. F., Ledentsov, N. N., Alferov Z. I., Soshnikov, I. P., Gerthsen, D., Plaut, A. C., Holst, J., Hoffmann, A., and Bimberg, D., Lasing in Vertical Direction in Structures with InGaN Quantum Dots, *Phys. Stat. Sol. A Appl. Res.* **180**, 91–96 (2000b).

Lal, M., Kumar, N. D., Joshi, M. P., and Prasad, P. N., Polymerization in Reverse Micelle Nanoreactor: Preparation pf Processable Poly(p-phenylenevinylene) with Controlled Conjugation Length, *Chem. Matter* **10**, 1065–1068 (1998).

Levy, R., Honerlage, B., and Grun, J. B., Optical Nonlinearities Due to Biexcitons, in *Optical Nonlinearities and Instabilities in Semiconductors,* H. Haug, ed., Academic Press, San Diego, 1988, pp. 181–216.

Li, L.-S., Hu, J. T., Yang, W. D., and Alivisatos, A. P., Band Gap Variation of Size- and Shape-Controlled Colloidal CdSe Quantum Rods, *Nano Lett.* **1**, 349–351 (2001).

Li, X., He, Y., Talukdar, S. S., and Swihart, M. T., A Process for Preparing Macroscopic Quantities of Brightly Photoluminescent Silicon Nanoparticles with Emission Spanning the Visible Spectrum, *Langmuir* **19**, 8490–8496 (2003).

Lieber, C. M., The Incredible Shrinking Circuit, *Sci. Am.* **9**, 59–64 (2001).

Littau, K. A., Szajowski, P. J., Muller, A. J., Kortan, A. R., and Brus, L., A Luminescent Silicon Nanocrystal Colloid via a High-Temperature Aerosol Reaction, *J. Phys. Chem.* **97**, 1224–1230 (1993).

Little, R. B., El-Sayed, M. A., Bryant, G. W., and Burke, S., Formation of Quantum-Dot Quantum-Well Heterostructures with Large Lattice Mismatch: ZnS/CdS/ZnS, *J. Chem. Phys.* **114**, 1813–1822 (2001).

Lockwood, D. J., ed., *Light Emission in Silicon from Physics to Devices,* Academic Press, New York, 1998.

Lucey, D., and Prasad, P. N., unpublished results, 2003.

McCombe, B. D., and Petrou, A., Optical Properties of Semiconductor Quantum Wells and Superlattices, in *Handbook of Semiconductors,* Vol. 2, *Optical Properties,* M. Balkanski, ed., Elsevier Science Publishers, Amsterdam, 1994, Chapter 6, pp. 285–384.

Mews, A., Kadavanich, A. V., Banin, U., and Alivisatos, A. P., Structural and Spectroscopic Investigation of CdS/HgS/CdS Quantum-Dot Quantum Wells, *Phys. Rev. B* **53**, 13242–13245 (1996).

Micic, O. I., Smith, B. B., and Nozik, A. J., Core-Shell Quantum Dots of Lattice-Matched ZnCdSe$_2$ Shells on InP Cores: Experiment and Theory, *J. Phys. Chem. B* **140**, 12149–12156 (2000).

Miller, D. A. B., Chemla, D. S., Damen, T. C., Gossard, A. C., Wiegmann, W., Wood, T. H., and Burrus, C. A., Electric Field Dependence of Optical Absorption Near the Band-Gap of Quantum-Well Structures, *Phys Rev. B* **32**, 1043–1060 (1985).

Mirkin, C. A, Letsinger, R. L., Mucic, R. C., and Storhoff, J. J., A DNA-Based Method for Rationally Assembling Nanoparticles into Macroscopic Materials, *Nature* **382**, 607–609 (1996).

Murray, C. B., Kagan, C. R., and Bawendi, M. G., Synthesis and Characterization of Monodisperse Nanocrystals and Close-Packed Nanocrystal Assemblies, *Annu. Rev. Mater. Sci.* **30**, 545–610 (2000).

Nirmal, M., and Brus, L., Luminescence Photophysics in Semiconductor Nanocrystals, *Acc. Chem. Res.* **32,** 407–414 (1999).

Olbright, G. R., and Peyghambarian, N., Interferometric Measurement of the Nonlinear Index of Refraction of CdS_xSe_{1-x} Doped Glasses, *Appl. Phys. Lett.* **48,** 1184–1186 (1986).

Peng, X. G., Schlamp, M. C., Kadavanish, A. V., and Alivisatos, A. P., Epitaxial Growth of Highly Luminescent CdSe/CdS Core/Shell Nanocrystals with Photostability and Electronic Accessibility, *J. Am. Chem. Soc.* **119,** 7019–7029 (1997).

Pettersson, H., Warburton, R. J., Lorke, A., Karrai, K., Kotthaus, J. P., Garcia, J. M., and Petroff, P. M., Excitons in Self-Assembled Quantum Ring-Like Structures, *Physica* **E6,** 510–513 (2000).

Prasad, P. N., Design, Ultrastructure, and Dynamics of Nonlinear Optical Effect in Polymeric Thin Films, in *Nonlinear Optical and Electroactive Polymers,* P. N. Prasad and D. R. Ulrich, eds., Plenum Press, New York, 1988, pp. 41–67.

Prasad, P. N., *Introduction to Biophotonics,* John Wiley & Sons, Hoboken, NJ, 2003.

Prasad, P. N., and Williams, D. J., *Introduction to Nonlinear Optical Effects in Molecules and Polymers,* John Wiley & Sons, New York, 1991.

Schmitt-Rink, S., Chemla, D. S., and Miller, D. A. B., Linear And Nonlinear Optical-Properties of Semiconductor Quantum Wells, *Adv. Phys.* **38,** 89–188 (1989).

Schooss, D., Mews, A., Eychmuller, A., and Weller, H., Quantum-Dot Quantum-Well CdS/HgS/CdS—Theory and Experiment, *Phys. Rev. B* **49,** 17072–17078 (1994).

Siegman, A. E., *Lasers,* University Science Books, Mill Valley, CA, 1986.

Singh, J., *Physics of Semiconductors and Their Heterostructures,* McGraw-Hill, New York, 1993.

Svelto, O., *Principles of Lasers,* 4th edition, Plenum Press, New York, 1998.

Swihart, M. T., Li, X., He, Y., Kirkey, W., Cartwright, A., Sahoo, Y., and Prasad, P. N., "High-Rate Synthesis and Characterization of Brightly Luminescent Silicon Nanoparticles with Applications in Hybrid Materials for Photonics and Biophotonics," *SPIE Proceedings on Organic and Hybrid Materials for Nanophotonics,* A. N. Cartwright ed., SPIE, Wellingham 2003, in press.

Takagahara, T., Effects of Dielectric Confinement and Electron–Hole Exchange Interaction on Excitonic States in Semiconductor Quantum Dots, *Phys. Rev. B* **47,** 4569–4584 (1993).

Tian, Y. C., Newton T, Kotov, N. A., Guldi, D. M., and Fendler, J. H., Coupled Composite CdS-CdSe and Core-Shell Types of (CdS)CdSe and (CdSe)CdS Nanoparticles, *J. Phys. Chem.* **100,** 8927–8939 (1996).

Tredicucci, A., Gmachl, C., Capasso, F., Sivco, D. L., Hutchinson, A. L., and Cho, A. Y., A Multiwavelength Semiconductor Laser, *Nature* **396,** 350–353 (1998).

Wang, J., Gudiksen, M. S., Duan, X.; Cui, Y., and Lieber, C. M., Highly Polarized Photoluminescence and Photodetection from Single Indium Phosphide Nanowires, *Science* **293,** 1455–1457 (2001).

Wang, Y., and Herron, N., Nanometer-Sized Semiconductor Clusters: Materials Synthesis, Quantum Size Effects, and Photophysical Properties, *J. Phys. Chem.* **95,** 525–532 (1991).

Warburton, R. J., Scholein, C., Haft, D., Bickel, F., Lorke, A., Karrai, K., Garcia, J. M., Schoenfeld, W., and Petroff, P. M., Optical Emission from a Charge-Tunable Quantum Ring, *Nature* **45,** 926–929 (2000).

Weisbuch, C., and Vinter, B., *Quantum Semiconductor Structures,* Academic Press, San Diego, 1991, pp. 87–100.

Williamson, A. J., Energy States in Quantum Dots, in *Quantum Dots,* Borovitskaya, E., and Shur, M. S., eds., World Scientific, Singapore, 2002, pp. 15–43.

Wilson, W. L., Szajowski, P. J., and Brus, L., Quantum Confinement in Size-Selected Surface-Oxidized Silicon Nanocrystals, *Science* **262,** 1242–1244 (1993).

Zhong, Z. H., Qian, F., Wang, D. L., and Lieber, C. M., Synthesis of *p*-Type Gallium Nitride Nanowires for Electronic and Photonic Nanodevices, *Nano Lett.* **3,** 343–346 (2003).

CHAPTER 5

Plasmonics

This chapter deals with metallic nanostructures and their applications, an area that has rapidly expanded into a major field called *plasmonics*. In the case of metallic nanostructures, the changes in the optical properties compared to the bulk form are not derived from quantum confinement of electrons and holes as discussed in Chapter 4. Rather, the optical effects in metallic nanostructures result from electrodynamics effects and from modification of the dielectric environment.

Examples of metallic nanostructures discussed in this chapter are: metallic nanoparticles, nanorods, and metallic nanoshells. A metal–dielectric boundary on the nanoscale produces considerable changes in the optical properties, making them size- and shape-dependent. A new type of resonance called plasmon or surface plasmon resonance, localized near the boundary between the metal nanostructure and the surrounding dielectric, also produces an enhanced electromagnetic field at the interface. This enhanced field can be used for metal–dielectric interface-sensitive optical interactions that form a powerful basis for optical sensing as well as for nanoscale localized optical imaging. The latter application also provides a method for apertureless near-field imaging, already discussed in Chapter 3. A new application of plasmonics is to use a close-packed array of metallic nanoparticles for confinement and guiding of an electromagnetic wave as plasmons through a waveguide much smaller in cross section than the wavelength of light. This topic is also covered in this chapter.

Section 5.1 introduces optical properties of metallic nanoparticles and nanorods. The origin of size and shape dependence of their optical properties is explained. Section 5.2 describes metallic nanoshells composed of metallic shell coated on a dielectric core such as a silica particle. Section 5.3 explains the local field enhancement effect near the metallic surface, produced by plasmon oscillations. Section 5.4 covers the properties of light emerging from apertures of size less than its wavelengths. Section 5.5 introduces the concept of plasmonic wave guiding where light is guided as coupled plasmon oscillations through channels of cross-section significantly smaller than those of optical wave guides. Section 5.6 lists some important features of metallic nanostructures in relation to their applications. Section 5.7 introduces the concept of radiative decay engineering in which the close proximity of a metallic interface is used to manipulate the radiative decay properties of a fluorescent molecule. Section 5.8 provides highlights of the chapter.

Nanophotonics, by Paras N. Prasad
ISBN 0-471-64988-0 © 2004 John Wiley & Sons, Inc.

For further reading the suggested references are:

Kreibig and Vollmer (1995): *Optical Properties of Metal Clusters*
Link and El-Sayed (2003):*Optical Properties and Ultrafast Dynamics of Metallic Nanocrystals*

5.1 METALLIC NANOPARTICLES AND NANORODS

Metallic nanostructures have been a subject of considerable interest in recent years (Link and El-Sayed, 1999; Kelly et al., 2003; Jackson and Halas, 2001). The field of metallic nanostructures is now more popularly called plasmonics, since the major manifestation produced by optical excitations is the collective oscillation of electrons, which are localized along the interface. Hence, this wave is also called a *surface plasmon wave,* which has been already introduced in Chapter 2 as a mode of excitation localized at the interface between a metal film and the surrounding dielectric. The majority of recent work has focused on the application of metallic nanoparticles and nanoshells.

In comparison to the semiconductor nanoparticles where quantum confinement produces quantization of the electron and hole energy states to produce major modifications of their optical spectra, metallic nanoparticles exhibit major changes in their optical spectra derived from effects that can be explained using a classical dielectric picture. The light absorption by metallic nanoparticles is described by coherent oscillation of the electrons, which is induced by interaction with the electromagnetic field. These oscillations produce surface plasmon waves. It should be noted that the term "surface plasmons" is used to describe the excitations at the metal–dielectric interface in the case of flat surfaces, where the plasmons can only be excited by using special geometries (e.g., the Kretschmann geometry as described in Chapter 2) required for matching of the wavevector, k_{sp}, of the surface plasmon wave with that of light producing it. In the case of metal nanostructures (e.g., nanoparticles), plasmon oscillations are localized and thus not characterized by a wavevector k_{sp}. To make a distinction, the plasmon modes in metallic nanoparticles are also sometimes referred to as localized surface plasmons. These localized plasmons are excited by light absorption in the nanoparticles, with the specific absorption bands being referred to as *plasmon bands*. To excite these localized plasmons in metallic nanostructures, no special geometry, such as those required for plasmon excitation along a planar metal–dielectric interface, is required. The specific wavelengths of light absorption producing plasmon oscillations are called surface plasmon bands or simply plasmon bands.

The main photonic applications of the metallic nanoparticles are derived from the local field enhancement under the resonance plasmon generation condition, which leads to enhancement of various light-induced linear and nonlinear optical processes within nanoscopic volume of the media surrounding the nanoparticles. Such field enhancement has been used for apertureless near-field microscopy. Another application presented by the metallic nanoparticles is that using an array of interacting metallic nanoparticles; light can be coupled and propagated as an electromagnetic wave through a dimension of nanometers in cross section, much smaller than the optical waveguiding dimension.

A systematic study of the optical properties of metallic nanoparticles reveals the following features (Link and El-Sayed, 1999):

- For metallic nanoparticles significantly smaller than the wavelength of light, light absorption is within a narrow wavelength range. The wavelength of the absorption peak maximum due to the surface plasmon absorption band is dependent on the size and the shape of the nanocrystals, as well as on the dielectric environment surrounding the particles.
- For extremely small particles (<25 nm for gold), the shift of the surface plasmon band peak position is rather small. However, a broadening of the peak is observed.
- For larger nanoparticles (>25 nm for gold), the surface plasmon peak shows a red shift. Figure 5.1 illustrates these features for a series of gold nanoparticles of different sizes.
- For a nanorod-shaped metallic nanoparticle, the plasmon band splits into two bands corresponding to oscillation of the free electrons along (longitudinal) and perpendicular (transverse) to the long axis of the rod. Figure 5.2 shows this splitting for a gold nanorod.
- The transverse mode resonance is close to that observed for spherical particles, but the longitudinal mode is considerably red-shifted, depending strongly on the aspect ratio, which is the length divided by the width of the rod.

The origin of these shifts is not due to quantum confinement. The quantum confinement does affect the energy spacing of the various levels in the conduction band.

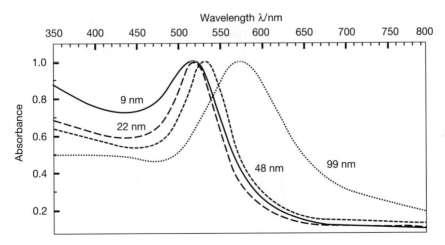

Figure 5.1. Optical absorption spectra of gold nanoparticles of different sizes. From Link and El-Sayed (1999), reproduced with permission.

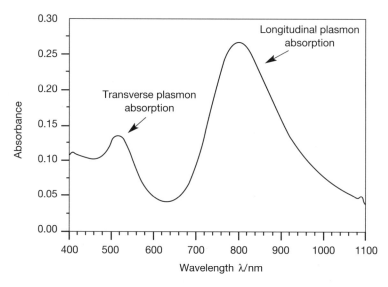

Figure 5.2. Absorbance of gold nanorods. From Link and El-Sayed (1999), reproduced with permission.

However, the quantization, derived from the confinement, affects the conductive properties of the metal and is often used to describe the metal-to-insulator transition occurring as the particle size is reduced from microscopic to nanoscopic size. When the dimensions of the metallic nanoparticles are large, the spacing of levels within the conduction band is significantly less than the thermal energy, kT (k is Boltzmann's constant and T is the temperature in kelvin), and the particle exhibits a metallic behavior. When the nanoparticles approach a size at which the increased energy separation due to the quantum confinement effect (smaller length of the box for the free electron) is more than the thermal energy, an insulating behavior results because of the presence of these discrete levels. However, the energy level separations are still too small to affect the optical properties of metals in the UV to the IR range.

Although a number of theoretical models have been proposed (Kelly et al., 2003), the original classical model of Mie (Born and Wolf, 1998) is often used to

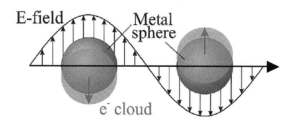

Figure 5.3. Schematic of plasmon oscillation in a metal nanosphere. From Kelly et al. (2003), reproduced with permission.

describe the optical properties of the metal nanoparticles. Often one utilizes a dipole approximation in which the oscillation of conduction electrons (plasmon oscillations), driven by the electromagnetic field of light, produces oscillating dipoles along the field direction where the electrons are driven to the surface of the nanoparticles as shown in Figure 5.3. A more rigorous theory (Kelly et al., 2003) shows that this dipolar-type displacement is applicable to smaller-size particles and gives rise to an extinction coefficient k_{ex} (measure of absorption and scattering strengths collectively) by the following equation (Kreibig and Vollmer, 1995):

$$k_{ex} = \frac{18\pi NV\varepsilon_h^{3/2}}{\lambda} \frac{\varepsilon_2}{[\varepsilon_2 + 2\varepsilon_h]^2 + \varepsilon_2^2} \tag{5.1}$$

Here λ is the wavelength of light, and ε_h is the dielectric constant of the surrounding medium. The terms ε_1 and ε_2 represent the real and the imaginary parts of the dielectric constant, ε_m, of the metal ($\varepsilon_m = \varepsilon_1 + i\varepsilon_2$) and are dependent on the frequency ω of light. If ε_2 is small or weakly dependent on ω, the absorption maximum corresponding to the resonance condition is produced when $\varepsilon_1 = -2\varepsilon_h$, leading to a vanishing denominator. Hence, a surface plasmon resonance absorption is produced at optical frequency ω at which the resonance condition $\varepsilon_1 = -2\varepsilon_h$ is fulfilled. The size dependence of the surface plasmon resonance comes from the size dependence of the dielectric constant ε of the metal. This is often described as the intrinsic size effect (Link and El-Sayed, 2003). In the case of noble metals such as gold, there are two types of contributions to the dielectric constant of the metal: One is from the inner d electrons, which describes interband transition (from inner d orbitals to the conduction band), and the other is from the free conduction electrons. The latter contribution, described by the Drude model (Born and Wolf, 1998; Kreibig and Vollmer, 1995), is given as

$$\varepsilon_D(\omega) = 1 - \frac{\omega_p^2}{\omega^2 + i\gamma\omega} \tag{5.2}$$

where ω_p is the plasmon frequency of the bulk metal and γ is the damping constant relating to the width of the plasmon resonance band. It relates to the lifetime associated with the electron scattering from various processes. In the bulk metal, γ has main contributions from electron–electron scattering and electron–phonon scattering, but in small nanoparticles, scattering of electrons from the particle's boundaries (surfaces) becomes important. This scattering produces a damping term γ that is inversely proportional to the particle radius r. This dependence of γ on the particle size introduces the size dependence in $\varepsilon_D(\omega)$ [thus ε_1 in Eq. (5.1)] and, consequently, in the surface plasmon resonance condition.

For larger-size nanoparticles (>25 nm for gold particles), higher-order (such as quadrupolar) charge cloud distortion of conduction electrons becomes important, as shown in Figure 5.4. These contributions induce an even more pronounced shift of the plasmon resonance condition as the particle size increases. This effect for the larger size particle is referred to as the *extrinsic size effect* (Link and El-Sayed, 2003). The position and the shape of the plasmon absorption band also depends on

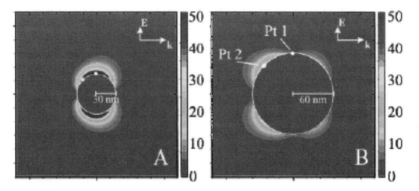

Figure 5.4. Calculated higher-order charge cloud distorsion around gold nanoparticles of sizes > 25 nm. From Kelly et al. (2003), reproduced with permission.

the dielectric constant ε_h of the surrounding medium as the resonance condition is described by $\varepsilon_1 = -2\varepsilon_h$ (see above). Hence, an increase in ε_h leads to an increase in the plasmon band intensity and band width, as well as produces a red shift of the plasmon band maximum (Kreibig and Vollmer, 1995). This effect of enhancing the plasmon absorption by using a higher dielectric constant surrounding medium forms the basis of what is known as *immersion spectroscopy*.

Metallic nanoparticles in different nonspherical shapes have been studied for some time. In recent years, shapes such as pyramidal have been made by Mirkin and co-workers (Jin et al., 2001). For small ellipsoidal particles the dielectric constant has a simple but illustrative expression. Let us consider the case of metal particles with dielectric constant ε_m and embedded in a host medium of dielectric constant ε_h. The volume fraction of particles (defined by the symbol f) is kept small in bulk composites due to the otherwise excessive losses. For oriented ellipsoidal particles and small volume fraction, (i.e., $f \ll 1$) of particles, the dielectric tensor has three values along the principal axes. The form of the result is (Bohren and Huffman, 1983)

$$\varepsilon_i = \varepsilon_h + f \frac{\varepsilon_h(\varepsilon_h - \varepsilon_m)}{\Gamma_i \varepsilon_m + (1 - \Gamma_i)\varepsilon_h} \quad (5.3)$$

Γ_i is a set of three parameters, defined along the principal axes of the particle, characterizing its shape. The parameters lie in the range (0,1) and their sum is restricted to be $\Gamma_1 + \Gamma_2 + \Gamma_3 = 1$. For the degenerate case of a sphere, $\Gamma_I = 1/3$ and the birefringence disappears. The denominator is resonant at the surface plasmon resonance as discussed above. The new resonance position is $\varepsilon_m = -(1 - \Gamma_i)\varepsilon_h/\Gamma_i$. By changing the shape, the resonance frequency can be moved by hundreds of nanometers. This has been demonstrated in a number of papers with prolate or oblate spheroidal particles in a waveguide geometry (Bloemer et al., 1988, Bloemer and Haus, 1996).

Schatz and co-workers have developed numerical models such as discrete dipole approximation (DDA) that can be applied to situations where the particle size is large enough that the simple expression is no longer valid. Their approach incorporates multipoles in the electromagnetic response (Hao and Schatz, 2003).

5.2 METALLIC NANOSHELLS

Another type of metallic nanoparticles receiving attention recently are nanoshelled particles containing two materials called a *core* and a *shell* (Haus et al., 1993; Zhou et al., 1994; Halas, 2002; Jackson and Halas, 2001). The core and shell materials are different and sometimes referred to as *heterostructured nanoparticles*. When the shell material is metallic, the changes of the surface plasmon resonance can be dramatic. Zhou et al. (1994) synthesized particles with a AuS core surrounded by a gold shell. The particle sizes are in the range of 30 nm in diameter, and the surface plasmon resonance absorption feature was shifted by more than 500 nm. In Halas's papers the metallic nanoshells are larger and composed of a dielectric core such as silica, of radius ~40–250 nm, which is surrounded by a thin metallic shell of thickness ~10–30 nm. Like the metallic nanoparticles and nanorods discussed in the previous section, the optical properties of these nanoshells are governed by plasmon resonances. However, the plasmon resonances are typically shifted to much longer wavelengths than those in the corresponding solid metal nanoparticles.

The core–shell particle has a core dielectric ε_c and a shell dielectric constant ε_s, and the particles are embedded in a host medium ε_h. The dielectric function for the coated spherical particles has the form (Bohren and Huffman, 1983)

$$\varepsilon = \varepsilon_h + f \frac{\varepsilon_h[(\varepsilon_s - \varepsilon_h)(\varepsilon_c + 2\varepsilon_s) + \delta(\varepsilon_c - \varepsilon_s)(\varepsilon_h + 2\varepsilon_s)]}{[(\varepsilon_s + 2\varepsilon_h)(\varepsilon_c + 2\varepsilon_s) + 2\delta(\varepsilon_c - \varepsilon_s)(\varepsilon_s - \varepsilon_h)]} \quad (5.4)$$

where δ is the ratio of core volume to the volume of the particle.

The local field enhancement factor extracted from the second term in Eq. (5.4) is

$$\gamma = \frac{[(\varepsilon_s - \varepsilon_h)(\varepsilon_c + 2\varepsilon_s) + \delta(\varepsilon_c - \varepsilon_s)(\varepsilon_h + 2\varepsilon_s)]}{[(\varepsilon_s + 2\varepsilon_h)(\varepsilon_c + 2\varepsilon_s) + 2\delta(\varepsilon_c - \varepsilon_s)(\varepsilon_s - \varepsilon_h)]} \quad (5.5)$$

The plasmon resonance frequencies and the spectra can again be described by Mie scattering using dielectric functions that incorporate changes due to enhanced electron scattering, derived from the ultrathin metallic layer structure (Averitt et al., 1997).

In the metallic shells, the following features are observed.

- As the metal shell thickness is decreased, keeping the dielectric core size constant, the optical resonance shifts to a longer wavelength. Figure 5.5 shows the calculated plasmon resonance spectra for different shell thicknesses of gold on a silica core of 60-nm radius. The spectral shift produced by varying the metallic shell thickness can cover a broad spectral range from visible to

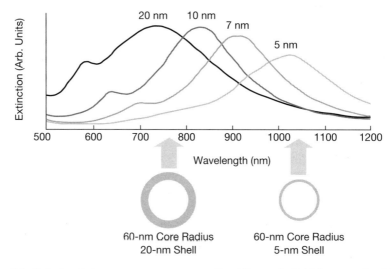

Figure 5.5. Calculated plasmon resonance spectra for different shell thicknesses of gold on a silica core of 60-nm radius. From Halas (2002), reproduced with permission.

IR. Theoretically, it is even possible to shift the plasmon resonance to beyond 10 μm in the IR.
- If the core/shell size ratio is kept constant and the absolute size of the particle is varied, the small particles observing the dipole limit (similar to the metallic nanoparticles) produce predominantly light absorption. As the particle size increases, the contribution of scattering relative to absorption increases.
- If the particle size increases beyond the dipole limit, multipole plasmon resonances appear in the extinction spectrum of the particles.

Methods have been reported recently, providing control to fabricate the core–shell structures with precision. An example of the approach to fabricate metal nanoshells is shown in Figure 5.6 (Hirsch et al., 2003). In this scheme, the silica nanoparticle is chemically modified to have amine groups on the surface. Small gold particles from a gold colloid suspension are absorbed onto the amine groups

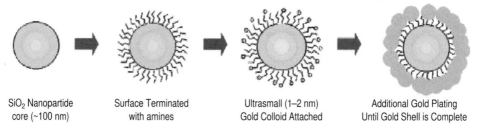

Figure 5.6. Schematics of fabrication of a gold shell around a silica nanoparticle. From Halas (2002), reproduced with permission.

on the surface. The SiO$_2$ nanoparticles decorated with the metal nanoparticles are then treated with HAuCl$_4$ in the presence of formaldehyde, whereby additional gold is formed by the reduction reaction, eventually leading to coalescing of the various metallic nanoparticles. This results in formation of a complete metal shell (Hirsch et al., 2003). This approach yielded a minimal metal thickness of ~5 nm. Other methods of producing silver nanoshells are described by Jackson and Halas (2001).

These precisely fabricated metal nanoshells, with judiciously chosen plasmon resonances, have been proposed for a number of applications. They range from application of nanoshells to prevent photo-oxidation of semiconducting polymer-based devices, to whole-blood immunoassay. The photo-oxidation of polymers generally occurs from the long-lived triplet state. Halas and co-workers showed that by tuning the nanoshell plasmon resonance to coincide with the energy of the triplet exciton state, the photo-oxidation process was considerably reduced. For the whole-blood immunoassay, gold nanoshells were used (Hirsch et al., 2003). In this method, appropriate antibodies are conjugated to the metal surface of the metallic shell, using standard surface coupling chemistry. A multivalent analyte binds to more than one such nanoshell–antibody conjugate, forming particle dimers and higher-order aggregates. Thus, the analyte (antigen) acts as a linker between nanoshells. The plasmon resonances of the nanoshell dimers and larger-size aggregates are red-shifted, compared to that of a single nanoshell. Thus, the dimerization and higher-order aggregation leads to a decrease in the extinction coefficient at a wavelength corresponding to the single-shell plasmon resonance. This decrease in the extinction was demonstrated to provide a simple immunoassay capability detecting subnanograms per milliliter quantities of analytes.

5.3 LOCAL FIELD ENHANCEMENT

Metallic films have been utilized in a surface plasmon geometry to produce enhanced electromagnetic field near the surface (evanescent wave) as described in Chapter 2. The metallic nanoparticles and nanoshells also produce significant enhancement of electric field near the particle surface. This enhancement has long been known and used in surface-enhanced Raman spectroscopy (SERS) (Lasema, 1996; Chang and Furtak, 1982). The enhancement is large enough to observe Raman spectra, even from a single molecule. More recently, this field enhancement using a metallic nanostructure has been used for apertureless near-field microscopy. The electromagnetic enhancement is derived from plasmon excitation in the particle. As discussed in Section 5.1, the metallic nanoparticles exhibit distinct dipole, quadrupole, and even higher multipole plasmon resonances, depending on their size and shape. Excitation of these resonances creates a considerably enhanced local electric field external to the particles. This field determines the normal and single molecule SERS intensities. Schatz and co-workers (Hao and Schatz, 2003) have conducted a detailed analysis of the local field enhancement in metallic nanoparticles using the discrete dipole approximation (DDA) described briefly in Section 5.1.

The DDA calculation by Hao and Schatz shows that for spherical Ag particles of radius less than 20 nm, the maximum field enhancement is less than 200 near the plasmon resonance at the wavelength of 410 nm. This electric field enhancement is wavelength-dependent. As the particle size increases, the field enhancement reduces, with a shift of the plasmon resonance to a longer wavelength. For example, for a Ag sphere of 90-nm radius, the plasmon resonance is at 700 nm with field enhancement only of 25.

An interesting observation is the SERS enhancement to the level where a single molecule can be detected from the hot site (considerably enhanced electric field) between two spherical particles (Nie and Emory, 1997; Michaels et al., 1999). Hao and Schatz (2003) used DDA calculation to show that for a dimer of 36-nm silver particles separated by 2 nm, there is a plasmon resonance at 520 nm, with polarization predominantly dipolar in character, and another plasmon resonance at 430 nm, quadrupolar in character. Maximum electric field enhancements for both dipole and quadrupole resonances are found to occur at the midpoint between the two spheres. The field enhancement is 3500 times at 430 nm (quadrupole resonance) and 11,000 times at 520 nm (dipole resonance). Hence these enhancements are substantially larger than those found for isolated Ag spherical nanoparticles.

Schatz and co-workers (Hao and Schatz, 2003) also show that in the case of non-spherical nanoparticles, such as a triangular prism, the field enhancement is considerably larger than that for a comparable-sized spherical particle.

5.4 SUBWAVELENGTH APERTURE PLASMONICS

Light, emerging from an aperture of size less than its wavelength (subwavelength aperture), is diffracted in all directions. Furthermore, the transmission of light through a subwavelength hole is normally extremely low. It has been reported that the use of a plasmonic structure in the form of a periodic array of subwavelength holes in a metal film produces extraordinarily high optical transmission (Ebbesen et al., 1998; Ghaemi et al., 1998; Lezec et al., 2002).

Ebbesen et al. (1998) reported that optical transmission through subwavelength hole arrays in an optically thick metal film can be enhanced by orders of magnitude by coupling with the surface plasmon resonance of the metal film. As already discussed in Chapter 2 and restated in Section 5.1, light cannot be coupled directly to the surface plasmon mode at a flat metal surface. This is because the wavevector k_{sp} of a surface plasmon wave is significantly larger than that of light propagating in vacuum or air. For this reason, various geometries are used to match the wavevectors of surface plasmon and light to simultaneously conserve energy E and wavevector (momentum) **k**. One such geometry discussed in Chapter 2 is the Kretschmann geometry using a prism. Another geometry utilizes a periodic corrugated (grating) structure on the metal film. The grating wavevector thus provides the additional wavevector to match the wavevectors of light and surface plasmon, thus providing the coupling between them.

In their study, Ebbesen et al. (1998) used a silver film of thickness ~ 200 nm in which a square array of cylindrical cavities was produced using a focused ion beam (FIB) system. The period of the array was ~ 900 nm and the hole diameter was 150 nm. They found that interaction of the incident light with the surface plasmon mode

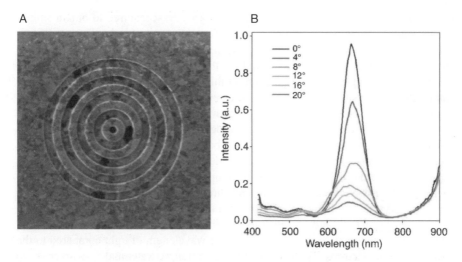

Figure 5.7. (A) Focused ion beam micrograph image of a bull's-eye structure surrounding a cylindrical hole in a suspended Ag film (groove periodicity, 500 nm; groove depth, 60 nm; hole diameter, 250 nm; film thickness, 300 nm). (B) Transmission spectra recorded at various collection angles for a bull's-eye structure on both sides of a suspended Ag film (groove periodicity, 600 nm; groove depth, 60 nm; hole diameter, 300 nm; film thickness, 300 nm). The tail above 800 nm is an artifact of the spectral measurement. The structure is illuminated at normal incidence with unpolarized collimated light. From Lezec (2002), reproduced with permission.

leads to wavelength-selective transmission, with efficiencies that are about 1000 times higher than that expected for subwavelength holes. The transmission enhancement is mediated by coupling to surface plasmons of the metal–dielectric interface on both sides of the metal film through the holes.

Lezec et al. (2002) showed that a single aperture, surrounded by a periodic corrugation in a metal surface, as shown in Figure 5.7, also displays surface plasmon-enhanced transmission. According to them, light exiting a single aperture may follow the reverse process described above which takes place in the presence of a periodic array of holes. Here the presence of a periodic structure on the exit surface can produce light emerging at certain angles for certain wavelengths. Lezec et al. observed that the transmitted light had a small angular divergence whose directionality can be controlled. Due to the very high transmission efficiencies, together with a subwavelength scattering, the subwavelength aperture plasmonics is attractive for applications in nanofabrication.

5.5 PLASMONIC WAVE GUIDING

A conventional optical waveguide that traps light within its boundaries has been briefly described in Chapter 2. Such a waveguide is a dielectric medium whose dimensions are controlled by the refractive index contrast required to obtain single-mode guiding, but are also limited by the diffraction limit of light. Thus, the minimum size of light confinement (transverse dimension of a waveguide) is of the order of $\lambda/2n$, where λ is the wavelength of the light guided and n is the effective refractive index of the guiding medium at this wavelength. For light in the visible range, this dimension is of the order of a few hundred nanometers. Furthermore, conventional channel waveguides cannot provide sharp bends for light guiding, due to considerable leakage of light at large angle bends. For this reason, other mechanisms of light confinement and guiding are being investigated so that an integrated and highly compact photonic circuit can be developed.

As shall be discussed in Chapter 9, photonic crystal structures overcome the bend problem and provide light guiding through sharp 90° corners. However, a photonic crystal, involving a dielectric medium for guiding of light, has the same diffraction limitation, and the thickness of the guiding dimension has to be large (hundreds of nanometers) because of the long wavelength of light compared to the wavelength of electrons. To overcome this limitation, Atwater and co-workers have proposed plasmonic guiding (Atwater, 2002; Maier et al., 2003) in which a periodic array of metallic nanostructures, embedded in a dielectric medium, guides and modulates light transmission in a regime dominated by near-field coupling. This plasmon waveguide consists of an array of metallic nanoparticles, nanorods, or nanowires, with their plasmon resonances in the region of optical waveguiding.

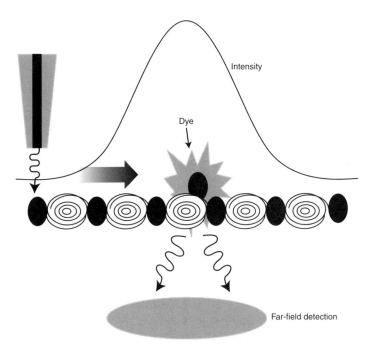

Figure 5.8. Schematics of plasmonic guiding and its use to generate fluorescence excitation in a dye. From Maier et al. (2003), reproduced with permission.

A schematic of the plasmon waveguide is shown in Figure 5.8. The plasmon excitations in the nanoparticles are coupled so that in the case of gold or silver particles <50 nm, the dominant dipole field (compared to higher multipole contributions), resulting from a plasmon oscillation in a single metallic nanoparticle, can induce plasmon oscillation in a closely spaced neighboring particle. This plasmon oscillation can propagate through the array as a coherent mode with a wavevector **k** along the nanoparticle array. In Figure 5.8, light emerging from the tip of a tapered fiber of a near-field microscope (see Chapter 3) locally excites the plasmon oscillation (Maier et al., 2003). The light wave is thus converted into electromagnetic energy, now in the form of plasmon oscillation, and is transported through the array of metallic nanoparticles. This guided electromagnetic energy can excite a dye, shown here as a fluorescent nanosphere, in the guided path. The resulting fluorescence can be collected in the far field, as shown in the figure. Atwater (2002) has shown that energy can be coherently transported in these subwavelength guiding structures of cross section of the order of $\lambda/20$, with the possibility of propagation through corners and tee structures. Their calculation for 50-nm silver nanospheres at a center-to-center distance of 75 nm shows that energy propagation velocities of about 10% of the speed of light can be achieved. Internal damping of the surface plasmon mode, resulting from resistive heating, was shown to induce transmission losses of about 6 dB/μm. Hence, the loss is high, but the attractive feature offered by a plasmon waveguide is its smaller dimension compared to a dielectric optical waveguide. Maier et al. (2003) have reported propagation distances of ~0.5 μm using closely spaced silver rods.

5.6 APPLICATIONS OF METALLIC NANOSTRUCTURES

A well-established application of use of metallic nanostructures, having a long history, is surface-enhanced Raman spectroscopy (SERS), mentioned in Section 5.3 (Chang and Furtak, 1982). More recently, a number of applications of metallic nanostructures have emerged. The three features utilized in these applications are:

- Local field enhancement around the surface of a metallic nanostructure
- Evanescent wave emanating from the surface, when exciting a surface plasmon resonance
- Sensitivity of the surface plasmon resonance to the dielectric medium surrounding the metallic nanostructure

These three types of applications are briefly outlined here. Some of them are discussed in detail in other sections of this book.

Local Field Enhancement. The local field enhancement has been utilized for a number of applications. The application of metallic nanoparticles to enhance the field in the neighboring nanoscopic region has formed the basis for apertureless near-field microscopy and spectroscopy. This approach for near-field microscopy has been discussed in Chapter 3. The local field enhancement—and, therefore, confinement of electromagnetic radiation around the metallic nanostructure—has been proposed for a new method called *plasmonic printing* for photofabrication of nanostructures. Plasmonic printing is described in Chapter 11, which deals with nanolithography.

The local field enhancement also contributes to the enhancement of the linear and nonlinear optical transitions in molecules within nanoscopic distance from a metallic nanostructure. Plasmonic enhancement of fluorescence by metallic nanoparticles is one example that has been utilized to increase fluorescence detection sensitivity, particularly for biological applications and for nanosensors. There are other plasmonic interaction factors also affecting the fluorescence. These effects are discussed in detail in Section 5.7. More recently, dramatic enhancement has been reported by Marder, Perry, and co-workers (Wenseleers et al., 2002) for two-photon excitation (discussed in Chapter 2), now extensively used for two-photon microscopy and two-photon three-dimensional microfabrication (see Chapter 11).

Nonlinear optical properties of nanoshelled particles were discussed and calculated in a number of papers (Kalyaniwalla et al. 1990; Haus et al. 1993). The local-field effect in the particles leads to an unusual feature called *intrinsic optical bistability*. This effect was reportedly observed in a silver-coated particle with a CdS core (Neuendorf et al., 1996).

Evanescent Wave Excitation. As discussed in Chapter 2, when light is coupled as a surface plasmon excitation in a metallic film or a nanostructure, it produces an evanescent electromagnetic field emanating from the surface and exponentially decaying into the surrounding dielectric field. This evanescent wave can be utilized to preferentially excite an optical transition in fluorescent molecules or fluorescent nanospheres near the surface of a metallic nanostructure. This evanescent wave excitation has been utilized in various fluorescence-based optical sensors. It has also been utilized to convert a propagating plasmon wave in a plasmon waveguide to a fluorescent optical signal, as discussed in Section 5.5.

Dielectric Sensitivity of Plasmon Resonance. The surface plasmon resonance frequency and its width are both sensitive to the surrounding dielectric. This feature has been discussed in Section 5.1 for metallic nanostructures and in Section 5.2 for metallic nanoshells. This sensitivity forms a basis for detection of biological analytes, which can bind to the surface of the metallic nanostructures through a typical antigen–antibody type coupling chemistry. Some of the biological applications are discussed in Chapter 13.

5.7 RADIATIVE DECAY ENGINEERING

Plasmonics can be used to manipulate the radiative decay properties of a fluorescent molecule (fluorophore). This approach is sometimes referred to as *radiative decay engineering* (Lakowicz, 2001). By using different sizes and shapes of the particles, and varying the distance of the fluorophores relative to the surface, one can achieve enhanced quantum yield or increased quenching of fluorescence. Furthermore, an appropriate fluorophore–metallic structure geometry can also produce directional emission rather than isotropic emission, which is observed in all directions around the fluorophore.

The interactions of a fluorophore with a metal nanostructure have been described by Barnes to be of three types (Barnes, 1998; Worthing and Barnes, 1999):

Local Field Enhancement. This effect has been discussed in detail in Section 5.3. This enhanced field concentrates the local excitation density around the nearby fluorophore molecule, producing enhancement of fluorescence.

Metal–Dipole Interaction. This interaction, already discussed in Chapter 3, Section 3.6, introduces an additional channel of nonradiative decay which is strongly dependent on the distance between the fluorophore molecule and the metal surface. Thus emission of fluorophores within 5 nm of a metal surface is often quenched.

Enhancement of Radiative Rate. Interaction between a fluorophore and a metal nanostructure can increase the intrinsic radiative decay rate of the fluorophore. To understand it, one needs to look at the microscopic theory of radiative transition. The radiative rate is defined by the quantum mechanical transition probability per unit time, W_{ij}, which according to the Fermi's Golden Rule is given as

$$W_{ij} = \frac{2\pi}{h}|\mu_{ij}|^2 \rho(\nu_{ij}) \tag{5.6}$$

where μ_{ij} is the transition dipole moment connecting the initial state, i, and the final state, f. The term $\rho(\nu_{ij})$ is the photon mode density at transition frequency ν_{ij} corresponding to the energy gap between the initial and the final states. It is this density term ρ which can be substantially increased by localizing a fluorophore within nanoscopic distances from a metal surface (Drexhage, 1974; Barnes, 1998; Worthing and Barnes, 1999). For this interaction to manifest, the fluorophore does not have to be in direct contact with the metallic surface. It is a microcavity (described in Chapters 2 and 9) effect (quantum electrodynamics), whereby localization of photon interactions in nanoscopic domains enhances the density of photon states.

Now the combined action of all three effects can be described by an apparent quantum yield Y, a net quantum yield Q_m, and lifetime τ_m for a fluorophore near the metal surface as

$$Y = |L(\omega_{em})|^2 Q_m \tag{5.7}$$

where

$$Q_m = \frac{(\Gamma + \Gamma_m)}{(\Gamma + \Gamma_m + k_{nr})} \tag{5.8}$$

$$\tau_m = (\Gamma + \Gamma_m + k_{nr})^{-1} \tag{5.9}$$

In Eq. (5.7) $L(\omega_{em})$ is the local field enhancement at the emission frequency ω_{em}. Γ_m is the enhancement of the radiative transition because of interactions with the metal surface. Γ is the radiative rate in the absence of the metal. The sum $\Gamma_m + \Gamma$ is the total radiative rate that is directly given by the quantum mechanical transitional probability rate described by Eq. (5.6). The enhancement Γ_m is thus given by the enhancement of the photon mode density $\rho(\nu_{ij})$ of Eq. (5.6). The term k_{nr} represents the nonradiative decay rate, including quenching due to metal–dipole interactions. The apparent quantum yield, Y, signifies the effect of local field enhancement (concentration of the field intensity near the metallic particle), and it can even be larger than 1. In contrast, a true quantum yield cannot be larger than 1.

If the local field $L(\omega_{em})$ is significantly enhanced or if Γ_m is significantly larger than k_{nr}, one can observe a net increase in the quantum yield, yet there is a concurrent decrease of the lifetime τ_m. In the opposite limit, when k_{nr} dominates due to significantly increased value of metal-induced quenching, both the quantum yield and the lifetime are reduced by the presence of the metal nanoparticles (or metal surfaces).

The radiative decay rate enhancement has been calculated for metallic particles, which are elongated spheroids of different aspect ratios. For an elongated silver spheroid, enhancement as large as 2000-fold has been predicted (Gresten and Nitzan, 1981). Let us take an example of a fluorophore for which the free-space quantum yield is 0.001, implying k_{nr} to be ~1000 times larger than the radiative rate, Γ [Eq. (5.8)]. Now if the radiative rate increases by a factor of 1000 due to a nearby metallic particle, the quantum yield will be 0.5, which is a 500-fold increase. Naturally, the effect of radiative rate enhancement is more pronounced for extremely low quantum yield or nearly "nonfluorescent" molecules, in which case nearly a million-fold increase in the number of emitted photons can be realized. This increase in the quantum yield is of significance to biology and biotechnology for both bioimaging and fluorescence-based biosensing.

The metal-induced enhancement of fluorescence has been more frequently investigated using fluorophores deposited on a silver island film with a spacer. These silver island films contain ~2-nm-diameter circular domains (islands) of silver metal and have frequently been used for surface-enhanced Raman spectroscopy (SERS) (Lasema, 1996). An example presented here is a thin film of the organometallic $Eu(ETA)_3$ where the electron lone pair containing ligand ethanol trifluoroacetonate (ETA) chelates europium (Eu) (Weitz et al., 1982). When deposited on silver island films, the fluorescence intensity of this chelate increased by a factor of 5 while the lifetime decreased 100-fold, from 280 μs to 2 μs as shown in Figure 5.9.

The magnitudes of the three effects described above depend on the location of the fluorophore around the particle and the orientation of its dipole relative to the metallic surface. For fluorophore distances <5 nm from the metal surface, the metal-induced quenching (increase of k_{nr}) dominates. However, for distances between 5 nm to 20 nm, enhancement of fluorescence due to either field enhancement or radiative rate increase can be realized.

As mentioned in Section 5.3, surface-enhanced Raman scattering (SERS) has proved to be an extremely valuable technique for obtaining Raman spectra of mole-

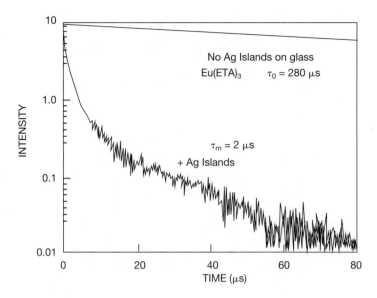

Figure 5.9. Fluorescence decay of Eu(ETA)$_3$ on silver-island films. Eu^{3+} was complexed with thenolytrifluoroacetonate (ETA). From Weitz et al. (1982), reproduced with permission.

cules at very low concentration, even to the single-molecule level (Lasema, 1996). SERS is most pronounced when there is a direct contact between the molecule and the metallic nanostructure. It is indeed a surface enhancement produced by the molecules localized on the surface of a metallic nanoparticle or an array of a metallic structure (such as an array of silver nanoparticles or silver island films). In contrast, fluorescence enhancement is observed at distances >5 nm at which the metal quenching effect is significantly reduced.

The comparison between SERS and plasmonic-enhanced fluorescence is nicely illustrated by the work of Cotton and co-workers (Sokolov et al., 1998). They used octadecanoic acid (ODA) Langmuir–Blodgett films (see Chapter 8), transferred from one monolayer (2.5 nm thick) to increasing number of multilayers on the top of a film of colloidal silver nanoparticles deposited on a glass surface that is derivatized to contain sulfhydryl groups. Thus, they produced separating layers of varying thickness between the metal surface and a fluorescein dye-labeled lipid (Fl-DPPE) which was then deposited on the top of the intervening layer Langmuir–Blodgett film. To complete the systematic study, they also deposited the dye-labeled lipid film directly on the metal surface. Figure 5.10 shows the results of emission obtained by excitation with the 488-nm laser line from an argon-ion laser. It also includes the result on the Fl-DPPE film deposited on a base glass slide. No fluorescence or Raman spectrum is observed in the absence of the metal. When Fl-DPPE is in the vicinity of the metal, both the fluorescence (broad emission) and the Raman peaks (sharp transitions) are significantly enhanced. When the Fl-DPPE is separated from the metal surface by the increasing the number of ODA monolayers, the

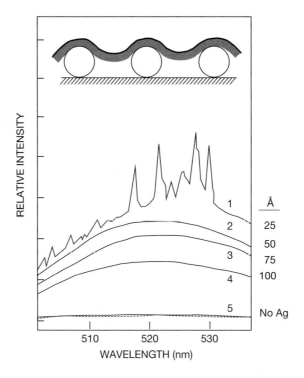

Figure 5.10. Emission spectra of mixed monolayers of fluorescein-labeled phospholipid FIPPE)/phospholipid (DPPE) with mole ratio 1:3 for different numbers of octadecanoic acid spacer layers (2.5 nm thick) on (1) mixed monolayer directly transferred on a colloidal film, (2) one spacer layer, (3) three spacer layers, (4) five spacer layers, and (5) mixed monolayer with three spacer layers on a bare glass slide. The dashed line epresents the instrument background. The spectra were taken with 488.0-nm laser excitation. The inset shows the sample configuration. From Sokolov et al. (1998), reproduced with permission.

Raman peaks disappear but the enhancement of fluorescence is still manifested, even at a distance of 10 nm from the metal surface.

An application of plasmonic-induced lifetime shortening is in enhancing the photostability of dyes. Photobleaching of dyes is a major problem in fluorescence-based sensing and imaging. When a molecules spends less time in the excited state, due to a shorter lifetime, possibility of its photochemical destruction is significantly reduced. Hence a significant reduction in the lifetime together with an enhanced quantum yield, achieved by using plasmonics, can greatly reduce photobleaching and, at the same time, greatly increase the detection limit.

Another feature offered by plasmonics for fluorescence detection is to produce highly directional emission. Normal fluorescence is emitted isotropically, of which only a small fraction can be captured by typical collection optics. The use of a continuous metallic film in a Kretschman-type geometry, discussed in Chapter 2, or a metallic grating structure allows the light to be coupled to the surface plasmon

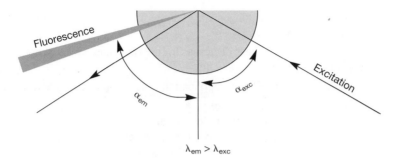

Figure 5.11. Biological assays based on evanescent wave surface plasmon excitation and directional emission. From Lakowicz et al. (2003), reproduced with permission.

mode (in this case a surface plasmon with a wavevector k_{sp}, see Chapter 2) at a specific plasmon coupling angle that produces wavevector matching between the surface plasmon wave and the optical wave (Weber et al., 1979; Benner et al., 1979; Kitson et al., 1996). In a geometry shown in Figure 5.11, which is similar to the one described in Chapter 2, the generated fluorescence can couple back into the metal surface and emerge at the plasmon coupling angle for the emission wavelength (Lakowicz et al., 2003; Weber et al., 1979; Benner et al., 1979).

5.8 HIGHLIGHTS OF THE CHAPTER

- Plasmonics encompasses the optical science and technology of nanostructures such as nanoparticles, nanorods, nanoshells, nanoscale thick films, and nano-size metallic islands.
- Interaction of metallic nanostructures with light generates resonances, called plasmons or surface plasmons, that give rise to a plasmon absorption band in a specific wavelength range. The plasmon band is both size- and shape-dependent.
- The size-dependent optical properties are not a result of quantum confinement, but are due to dielectric and electrodynamic effects.
- The absorption of light produces coherent, collective oscillation of electrons that are localized near the surface of the metallic nanostructure.
- Resonance coupling of light into a metallic nanostructure generating surface plasmon waves produces a large enhancement of the local electric field near the metal surface.
- For a nanorod-shaped metallic nanoparticle, the plasmon band splits into two bands corresponding to oscillations of free electrons along (longitudinal) and perpendicular (transverse) to the long axis of the rod.
- Metallic nanoshells are comprised of a dielectric core such as silica, surrounded by a metallic shell. The plasmon resonances in the nanoshells are

typically red-shifted to longer wavelengths compared to the corresponding solid metal nanoparticles.
- As the metal shell thickness decreases for the same core size, the optical resonance shifts to a longer wavelength.
- Nanoshells have a number of applications such as reducing photo-oxidation and in whole blood immunoassay.
- Local field enhancement near a metallic surface has found a number of applications such as in surface-enhanced Raman spectroscopy (SERS) and in electronic linear and nonlinear optical spectroscopy.
- A convenient numerical method to evaluate the local field enhancement is provided by the discrete dipole approximation (DDA).
- In the case of interacting metallic nanoparticles, the maximum field enhancement occurs at the midpoint between the two spheres.
- The field enhancement for a nonspherical nanoparticle is considerably larger than that for a comparable size spherical particle.
- Metallic films containing a periodic array of subwavelength holes produce extraordinarily high optical transmission. This enhanced transmission through the holes is derived from coupling of light to surface plasmons on both sides of the metallic film.
- Plasmonic guiding refers to guiding of electromagnetic wave via coupling of surface plasmon excitation among an array of closely spaced metallic nanostructures. In this array, a plasmon oscillation in one metallic nanoparticle induces another plasmon oscillation in the neighboring particle.
- Applications of plasmonics principally utilize three features. These are: (a) local field enhancement near the metallic surface, (b) evanescent wave emanating from the surface under the surface plasmon resonance condition, and (c) sensitivity of the surface plasmon resonance to the dielectric medium surrounding the metal.
- Radiative decay engineering refers to the manipulation of the radiative properties of a fluorescent molecule by using metallic nanostructures of different size and shape and also by varying the distance of the fluorophore from the metal surface.
- The three different types of plasmonic effects determining the radiative and nonradiative properties of a fluorophore are (a) local field enhancement producing enhancement of the local excitation density and consequently fluorescence, (b) metal–dipole interaction leading to enhanced nonradiative process and hence quenching of fluorescence, and (c) enhancement of radiative rate derived from increased photon mode density (density of photons at the energy gap of the fluorophore).
- Through radiative decay engineering, one can enhance fluorescence quantum yield, even though lifetime of fluorescence shortens due to increased metal-induced nonradiative processes. This feature is useful for enhancing the photostability of dyes.

- Surface plasmonic coupling in a metallic film can produce directional emission and thus provides a significant enhancement in collection of emitted photons.

REFERENCES

Atwater, H., Guiding Light, *SPIE's OE Mag.* **July,** 42–44 (2002).

Averitt, R. D., Sarkar, D., and Halas, N. J., Plasmon Resonance Shifts of Au-Coated Au_2S Nanoshells: Insight into Multicomponent Nanoparticle Growth, *Phys. Rev. Lett.* **78,** 4217–4220 (1997).

Barnes, W. L., Fluorescence Near Interfaces: The Role of Photonic Mode Density, *J. Mod. Opt.* **45,** 661–699 (1998).

Benner, R. E., Dornhaus, R., and Chang, R. K., Angular Emission Profiles of Dye Molecules Excited by Surface Plasmon Waves at a Metal Surface, *Opt. Commun.* **30,** 145–149 (1979).

Bloemer, M. J. and Haus, J. W., Broadband Waveguide Polarizers Based on the Anisotropic Optical Constants of Nanocomposite Films, *J. Lightwave Tech.* **14,** 1534–1540 (1996).

Bloemer, M .J., Budnick, M. C., Warmack, R. J., and Farrell, T. L., Surface Electromagnetic Modes in Prolate Spheroids of Gold, Aluminum and Copper, *J. Opt. Soc. Am. B* **5,** 2552–2559 (1988).

Bloemer, M. J., Haus, J. W., and Ashley, P. R., Degenerate Four-Wave Mixing in Colloidal Gold as a Function of Particle Size, *J. Opt. Soc. Am. B,* **7,** 790–795 (1990).

Bohren, C. F., and Huffman, D. R., *Absorption and Scattering of Light by Small Particles,* John Wiley & Sons, New York, 1983.

Born, M., and Wolf, E., *Principles of Optics,* 7th edition, Pergamon Press, Oxford, 1998.

Chang, R. K. and Furtak, T. A., eds., *Surface-Enhanced Raman-Scattering,* Plenum, New York, 1982.

Drexhage, K. H., Interaction of Light with Monomolecular Dye Layers, *Prog. Opt.* **12,** 163–232 (1974).

Ebbesen, T. W., Lezec, H. J., Ghaemi, H. F., Thio, T., and Wolff, P. A., Extraordinary Optical Transmission Through Sub-wavelength Hole Arrays, *Nature* **391,** 667–669 (1998).

El-Sayed, M. A., Some Interesteresting Properties of Metal Confined in Time and Nanometer Space of Different Shapes, *Acc. Chem. Res.* **34,** 257–264 (2001).

Ghaemi, H. F., Thio, T., Grupp, D. E., Ebbesen, T. W., and Lezec, H. J., Surface Plasmons Enhance Optical Transmission Through Subwavelength Holes, *Phys. Rev. B* **58,** 6779–6782 (1998).

Gresten, T., and Nitzan, A., Spectroscopic Properties of Molecules Interacting with Small Dielectric Particles, **75,** 1139–1152 (1981).

Halas, N., The Optical Properties of Nanoshells, *Opt. Photon. News* **August,** 26–31 (2002).

Hao, E., and Schatz, G. C., Electromagnetic Fields Around Silver Nanoparticles and Dimers, *J. Chem. Phys.* (2003).

Haus, J. W., Inguva R., and Bowden, C. M., Effective Medium Theory of Nonlinear Ellipsoidal Composites, *Phys. Rev. A,* **40,** 5729–5734 (1989).

Haus, J. W., Kalyaniwalla, N., Inguva, R., Bloemer, M. J., and Bowden, C. M., Nonlinear optical properties of conductive spheroidal composites, *J. Opt. Soc. Am. B* **6**, 797–807 (1989).

Haus, J. W., Zhou, H. S., Takami, S., Hirasawa, M., Honma, I., and Komiyama, H., Enhanced Optical Properties of Metal-Coated Nanoparticles, *J. Appl. Phys.* **73**, 1043–1048 (1993).

Hirsch, L. R., Jackson, J. B., Lee, A., and Halas, N. J., A Whole Blood Immunoassay Using Gold Nanoshells, *Anal. Chem.* **75**, 2377–2381 (2003).

Jackson, J. B., and Halas, N. J., Silver Nanoshells: Variation in Morphologies and Optical Properties, *J. Phys. Chem. B* **105**, 2743–2746 (2001).

Jin, R., Cao, Y. W., Mirkin, C. A., Kelly, K. L., Schatz, G. C., and Zheng, J. G., Photoinduced Conversion of Silver Nanospheres to Nanoprisms, *Science* **294**, 1901–1903 (2001).

Kalyaniwalla, N., Haus, J. W., Inguva, R., and Birnboim, M. H., Intrinsic optical bistability for coated particles, *Phys. Rev. A* **42**, 5613–5621 (1990).

Kelly, K. L., Coronado, E., Zhao, L. L., and Schatz, G. C., The Optical Properties of Metal Nanoparticles: The Influence of Size, Shape, and Dielectric Environment, *J. Phys. Chem. B* **107**, 668–677, 2003.

Kitson, S. C., Barnes, W. L., and Sambles, J. R., Photoluminescence from Dye Molecules on Silver Gratings, *Opt. Commun.* **122**, 147–154 (1996).

Kreibig, U., and Vollmer, M., *Optical Properties of Metal Clusters,* Springer Series in Materials Science 25, Springer, Berlin, 1995.

Lakowicz, J. R., Radiative Decay Engineering: Biophysical and Biomedical Applications, *Anal. Biochem.* **298**, 1–24 (2001).

Lakowicz, J. R., Malicka, J., Gryczynski, I., Gryczynski, Z., and Geddes, C. D., Radiative Decay Engineering: The Role of Photonic Mode Density in Biotechnology, *J. Phys. D: Appl. Phys.* **36**, R240–R249 (2003).

Lasema, J. J., ed., *Modern Techiniques in Raman Spectroscopy,* John Wiley & Sons, New York, 1996.

Lezec, H. J., Degiron, A., Devaux, E., Linke, R. A., Martin-Moreno, L., Garcia-Vidal, F. J., and Ebbesen, T. W., Beaming Light from a Subwavelength Aperture, *Science* **297**, 820 822 (2002).

Link, S., and El-Sayed, M., Spectral Properties and Relaxation Dynamics of Surface Plasmon Electronic Oscillations in Gold and Silver Nanodots and Nanorods, *J. Phys. Chem. B* **103**, 8410–8426 (1999).

Link, S., and El-Sayed, M. A., Optical Properties and Ultrafast Dynamics of Metallic Nanocrystals, *Annu. Rev. Phys. Chem.* **54**, 331–366 (2003).

Maier, S. A., Kik, P. G., Atwater, H. A., Meltzer, S., Harel, E., Koel, B. E., and Requicha, A. A. G., Local Detection of Electromagnetic Energy Transport Below the Diffraction Limit in Metallic Nanoparticle Plasmon Waveguides, *Nature Materials/Advanced Online Publication,* 2 March 2003, p. 1–4.

Michaels, A. M., Nirmal, M., and Brus, L. E., J., Surface Enhanced Raman Spectroscopy of Individual Rhodamine 6G Molecules on Large Ag Nanocrystals, *J. Am. Chem. Soc.* **121**, 9932–9939 (1999).

Neuendorf, R., Quinten, M., and Kreibig, U., Optical Bistablility of Small Heterogeneous Cluster, *J. Chem. Phy.* **104**, 6348–6354 (1996).

Nie, S., and Emory, S. R., Probing Single Molecules and Single Nanoparticles by Surface-Enhanced Raman Scattering, *Science,* **275**, 1102–1106 (1997).

Perenboom, J. A. A. J., Wyder, P., and Meier, F., Electronic Properties of Small Metallic Particles, *Phys. Rep.* **78,** 173–292 (1981).

Shalaev, V. M., *Nonlinear Optics of Random Media: Fractal Composites and Metal Dielectric Films, Springer Tracts in Modern Physics,* Vol. 158, Springer, Berlin, 2000.

Sokolov, K., Chumanov, G., and Cotton, T. M., Enhancement of Molecular Fluorescence Near the Surface of Colloidal Metal Films, *Anal. Chem.* **70,** 3898–3905 (1998).

van Beek, L. K. H., Dielectric Behavior of Heterogeneous Sytems, in *Progress in Dielectrics,* J. B. Birks, ed., CRD, Cleveland, OH, 1967.

Weber, W. H., and Eagen, C. F., Energy Transfer From an Excited Dye Molecule to the Surface Plasmons of an Adjacent Metal, *Opt. Lett.* **4,** 236–238 (1979).

Weitz, D. A., Garoff, S., Hanson, C. D., and Gramila, T. J., Fluorescent Lifetimes of Molecules on Silver-Island Films, *Opt. Lett.* **7,** 89–91 (1982).

Wenseleers, W., Stellacci, F., Meyer-Friedrichsen, T., Mangel, T., Bauer, C. A., Pond, S. J. K., Marder, S. R., and Perry, J. W., Five Orders-of-Magnitude Enhancement of Two-Photon Absorption for Dyes on Silver Nanoparticle Fractal Clusters, *J. Phys. Chem. B* **106,** 6853–6863 (2002).

Worthing, P. T., and Barnes, W. L., Spontaneous Emission within Metal-Clad Microcavities, *J. Opt. A: Pure Appl. Opt.* **1,** 501–506 (1999).

Zhou, H. S., Honma, I., Haus, J. W., and Komiyama, H., The Controlled Synthesis and Quantum Size Effect in Gold Coated Nanoparticles, *Phys. Rev. B* **50,** 12052–12056 (1994).

CHAPTER 6
Nanocontrol of Excitation Dynamics

Chapter 6 deals with the manipulation of a nanoscale dielectric environment to control excitation dynamics. Even though this may not have a quantum confinement effect on electronic states (such as in insulating materials), the dynamics of electronic excitation can still exhibit important manifestations derived from nanostructure control. The nanostructure dependence of dynamics can be derived from the interaction of the electronic excitation with the surroundings and its phonon modes.

The factors that control the dynamics are discussed in Section 6.1. The use of dielectric modification to control excited-state dynamics and thus enhance a particular optical manifestation (such as emission at a specific wavelength) is illustrated by two examples:

1. The rare-earth ion containing nanostructures in which the phonon interactions with the surrounding lattice and subsequently the excited-state relaxation is controlled to produce enhancement of a particular emission. An important type of emission is that of the Er^{3+} ions at 1550 nm, which forms the basis for optical amplification. It is shown in Section 6.2 that the nanostructure control produces significant increase of the emission lifetime to provide improved optical amplification ability. Another important type of emission, discussed in Section 6.3, is up-conversion where excitation by a CW-IR laser beam at 974 nm produces up-converted emission in the red, green, or blue regions, depending on the nanostructure composition and interactions of the rare-earth ion containing nanoparticles. This up-conversion process finds important applications in up-conversion lasing (Scheps, 1996). Another application is in display technology (Downing et al., 1996). Other applications are for infrared quantum counters and temperature sensors (Joubert, 1999). The up-conversion process also finds biological applications in bioimaging and light-activated therapy, which are discussed in Chapter 13. Another important type of up-conversion process is photon avalanche that occurs above a certain pump threshold power. It is discussed in Section 6.4. Section 6.5 describes the converse process of down-conversion, also called *quantum cutting*.

2. Organic–inorganic hybrid core-shell nanoparticles where an organic fluorescent molecule is encapsulated within an inorganic shell (silica shell in the present case). This example is presented in Section 6.6. The shell shields the organic molecule from interactions with the external environments (such as aqueous environments) which act as a quencher of emission. The enhanced emission efficiency, to-

gether with the capabilities of the silica surface to be functionalized to target a specific biological site, provides a powerful tool for bioimaging. The highlights of the chapter are provided in Section 6.7.

Some further reading references are:

Dieke (1968): *Spectra and Energy Levels of Rare Earth Ions in Crystals*
Gamelin and Güdel (2001): *Upconversion Processes in Transition Metal and Rare Earth Metal Systems*
Jüstel et al. (1998): *New Developments in the Field of Luminescent Materials for Lighting and Displays*

6.1 NANOSTRUCTURE AND EXCITED STATES

Even in the case of insulating materials for which electronic wavefunctions are localized within an atom, ion, or molecule, the immediate surrounding (within nanoscopic region) plays an important role in determining the fate of an electronic excited state. Thus a nanostructure control either by judiciously choosing the nanoenvironment of the species to be excited or by utilizing a nanoconfined structure (such as a nanoparticle) can be utilized to manipulate the excitation dynamics. In this section, some important factors derived from this nanocontrol are discussed. Examples selected are rare-earth ions and molecules for which the electronic wavefunctions are fairly localized within the ion (or the molecule). For molecules, the electronic transitions can be classified on the basis of the nature of molecular orbitals involved (Prasad, 2003). An example is a $\pi \rightarrow \pi^*$ transition involving promotion of an electron from a filled bonding π-molecular orbital (e.g., highest occupied π-molecular orbital, abbreviated as HOMO) to an empty anti-bonding π^*-molecular orbital (e.g., lowest unoccupied molecular orbital, abbreviated as LUMO). Other examples are $n \rightarrow \pi^*$, $\sigma \rightarrow \pi^*$, $\sigma \rightarrow \sigma^*$ transitions as well as metal–ligand charge transfer bands in organometallics (Prasad, 2003). These transitions are further characterized by molecular symmetry of the molecule.

In the case of rare-earth ions, important transitions are $f \rightarrow f$ and $f \rightarrow d$ involving the f and d orbitals of the rare-earth ions. The states in these ions split into levels by many-electron and spin–orbit interactions. These levels are characterized by term symbols, represented as $^{2S+1}L_J$, where S is the total spin represented in its numerical value, L is the overall orbital angular momentum represented by a capital letter (S for $L = 0$, P for $L = 1$, etc.); the letter J in numerical value represents the overall angular momentum (Dieke, 1968).

For simplicity we shall consider a dilute guest–host system where an ion or a molecule is dispersed as a guest in a host matrix (or a nanoparticle). Thus, exciton interactions, whereby nearest-neighbor molecules interact to produce a spread of an excited electronic state into an exciton band, are not considered here.

For the case of the localized electronic states discussed above, the following nanoscopic interactions play key roles in controlling the excitation dynamics:

Local Field Interactions. Even though the electronic wavefunctions may be localized within the impurity (molecular or ionic), the electronic interactions may be different when the impurity is in ground or excited states. Thus, the excitation energy ΔE_{gf} between the ground state of the impurity and its excited state is dependent on the nanoscopic environment, and the change in the intermolecular interaction potential ΔV is given by

$$\Delta V = V_{ff} - V_{gg} \qquad (6.1)$$

Thus, the energy gap can be manipulated by changing the nanoscopic environment. A clear example is what is called crystal field effect (or ligand field effect) for transition metal and rare-earth ions where the symmetry (number and geometric arrangement of nearest neighbor host centers) surrounding the impurity center leads to a splitting of the d orbitals and thus shifting of the various energy levels. This has a profound effect on the excitation dynamics. For example, in the case of Pr^{3+} ion, the strength of crystal field interaction determines whether the intensely absorbing $4f \rightarrow 4f5d$ transition is above or below the S_0 level (derived from the $4f \rightarrow 4f$ transition). This has important consequences in determining the excited-state dynamics, as discussed below in Section 6.5 in description of the process of quantum cutting.

Electron–Phonon Coupling. Phonons are the vibrations of the lattice and, in the bulk, fall into two types:

1. *Low-Frequency Acoustic Phonons.* These form a band described by the dispersion of their frequency, ω, as a function of the wavevector. Their dispersion (ω_k versus k) and density of states are described by the Debye model, in which at low frequencies the dispersion is linear and the slope is the speed of sound. This band spans from $\omega_k = 0$ at $k = 0$.
2. *Optical Phonons.* These are high-frequency phonons and have a nonzero value of ω at $k = 0$.

For details of phonon modes in a crystal lattice, the readers are referred to the book by Kittel (2003). For phonons and phonon interactions in molecular solids, the readers are referred to the review by Hochstrasser and Prasad (1974). The electron–phonon interaction is produced by the variation of the electronic interaction (6.1) during lattice vibration of a given phonon mode (pattern of displacement). Under linear response theory, the electron–phonon interaction is determined by (Hochstrasser and Prasad, 1974)\

$$(\partial V_{ff}/\partial R_n^S)_0 \cdot \tilde{\Phi}_s^{(n)} \qquad (6.2)$$

where R_n^S is the linear lattice displacement for phonon mode S and Φ is the amplitude of the mode at the impurity site n. In other words, all phonon modes may not have the same amplitude at the impurity site. For example, certain optical modes may have very large amplitudes at the impurity site and very little at the host site. These modes are called *localized phonons*.

Electron–phonon interactions play an important role in determining two types of processes in electronic excitation:

1. *Dephasing of Electronic Transition.* Here the electron–phonon interactions cause fluctuation of the electronic excitation energy ΔE_{gf} and thus cause line broadening that is temperature-dependent. This is called *homogeneous line broadening*. Often, the electronic transitions are inhomogeneously broadened due to the variation of interaction ΔV, given by (6.1), from one site to another. The homogeneous broadening caused by electron–phonon interaction is determined by a hole-burning experiment, in which a very narrow excitation source (e.g., from a laser) is used to excite a selected site causing saturation or photobleaching, so that this site does not absorb any more. Consequently, a hole appears in the inhomogeneously broadened line shape of the transition. The linewidth of the hole is limited by the homogeneous broadening. The dephasing is characterized by a dephasing time T_2, given as

$$T_2 = \frac{1}{2\pi\Delta\nu}$$

where $\Delta\nu$ is the homogeneous linewidth.

2. *Relaxation of the Excited State.* Here the population of the excited state decreases by radiative or nonradiative transition from the excited state to a lower electronic state. The energy difference between the two electronic states is converted into phonon energy by creation of phonons due to electron–phonon interaction. This process of population relaxation is characterized by a population relaxation time T_1 which also contributes to line broadening due to the Heisenberg uncertainty principle ($\Delta\nu \times T_1 \approx \hbar$). This broadening is called *lifetime broadening*. Thus the total dephasing time T_2 includes the T_1 contribution.

In the case of the excess energy in nonradiative relaxation of electronic level 2 to electronic level 1 being dissipated by creating phonons, the process is known as a *direct process*. If it is accomplished by creation of only one phonon, it is called a *one-phonon process*. If many phonons are needed to make up for the energy difference ΔE between levels 2 and 1, the relaxation is termed as a *multiphonon process*. Another type of process is a Raman process in which a specific phonon mode interacts with the electronic excitation to accept the excess energy and create another phonon mode of higher frequency (for anti-Stokes Raman process). This type of process requires existing population of phonons and hence cannot occur at the temperature of 0 K. The anti-Stokes process exhibits a strong temperature dependence.

Nanostructure manipulation can be used to modify electron–phonon interaction and consequently influence the T_1 and T_2 processes. First, the phonon spectra and the phonon density of states can be modified by simply the selection of the host lattice that has a strong influence on the multiphonon relaxation process. Second, the electron–phonon coupling can also be influenced by the amplitude Φ_i^S of the highest-frequency phonon of the host lattice at the impurity site or coupling of the localized phonons of the impurity with the phonons of the host.

Multiphonon relaxation process is often described by using the theory developed by Huang and Rys (1950) in which the multiphoton relaxation is induced by a single-phonon mode. For a detailed description of the multiphonon relaxation and the expression for the nonradiative rate, a good reference is Yeh et al. (1987). The Huang and Rys model shows that the highest-frequency phonons provide a more efficient multiphonon relaxation pathway. For this reason the effective single-phonon frequency for multiphonon relaxation process is generally taken as the highest frequency (also called cut-off frequency) of the host (Englman, 1979). On this argument, hosts of low cut-off phonon frequencies are choices to reduce nonradiative decays. However, Auzel and co-workers (Auzel and Chen, 1996; Auzel and Pell, 1997) have shown that the actual phonon coupling mechanism for phonon-assisted processes in rare-earth ions may be more complicated. They have proposed an effective phonon mode concept that takes into account the strength of electron–phonon coupling together with the phonon density of states. The work of de Araujo and co-workers (Menezes et al., 2001, 2003) on phonon-assisted processes in fluoroindate glass appears to support this concept.

Finally, the phonon spectrum of the host lattice is strongly influenced by reducing the size of the host from the bulk to a nanocrystal size. Important modification is in the acoustic phonon range where there is an appearance of a low-frequency gap and the formation of discrete vibrational states. This low-frequency gap can be understood by the consideration that an acoustic phonon cannot exist in a nanoparticle of diameter a, if

$$\lambda/2 > a \tag{6.3}$$

Here λ is the wavelength of phonons and a is the diameter. Thus for a spherical particle of size a, the low-frequency cut-off, ω (angular frequency), for phonons is given by

$$\omega = \pi/av$$

where v is the sound velocity. For a nanocrystal of size $a = 2.5$ nm, the phonon cut-off frequency is 30.0 cm^{-1}. As we shall see from the example presented in Section 6.2, the relaxation between two levels separated by less than this value cannot occur by direct one-phonon process in a nanoparticle of size $a = 2.5$ nm. Disappearing acoustic phonons even in much larger nanoparticles of 10 to 20-nm size were recently postulated (Liu et al., 2003a,b) to explain anomalous thermalization effects induced by optical excitation in rare-earth nanophosphors. These effects are manifested in the appearance of hot emission bands, as was recently reported for Pr^{3+}-doped Y_2SiO_5 nanoclusters (Malyukin et al., 2003) and for Nd^{3+} and Yb^{3+} co-doped nanocrystalline YAG ceramics (Bednarkiewicz et al., 2003).

Another modification of the phonon spectrum is the introduction of surface phonon modes that are localized at the interface of the nanoparticle and the surrounding medium. The surface phonon mode is strongly influenced by the dielectric constant of the surrounding medium and can introduce new channels of surface-induced nonradiative relaxations (T_1 processes).

Impurity Pair Interactions (e.g., Ion–Ion Interactions). These interactions introduce new channels of excitation dynamics. The examples provided below are: (i) energy up-conversion where absorption of two photons produce emission of a photon of higher energy and (ii) energy down-conversion, also called quantum cutting, in which the absorption of a high-energy photon produces emission of two photons of lower frequencies. An important type of relaxation in both these processes is cross-relaxation, whereby an ion transfers part of its excitation energy to another ion. These ion-pair interactions show strong dependence on the ion-pair distance and thus on the nanostructure. Andrews and Jenkins (2001) have proposed a quantum mechanical theory of three-center energy transfer for up-conversion and down-conversion in rare-earth doped materials. These energy transfer processes will be dependent on ion–ion interactions and nanostructure.

6.2 RARE-EARTH DOPED NANOSTRUCTURES

An example of a nanostructure that provides a control of excited-state dynamics to enhance a specific radiative process is the rare-earth nanoparticles. The rare-earth ion containing nanoparticles are also examples that do not show quantum confinement effects, but they are useful for many different applications.

Rare-earth ion doped glasses have been used for applications such as in optical amplification (Digonnet, 1993), lasing (Weber, 1990), optical data storage (Nogami et al., 1995; Mao et al., 1996), and chemical sensing (Samuel et al., 1994). Most of the rare-earth ions used are in the 3+ oxidation state (with an effective charge of +3). In such a case, the lowest-energy electronic transitions are $4f \rightarrow 4f$ involving the inner f-atomic orbitals (Dieke, 1968). These transitions are very narrow and exhibit multiple structures derived from electronic interactions, as well as spin–orbit coupling. The energy levels in these ions are represented by their term symbols as discussed in Section 6.1. In fact, many ions emit from a number of excited energy levels, thus exhibiting multiple radiative channels. Under controlled conditions, 2+ oxidation states can be produced for some rare-earth ions such as Eu and Sm, where the extra electron occupies the $5d$ orbital and can provide a $5d$–$4f$ transition. This transition is dipole allowed and approximately 10^6 times more intense than the f–f transition of the corresponding trivalent ion. Furthermore, the d–f transition is considerably blue-shifted (shorter wavelength) and broad, similar to what is observed for organic molecules. Both the f–f and d–f transitions involve electrons that are localized in atomic orbitals of the ions. Therefore, no size-dependent quantization effect, as seen in the case of confinement of delocalized electrons in the valence and conduction bands of semiconductors (Chapter 4), is found for these transitions. However, a nanostructure control of the environment surrounding a rare-earth ion has important manifestations in influencing the optical properties, some examples of which are provided here:

- *Control of Local Interaction Dynamics.* The efficiency of emission from a specific excited energy level is determined by the extent of competing nonradiative decay due to coupling with phonon modes (lattice vibrations) of the surrounding dielectric host, as discussed in Section 6.1. The nonradiative decay probability is often determined by the rate of conversion of the excitation energy into multiple phonons of the surrounding dielectric host (multiphonon process). Qualitatively, the larger the number of phonons needed to convert the excitation energy into phonon energy, the lower is the efficiency of the nonradiative process. Hence, to enhance the emission efficiency by reducing nonradiative rate, it is desirable to have the rare-earth ion incorporated into a dielectric host of very low frequency phonons. The luminescence efficiencies of the nanoparticle of alumina doped by europium and terbium have been reported to be higher than that of microspheres (Patra et al., 1999). ZrO_2 and Y_2O_3 are examples of such hosts with their highest phonon frequencies of 470 cm^{-1} and 300–380 cm^{-1}, respectively, compared to Al_2O_3 (870 cm^{-1}) and SiO_2 (1100 cm^{-1}) (Patra et al., 2002). They thus provide much improved emission efficiency. Because these media are crystalline, one can use nanocrystals of Y_2O_3 and ZrO_2 containing rare-earth ions to provide low-frequency phonon media and then disperse these nanocrystals in SiO_2 or other glass bulk media that can easily be produced in the form of a film, a fiber, or a channel waveguide.
- *Control of Oxidation State.* The manipulation of local environment can produce stabilization of the 2+ oxidation state for a rare-earth ion. An example is the europium ion, which generally forms a 3+ oxidation state. Our work (Biswas et al., 1999) has shown that by co-doping with Al and sintering above 1000°C, a spontaneous reduction of Eu^{3+} ion occurs to produce Eu^{2+} whereby the sharp and narrow emissions in the range of ~600 nm (red) from the Eu^{3+} ions are replaced by a broad emission band of Eu^{2+} at ~440 nm (blue).
- *Generation of New Up-Conversion Processes.* Rare-earth ions also exhibit a number of up-conversion processes whereby an IR excitation produces an up-converted emission from a higher level (Scheps, 1996; Gamelin and Güdel, 2001; de Araujo et al., 2002). Many new channels of up-conversion involve interaction with another ion within a nanoscopic domain (i.e., ion-pair interactions). Thus, nanostructure design can be used to enhance these processes. The up-conversion is discussed in detail separately in Section 6.3. Photon avalanche is another type of up-conversion process which is discussed in Section 6.4.
- *Generation of Quantum Cutting Processes.* Quantum cutting refers to an optical down-conversion process in which the energy of an absorbed photon is split to produce emission of two photons of lower energies (Wegh et al.,

1999a). In the case of rare-earth ions, the absorption of vacuum UV photons has been shown to produce emission of two-photons in the visible, which is of considerable interest for producing mercury-free fluorescent tubes and plasma display panels. It was also demonstrated that quantum cutting can be observed in infrared by the absorption of visible photons (Strek et al., 2000). Nanostructure control can be utilized to produce quantum cutting by energy transfer between two different ions (Wegh et al., 2000). The efficiency of these quantum cutting two-photon down-conversion processes have been reported to be as high as 200%. This topic is discussed in Section 6.5.

- *Modification of Phonon Spectrum.* Even though the electronic spectra of rare-earth ions show no quantum confinement effect, the phonon spectrum of the lattice shows important manifestations in going from a bulk to nanocrystal environment, as discussed in Section 6.1. Gap appears in the acoustic-phonon spectrum with the formation of discrete vibrational states (Meltzer and Hong, 2000). This modification of the phonon spectrum produces important manifestations in the electron–phonon interaction and thus the dynamics of multiphonon relaxations and phonon-induced line broadening of electronic transitions. Thus for the $^7F_0 \rightarrow {^5D_0}$ transition of Eu^{3+} ions in Eu_2O_3, a different temperature dependence ($\sim T^3$) is found for the linewidth in nanoparticles, compared to $\sim T^7$ dependence in the bulk form. A quantitative study of the line broadening is facilitated by the narrow linewidth associated with the rare-earth $f \rightarrow f$ transitions. Similarly, Yang et al. (1999) observed that the decay rates from the upper levels of the $^5D_1$ manifold by a one-phonon process, which requires phonons of frequencies 3 and 7 cm^{-1} for certain sites in $Y_2O_3:Eu^{3+}$ nanocrystals, are dramatically reduced by up to two orders of magnitude compared to those observed in micron-sized materials. They interpretted this to arise from the fact that the nanocrystals do not have such low-frequency phonon modes.

- *Size Dependence of Emission Bands.* The size dependence of the electron–phonon coupling, derived from the change in the phonon spectrum, together with a change in the crystal field strength, can produce a shift of the emission peak, which is not due to quantum confinement of electrons. For the ZnS nanoparticles containing Eu^{2+} ions, a size dependence of the emission peak involving the $4d \rightarrow 4f$ transition in Eu^{2+} has been observed (Chen et al., 2001). The emission bands of the 4.2-, 3.2-, and 2.6-nm-size ZnS:Eu^{2+} nanoparticles peak respectively at 670, 580, and 520 nm. Chen et al. (2001) interpreted this to arise from decreases in both the electron–phonon coupling and the crystal field strength with the decreasing particle size. Variation of the charge-transfer band intensity was demonstrated in Eu^{3+} doped Lu_2O_3 nanocrystallites to be dependent on size of nanoparticles (Strek et al., 2002).

An example of the application of nanostructure control to enhance a particular optical transition is provided by the Er^{3+} ions. Erbium-doped fiber amplifiers (EDFA) are utilized as very efficient optical amplifiers, widely used for telecommunications at 1.55 μm. By nanostructure control, one can manipulate the excited-state dynamics to produce emission from a given level (thus emission of a specific color). Thus, one can derive different types of emission from the same ion. The 1.5-μm emission from erbium, used for amplification, is produced by pumping at 980 nm, with co-pumping at 1400 nm to enhance the efficiency. The lifetime of this transition in commercially sold optical amplifiers is typically of the order of 8 ms. Our Institute for Lasers, Photonics, and Biophotonics has made these materials by nanostructure processing. Nanostructure control by using a multicomponent environment to manipulate the local interactions around the erbium ion has produced nanoparticles with a much longer lifetime (up to 17 ms) (Biswas et al., 2003). This increased lifetime allows more efficient amplification. As the nonradiative process is reduced, an increased lifetime leading to the theoretical limit of radiative lifetime can be realized. This will ensure that all absorbed photons are used to produce emission (used for the radiative process).

The second factor that enters into material's figure of merit to enhance its characteristic as a gain medium for amplifiers is the number density of the emitter. The number density of erbium ions should be as high as possible without inducing "concentration quenching," which occurs when two ions communicating with each other produce nonradiative dissipation of excited state energy. Therefore, it is important to ensure that no ion clustering takes place at higher concentration. By nanostructure control, one can also increase the number density without concentration quenching caused by ion clustering. This goal was also accomplished by Biswas et al. (2003). Therefore, a nanostructure-controlled nanoparticle technology can produce efficient optical amplification.

6.3 UP-CONVERTING NANOPHORES

Another area of nanometer control is in producing efficient up-conversion. Some of the up-conversion processes that efficiently occur in a rare-earth or transition metal ion doped medium (Scheps, 1996; Gamelin and Güdel, 2001) are shown in Figure 6.1.

The more prominent up-conversion processes in these ions involve absorption of two photons to two distinct levels. Thus, these processes are different from the direct or simultaneous two-photon processes, discussed in Chapter 2, where the absorption of two photons occurs simultaneously without involving any real intermediate one-photon levels. The up-conversion process in the rare-earth ions can be divided into two broad classes, as described in Table 6.1. The nanostructure control, in the case of ESA, provides the ability to minimize nonradiative relaxation (loss of population) of the intermediate level generated by the first photon. For cooperative mechanism, the nanostructure design involves the control of distances between the ion pairs within nanoscopic region, as the dominant electronic interactions between the two interacting ions are electric multipole or exchange type, both of them being strongly dependent on the ion–ion separation.

The advantages offered by these types of up-conversion processes are that they can be induced even by a low-power CW laser. In contrast, a simultaneous two-

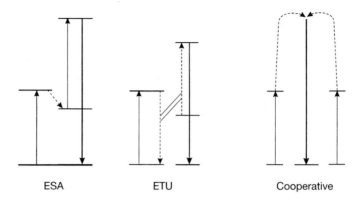

Figure 6.1. Up-conversion processes.

photon process requires a high-intensity pulse laser source, such as a mode-locked Ti:sapphire laser producing ~ 100-fs pulses (see Chapter 2).

The nanoparticles containing the up-converting rare-earth ions can be called *up-converting nanophores*. They offer the advantage that they can be dispersed in a glass or plastic medium for display. They can also be surface functionalized to be dispersable in an aqueous medium for bioimaging. A major challenge has been to produce small nanophores in the desired size range of <50 nm for these purposes, at the same time providing highly efficient up-conversion. We have succeeded in this goal by using low-frequency phonon host nanocrystals such as ZrO_2 and Y_2O_3, as discussed in Section 6.1 (Patra et al., 2002). We observed that the up-conversion luminescence intensity depends on the crystal structure and particle size (Patra et al., 2003a). The asymmetric structure of the low-symmetry site of

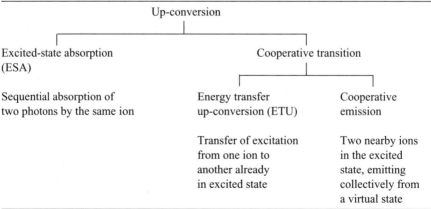

Table 6.1. Up-Conversion Processes in Rare-Earth Ions

Excited-state absorption (ESA)	Cooperative transition	
	Energy transfer up-conversion (ETU)	Cooperative emission
Sequential absorption of two photons by the same ion	Transfer of excitation from one ion to another already in excited state	Two nearby ions in the excited state, emitting collectively from a virtual state

the rare-earth ions allows intermixing of the *f* states with higher electronic configuration and as a result the optical transition probabilities increase. The up-conversion luminance value for same concentration of Er^{3+} ions in $BaTiO_3$ is much higher than that in the TiO_2 host (Patra et al., 2003b). Because the perovskite oxide matrix has low-frequency soft transverse optical (TO) mode, the low phonon energies reduce the probability of a multiphonon nonradiative process. The presence of a large energy gap between the emitting and the terminal levels also reduces the nonradiative decay. This reduction leads to an increase in the emission efficiency. The efficiency of these materials is dependent on the excited-state dynamics of the rare-earth ions and their interactions with the host matrix. This interaction can be a function of the host phase and the dopant concentration. The excited-state dynamics is also dependent on the energy migration between the active ions, the statistical distribution of the active ions, and the site symmetry of the active ions in the host matrix.

Figure 6.2 shows the emission from three different sets of yttria (Y_2O_3) nanoparticles, containing pairs of rare-earth ions. These nanoparticles are roughly 25–35 nm in size. They cover the visible spectrum from blue to red. The red emission results from a two-photon process. The green emission is also produced by a two-photon process. The blue emission is a three-photon process. If one disperses these nanoparticles in a homogeneous medium, like glass, or in some way proper for bioimaging in a solution, they should be transparent. These bulk media should not scatter light because the nanoparticles are ≤ 35 nm and yet they are very efficient upconverters. Figure 6.3 shows, as an example, a photograph of one such up-converting transparent glass. These nanoparticles can be used for a variety of ap-

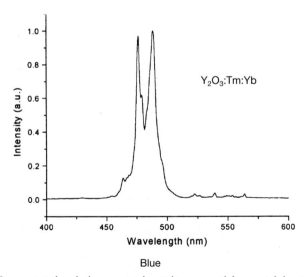

Blue

Figure 6.2. Up-converted emission spectra in yttria nanoparticles containing two rare-earth ions. (*continued*)

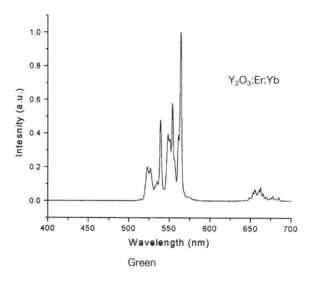

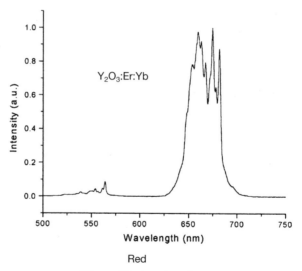

Figure 6.2. (*continued*)

plications—from imaging, to temperature monitoring, to large area displays—when incorporated in polymer sheets.

One can also disperse the nanoparticles in a charge-transporting medium to generate electroluminescence, which means a recombination of electron–hole pairs at a nanoparticle in a plastic matrix. This combination can provide the features of luminescence characteristic of the inorganics, but in a flexible physical form of a plastic matrix.

Figure 6.3. Photograph of an up-converting transparent glass laid over the writing to show its clarity.

6.4 PHOTON AVALANCHE

Photon avalanche is an up-conversion process which produces up-conversion above a certain threshold of excitation power (Joubert, 1999; Gamelin and Güdel, 2001). Below this threshold, very little up-converted fluorescence is produced and the medium is transparent to the pump beam (in the IR for the rare-earth ions). Above the pump threshold, the pump beam is strongly absorbed and the fluorescence intensity increases by orders of magnitude.

The mechanism of photon avalanche is illustrated by the energy diagrams of Figure 6.4. It involves a cross-relaxation. In this process the pump beam does not have sufficient energy to directly populate level 2 by ground-state absorption (GSA). However, there is a strong excited-state absorption (ESA) from level 2 to level 3 at the pump wavelength. Thus, if somehow the metastable (intermediate) level 2 becomes populated, it readily absorbs the pump photon to level 3, which produces an up-converted emission.

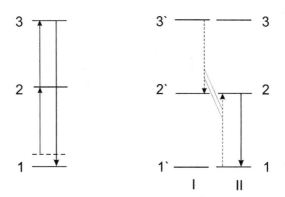

Population by phonon-assisted   Population by cross-relaxation

Figure 6.4. Mechanisms for populations of intermediate levels in photon avalanche up-conversion.

The population of level 2 is created by two mechanisms. Initially, population of level 2 is created by weak excitation from level 1 involving assistance of phonons. In the diagram it is shown as originating from a phonon state (dotted line) or it could be through the phonon side band. As soon as level 2 is populated, it readily absorbs another photon from the pump beam due to strong excited-state absorption and reaches level 3. In some ions, level 3 can then produce an up-converted emission. Some excited ions transfer part of its energy from level 3 to another ion by a nonradiative energy transfer mechanism, called *cross-relaxation*, which results in both ions being in level 2. This cross-relaxation process, abbreviated as CR, is represented in the figure by the dashed line. Now both ions are in state 2 from where they readily populate level 3 to further initiate cross-relaxation and more ESA to increase level 3 population, resulting in a dramatic increase of fluorescence as an avalanche process. The strong pump power dependence comes from the strong dependence of CR on the population of level 3, and hence on the pump power.

The requirement for photon avalanche can thus be summarized as follows:

- The excitation energy is not in resonance with any ground-state absorption,
- Excited-state absorption is very strong,
- Rare-earth ion concentration is high enough for ion–ion interactions to produce efficient cross-relaxation populating the intermediate state.

The photon avalanche process can readily be distinguished from other up-conversion processes, discussed in Section 6.2, by the observations of a threshold condition discussed above. An example of photon avalanche is provided by the Pr^{3+} ion in a $LaCl_3$ or $LaBr_3$ host. A pump beam from a continuous-wave (CW) dye laser that matched the ESA $^3H_5 \rightarrow {}^3P_1$ was used (Chivian et al., 1979; Kueny et al., 1989). When the pump power level exceeded a certain critical intensity, strong Pr^{3+} up-converted fluorescence from $^3P_1$ and $^3P_0$ was produced. Photon avalanche has also been reported for Nd^{3+}, Er^{3+}, and Tm^{3+} (Joubert et al., 1994; Bell et al., 2002; Jouart et al., 2002), and Os^{4+} (Wermuth and Gudel, 1999).

Nanoscale manipulation can have two advantages. First, the initial population of level 1 through phonon interactions can be judiciously controlled by optimizing the phonon density of states and electron–phonon coupling. Second, the ion–ion interaction to produce efficient cross-relaxation can also be optimized by a nanostructure control.

6.5 QUANTUM CUTTING

As described above, quantum cutting refers to a process in which a high-energy photon (such as in vacuum ultraviolet) absorption in a medium results in emission of two photons of lower energies (such as in the visible) (Jüstel et al., 1998). This is thus a down-conversion process, opposite of the up-conversion process discussed in Sections 6.3 and 6.4. There is a great deal of interest in this process for producing mercury-free fluorescent lamps and plasma display panels which require efficient conversion of vacuum ultraviolet radiation to visible light (Wegh et al., 199a, 2000; Jüstel et al., 1998; Kück et al., 2003; Vink et al., 2003).

Fluorescent mercury lamps contain phosphors coated on the inside wall of a glass tube that absorb the ultraviolet light radiation of mainly $\lambda = 254$ nm generated by mercury discharge. These phosphors subsequently emit in the blue, green, and red to produce white light. The quantum efficiency of the lanthanide rare-earth-based phosphors can be as high as 90%. Therefore, they are very efficient. For environmental reasons, there is a great deal of interest in replacing mercury discharge with other types of discharge. An attractive alternative is a xenon discharge that produces vacuum UV radiation at 147 nm (Xe fundamental line) and 172 nm (Xe excimer band), with their ratio depending on the gas pressure in the discharge cell. Another advantage offered by Xe discharge is that there is no delay in emission, whereas for mercury lamps, mercury has to first evaporate. The efficiency of the fluorescent lamp can be described as (Jüstel et al., 1998)

$$\phi_{lamp} = \phi_{discharge} [\lambda_{uv}/\lambda_{vis}]QE$$

where ϕ_{lamp} is the overall lamp efficiency, QE is the quantum efficiency of the employed phosphors, λ_{uv} and λ_{vis} are UV excitation and visible emission wavelengths' $\phi_{discharge}$ is the efficiency of mercury plasma. From this equation it is evident that the lamp efficiency is significantly reduced at shorter wavelength produced by a Xe discharge. Quantum cutting provides an exciting possible solution where the absorption of a UV photon can produce two photons in the visible range. Thus, theoretically, one can reach a maximum quantum efficiency of 200%. The materials producing quantum cutting action are called *quantum cutters*.

Some possible mechanisms of down-conversion (Wegh et al., 1999b) producing quantum cutting are illustrated in Figure 6.5. The mechanisms can involve a single

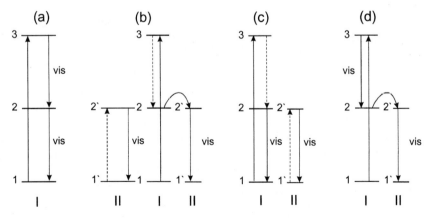

Figure 6.5. Energy level diagram for some possible quantum cutting processes exhibited by rare-earth ions.

type of ion, I, or two types of ions, I and II. Figure 6.5a represents the case where a single ion I produces quantum cutting by absorption producing excitation of its high energy level. In the energy level diagram scheme shown in this figure, the bold straight line refers to the radiative transitions (absorption or emission of photons). The absorption of a high-energy (VUV) photon produces excitation in ion I from level 1 to highly energetic excited level 3. The quantum cutting of this excitation energy produces photons in visible range, derived from successive emission from levels 3 (to 2) and 2 (to 1).

The mechanisms shown in Figures 6.5b and 6.5c involve cross-relaxation shown by a dashed curve. Again, as described in earlier sections, cross-relaxation here refers to a nonradiative process in which a part of the excitation energy is transferred from ion I to ion II. In the mechanism represented by Figure 6.5b, the cross-relaxation produces quantum cutting to leave ions I and II in excited states 2 and 2', respectively. The excited ion II returns to the ground state by emitting a photon in the visible range. The ion I in the excited state 2 can transfer its remaining excitation energy to another neighboring ion II, which also emits a photon in the visible.

The two mechanisms represented by Figures 6.5c and 6.5d involve only one energy transfer step from ion I to ion II. Hence in these cases, one of the photons in the visible region is produced by emission from ion I, the other one from ion II. In the case of Figure 6.5c, quantum cutting is produced by cross-relaxation from ion I to ion II, leaving ion I in state 2 and exciting ion II to state 2'. Subsequent emission of visible photons occurs from level 2 (to level 1) of ion I and level 2' (to level 1') of ion II. In Figure 6.5d, ion I emits a visible photon from level 3, ending up in level 2 from where energy transfer to level 2' of ion II produces the emission of another visible photon from level 2'.

Nanostructure control provides the following opportunities:

- Control of energy level structure (relative ordering and spacing of energy levels) by manipulation of the crystal field surrounding the ions. This is particularly important in the case of single-ion two-step emission, as exhibited by Figure 6.5a, which is also sometimes referred to as photon cascade emission. A widely studied rare-earth ion exhibiting this type of process is Pr^{3+}, which is discussed below in detail.

- Selection of lattice with low phonon frequency and density of phonon states (as discussed earlier) to minimize competing nonradiative multiphonon relaxation. For this reason, the work on quantum cutters has focused on the use of hosts such as fluorides, aluminates, and borates with smaller maximum phonon frequencies (~500 cm^{-1}) compared to many oxides such as silica (maximum phonon frequencies > 1000 cm^{-1}). Nanocrystal environments with discrete and controllable phonon spectrum may provide some advantages.

- Optimization of ion-pair interactions for mechanisms represented in Figures 6.5b–d. The ion pairs involve multipolar and/or exchange interactions that are strongly dependent on the ion-pair separation. Nanostructure control provides an opportunity to optimize the ion-pair interaction.

Here two examples of quantum cutting are discussed. One is by photon cascade emission in Pr^{3+} (Vink et al., 2003; Kück et al., 2003). In this case, the absorption of a vacuum UV photon (~185 nm) takes place via an intense $4f5d$ transition (involving a dipole-allowed interorbital $f \rightarrow d$ transition). The photon cascade emission starts from the $^1S_0$ state; the two predominant emissions involved in the two-step emission from a Pr^{3+} ion are $^1S_0 \rightarrow {}^1I_6$ (~400 nm) and $^3P_0 \rightarrow {}^3H_4$ (~490 nm). These energy level designations utilize their term symbols. The level $^3P_0$ is populated by nonradiative relaxation from $^1I_6$, which is populated following emission from $^1S_0$. For quantum cutting in Pr^{3+}, the $4f5d$ state has to be higher in energy than the $^1S_0$ level. This is the case for hosts exhibiting a weak crystal field. Examples are YF_3, $SrAl_{12}O_{19}$, LaB_3O_6, and so on. In the case of hosts exhibiting a strong crystal field, the $4f5d$ state moves close to or below the $^1S_0$ state and no quantum cutting can be observed. This is found to be the case for $LiYF_4$ and YPO_4 host. Another way to look at the crystal field effect is from the coordination number of the ion. A high coordination number of the Pr^{3+} ion leads to a small crystal field splitting of the $4f5d$ state which leads to the $4f5d$ state being above the $^1S_0$ level.

An example of quantum cutting involving two ions is provided by the Eu^{3+}, Gd^{3+} ion pairs in the Eu^{3+} ion-doped $LiGdF_3$ host (Wegh et al., 1999a) and BaF_2 containing both Gd and Eu (Liu et al., 2003a). Here a VUV photon is absorbed by the Gd^{3+} sensitizer absorption band ($^8S_{7/2} \rightarrow {}^6G_J$) which in one step transfers a part of this energy to an Eu^{3+} ion by cross-relaxation which then emits at ~612 nm from the ^5D_J manifold (set of levels corresponding to different J values) to the ^7F_J manifold. In the second step the resulting Gd^{3+} ion in a lower excited level transfers its energy to another Eu^{3+} ion that emits at the same wavelength. Hence the advantage offered by this process is that both visible photons are at the same wavelength (~612 nm), which is also in the useful range for fluorescent lamp applications. A quantum efficiency close to 200% for this process has been reported (Wegh et al., 1999a,b; Liu et al., 2003a).

Pr^{3+}-based system produces the 407-nm emission derived from the $^1S_0 \rightarrow {}^1I_6$ transition; this emission is close to UV, which prohibits good color rendering. In contrast, the Eu^{3+}-based quantum cutter produces both photons at ~612 nm in the more suitable visible region. However, a disadvantage of the Eu^{3+} quantum cutter is that the Gd^{3+} excitation by VUV photons is an $f \rightarrow f$ transition which is narrower and has a considerably smaller absorption cross section compared to the $f \rightarrow d$ transition involved in the Pr^{3+} ion.

Recent work in this area has been in finding other hosts and ions for quantum cutting (Liu et al., 2003a; Wegh et al., 2000). A problem, from the perspective of bulk forms, is that fluoride host materials, which have widely been used for quantum cutting, are in general brittle. Here again nanoparticle technology offers advantages. Nanoparticles can provide the desired local interaction with the lattice (crystal field, ion-pair interaction) and phonon bath (low-frequency phonons to reduce multiphonon nonradiative decay rates). At the same time, they can be dispersed in a bulk medium with more desirable mechanical and processing features.

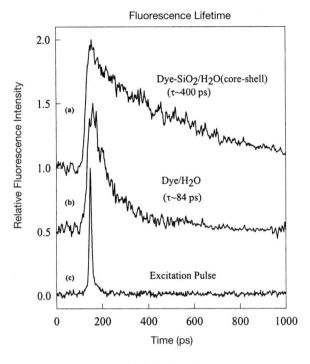
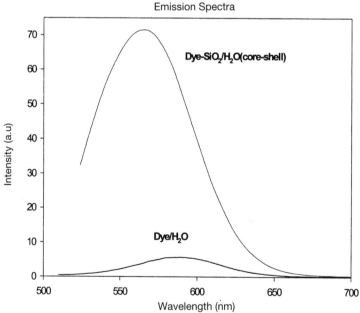

Figure 6.6. Emission spectra and fluorescence lifetime of a dye encapsulated in a silica bubbles (Dye-SiO$_2$/H$_2$O (core shell)) is compared with these for a free dye aqueous solution (Dye/ H$_2$O).

6.6 SITE ISOLATING NANOPARTICLES

A nanoparticle can be used to encapsulate an emitter (e.g., organic fluorophore) and isolate it from an environment that acts as a fluorescent quencher. An example is provided by a core-shell type of multilayer structure where a zinc sulfide (ZnS) core is made first as the core. Then one can place an organic dye on the top of the core. Subsequently, sol–gel chemistry is used to make a very thin shell of silica on the top to encapsulate the core. This approach is demonstrated by using an ionic organic fluorescent dye ASPI-SH, as shown in Figure 6.6. This shows ZnS nanocrystals with the dye around them and the silica shell encapsulating everything. The advantage of making this multilayered hierarchical structure is that, by using multistep nanochemistry, silica encapsulation can provide a shield from solvent (water) induced nonradiative quenching. Figure 6.6 shows the emission properties of this organic dye deposited on the zinc sulfide core and encapsulated with silica. If the dye is directly placed in water, it is in contact with the high vibration frequency -OH groups of water. These -OH groups act as quenchers, considerably reducing the emission efficiency. Comparing the two curves in Figure 6.6 (bottom), one can see that the intensity of fluorescence of the dye in water is significantly (two orders of magnitude) lower than that for a silica encapsulated dye.

The top curve of Figure 6.6 (top) shows a 400 ps fluorescence lifetime when we use the dye, zinc sulfide and a silica type core-shell encapsulated structure, as described above. When the dye is in direct contact with water, fluorescence lifetime obtained is 84 ps. This is an example where there is no quantum confinement effect on the dye emission, but nanostructure control allows one to control the dynamics of the excited state of the dye to enhance its luminescence efficiency and increase its lifetime by reducing the non-radiative contribution.

6.7 HIGHLIGHTS OF THE CHAPTER

- Even for those nanoscale materials that do not exhibit quantum confinement effects, excited-state dynamics can be controlled by manipulation of the nanostructure.
- The dynamics of excited states can be manipulated in a nanoparticle by the judicious choice of interionic (intermolecular) interactions and interaction with phonon modes (the lattice vibrations of the surrounding lattice).
- Rare-earth ion-containing nanoparticles are examples in which interionic coupling and coupling with the phonon modes can be controlled to enhance a particular emission as well as to produce new types of emission.
- The rare-earth ions exhibit a number of optical transitions that arise from promotion of electrons from an f to another f orbital or an f to a d orbital.
- An important optical emission, from telecommunications prospectives, is the 1.55-μm emission of an Er^{3+} ion which is used for optical amplification. It is shown here that nanostructure manipulation can significantly enhance this amplification.

- Phonon interactions determine the nonradiative decay from an excited electronic level to a lower energy level. This often involves a multiphonon process in which the excess energy is converted to many phonons.
- The multiphonon relaxation is determined by the phonon state distribution of the host lattice, the amplitude of the phonon vibration at the ionic site, and the strength of electron–phonon coupling. All of them depend on the nanoscale surroundings of the ion.
- The two important manifestations of electron–phonon interactions are (i) broadening of electronic transitions due to fluctuation of the electronic excitation energy derived from coupling to phonons, a process called dephasing of electronic excitation, and (ii) nonradiative relaxation of an excited state, leading to its depletion.
- Phonon spectrum (distribution of phonon energy states) of the host lattice is strongly influenced by reducing the size of the host from the bulk to a nanocrystal.
- An important modification of phonon spectrum is in the low-frequency acoustic phonon range where no phonons of wavelengths longer than twice the diameter of a nanoparticle can propagate. Thus there is a low-frequency gap in the phonon spectrum.
- In Huang and Rys model, the multiphonon relaxation is induced by a single-phonon mode, often of the highest frequency, also called the cut-off frequency of the host lattice.
- Hosts with lower cut-off phonon frequency produce significantly less nonradiative relaxation, and hence more efficient emission, compared to those with higher cut-off phonon frequencies.
- In the process of cross-relaxation, an ion transfers part of its excitation energy to another ion by ion–ion interactions that show strong dependence on their distance and thus on the nanostructure.
- The manipulation of local environment can produce stabilization of the 2^+ oxidation state (effective charge of 2^+) for a rare-earth ion, which has very different spectroscopic properties compared to the normal 3^+ oxidation state.
- Rare-earth ions exhibit a number of up-conversion processes whereby an IR excitation produces up-converted emission from a higher level in the visible spectral range.
- The prominent up-conversion processes are derived from three different mechanisms (i) excited-state absorption leading to sequentional absorption of two or more photons by the same ions, (ii) energy transfer up-conversion resulting from transfer of excitation from one ion to another already in the excited state, and (ii) cooperative emission whereby two neighboring ions in the excited state emit collectively from a higher-energy virtual state.
- Photon avalanche is an up-conversion process that produces up-conversion above a certain threshold of excitation power. The basic principle involves initial population of an intermediate level by some mechanism such as a

- phonon-assisted one. Another photon is readily absorbed from the intermediate level, which then populates the intermediate levels in other ions by cross-relaxation to start the avalanche process.
- Quantum cutting refers to a process in which the absorption of a high-energy photon (such as in vacuum ultraviolet) results in emission of two photons of lower energies (such as in visible). Hence, it is a down-conversion process.
- The mechanisms of quantum cutting may involve either a two-step emission from the same ion or a cross-relaxation from a highly excited ion to another ion whereby both ions emit.
- Nanoscale control can be used to enhance both up-conversion and quantum cutting.
- Another nanoscale control of excitation dynamics is provided by encapsulation of emitting species in a nanoparticle (or a nanosurrounding) to isolate it from the quenching induced by high-frequency phonons of the solvent or the host matrix.
- An example of site encapsulation is provided by shielding of an organic fluorophore in a silica bubble from water which has high-frequency phonons and is known to be a solvent that quench emission of dyes.

REFERENCES

Andrews, D. L., and Jenkins, R. D., A Quantum Electrodynamical Theory of Three-Center Energy Transfer for Upconversion and Downconversion in Are Earth Doped Materials, *J. Chem. Phys.* **114,** 1089–1110 (2001).

Auzel, F. and Chen, Y. H., The Effective Frequency in Multiphonon Processes: Differences for Energy Transfers or Side-Bands and Non-radiative Decay, *J. Lumin.* **66–67,** 224–227 (1996).

Auzel, F., and Pelle, F., Bottleneck in Multiphonon Nonradiative Transitions, *Phys. Rev. B* **55,** 11006–11009 (1997).

Bednarkiewicz, A., Hreniak, D., Deren, P., and Strek, W., Hot Emission in Nd^{3+}/Yb^{3+}:YAG Nanicrystalline Ceramics, *J. Lumin.* **102–103,** 438–444 (2003).

Bell, M. J. V., de Sousa, D. F., de Oliveira, S. L., Lebullenger, R., Hernandes, A. C., and Nunes, L. A. O., Photon Avalanche Upconversion in Tm^{3+}-Doped Fluoroindogallate glasses, *J. Phys. Condens. Mater.* **14,** 5651- 5663 (2002).

Biswas, A., Friend, C. S., and Prasad, P. N., Spontaneous Reduction of Eu^{3+} Ion in Al Co-Doped Sol–Gel Silica Matrix During Densification, *Mater. Lett.* **39,** 227–231 (1999).

Biswas, A., Maciel, G. S., Kapoor, R., Friend, C. S., and Prasad, P. N., Er^{3+}-Doped Multicomponent Sol–Gel Processed Silica Glass for Optical Signal Amplification at 1.5 μm, *Appl. Phys. Lett.* **82,** 2389–2391 (2003).

Chen, W., Malm, J.-O., Zwiller, V., Wallenberg, R., and Bovin, J.-O., Size Dependence of Eu^{2+} Fluorescence in ZnS:Eu^{2+} Nanoparticles, *J. Appl. Phys.* **89,** 2671–2675 (2001).

Chivian, J. S., Case, W. E., and Eden, D. D., The Photon Avalanche: A New Phenomenon in Pr^3-Based Infrared Quantum Counters, *Appl. Phys. Lett.* **35,** 124–125 (1979).

de Araüjo, C. B., Maciel, G. S., Menezes, L., Rakov, N., Falcao-Fieho, E. L., Jerez, V. A., and Messaddeq, Y., Frequency Upconversion in Rare-Earth Doped Fluorinated Glasses, *C. R. Chim.* **5**, 885–898 (2002).

Dieke, G.-H., *Spectra and Energy Levels of Rare Earth Ions in Crystals,* Interscience, New York, 1968.

Digonnet, M. J. F., ed., *Rare Earth Doped Fiber Lasers and Amplifiers,* Marcel Dekker, New York, 1993.

Downing, E., Hesselink, L., Ralston, J., and Macfarlane, R., A Three-Color, Solid-State, Three-Dimensional Display, *Science* **273**, 1185–1189 (1996).

Englman, R., *Non-Radiative Decay of Ions and Molecules in Solids,* North-Holland, Amsterdam, 1979.

Gamelin, D. R., and Güdel, H. V., Upconversion Processes in Transition Metal and Rare Earth Metal Systems, *Top. Cur. Chem.* **214**, 1–56 (2001).

Hochstrasser, R. M., and Prasad, P. N., Optical Spectra and Relaxation in Molecular Solids, in *Excited States,* Vol. 1, E. C.-Lin, ed., Academic Press, New York, 1974, pp. 79–128.

Huang and Rhys (1950).

Jouart, J. P., Bouffard, M., Duvaut, T., and Khaidukov, N. M., Photon Avalanche Upconversion in LiKYF$_5$ Crystal Doubly Doped with Tm^{3+} and Er^{3+}, *Chem. Phys. Lett.* **366**, 62–66 (2002).

Joubert, M. F., Photon Avalanche Upconversion in Rare-Earth Laser Materials, *Opt. Mater.* **11**, 181–203 (1999).

Joubert, M. F., Guy, S., Jacquier, B., and Linares, L. J., The Photon-Avalanche Effect: Review, Model and Application, *Opt. Mater.* **4**, 43–49 (1994).

Jüstel, T., Nikel, H., and Ronda, C., New Developments in the Field of Luminescent Materials for Lighting and Displays, *Angew. Chem. Int. Ed.* **37**, 3084–3103 (1998).

Kittel, C., *Introduction to Solid State Physics,* 7th edition, John Wiley & Sons, New York, 2003.

Kück, S., Sokolska, I., Henke, M., Doring, M., and Scheffler, T., Photon Cascade Emission in Pr^{3+}-Doped Fluorides, *J. Lumin.* **102–103**, 176–181 (2003).

Kueny, A. W., Case, W. E., and Koch, M. E., Nonlinear-Optical Absorption Through Photon Avalanche, *J. Opt. Soc. Am. B* **6**, 639–642 (1989).

Liu, B., Chen, Y., Shi, C., Tang, H., and Tao, Y., Visible Quantum Cutting in BaF$_2$:Gd, Eu via Downconversion, *J. Lumin.* **101**, 155–159 (2003a).

Liu, B., Chen, X.Y., Zhuang, H. Z., Li, S., and Niedbala, R. S., Confinement of Electron–Phonon Interaction on Luminescence Dynamics in Nanophosphors of Er^{3+}:Y$_2$O$_2$S, *J. Solid State Chem.* **171**, 123–132 (2003b).

Malyukin, Y. V., Maslov, A. A., and Zhmurin, P. N., Single-Ion Fluorescence of a Y$_2$SiO$_5$:Pr^{3+} Nanocluster, *Phys. Lett. A* **316**, 147–152 (2003).

Mao, Y., Gavrilovic, P., Singh, S., Bruce, A., and Grodkiewicz, W. H., Persistent Spectral Hole Burning at Liquid Nitrogen Temperature in Eu3+-Doped Aluminosilicate Glass, *Appl. Phys. Lett.* **68**, 3677–3679 (1996).

Meltzer, R. S., and Hong, K. S., Electron–Phonon Interactions in Insulating Nanoparticles: Eu$_2$O$_3$, *Phys. Rev. B* **61**, 3396–3403 (2000).

Menezes, L. de S., Maciel, G. S., and de Araüjo, C. B., Thermally Enhanced Frequency Upconversion in Nd^{3+} Doped Fluoroindate Glass, *J. Appl. Phys.* **90**, 4498–4501 (2001).

Menezes, L. de S., Maciel, G. S., and de Araüjo, C. B., Phonon-Assisted Cooperative Energy Transfer and Frequency Upconversion in a Yb^{3+}/Tb^{3+} Codoped Fluoroindate Glass, *J. Appl. Phys.* **94,** 863–866 (2003).

Nogami, M., Abe, Y., Hirao, K., and Cho, D. H., Room Temperature Persistent Spectra Hole Burning in Sm^{2+}-Doped Silicate Glasses Prepared by the Sol–Gel Process, *Appl. Phys. Lett.* **66,** 2952–2954 (1995).

Patra, A., Sominska, E., Ramesh, S., Koltypin, Y., Zhong, Z., Minti, H., Reisfeld, R., and Gedanken, A., Sonochemical Preparation and Characterization of Eu_2O_3 and Tb_2O_3 Doped in and Coated on Silica and Alumina Nanoparticles, *J. Phys. Chem. B* **103,** 3361–3365 (1999).

Patra, A., Friend, C. S., Kapoor, R., and Prasad, P. N., Upconversion in $Er^{3+}:ZrO_2$ Nanocrystals, *J. Phys. Chem. B* **106,** 1909–1912 (2002).

Patra, A., Friend, C. S., Kapoor, R., and Prasad, P. N., Effect of Crystal Nature on Upconversion Luminescence in $Er^{3+}:ZrO_2$ nanocrystals, *Appl. Phys. Lett.* **83,** 284–286 (2003a).

Patra, A., Friend, C. S., Kapoor, R., and Prasad, P. N., Fluorescence Upconversion Properties of Er^{3+}-Doped TiO_2 and $BaTiO_3$ Nanocrystallites, *Chem. Mater.* **15,** 3650–3655 (2003b).

Prasad, P. N., *Introduction to Biophotonics,* John Wiley & Sons, New York, 2003.

Samuel, J., Strinkowski, A., Shalom, S., Lieberman, K., Ottolenghi, M., Avnir, D., and Lewis, A., Miniaturization of Organically Doped Sol–Gel Materials: A Micron-Size Fluorescent pH Sensor, *Mater. Lett.* **21,** 431–434 (1994).

Scheps, R., Upconversion Laser Processes, *Prog. Quantum Electron.* **20,** 271–358 (1996).

Strek, W., Deren, P., and Bednarkiewicz, A., Cooperative Processes in $KYb(WO_4)_2$ Crystal Doped with Eu^{3+} and Tb^{3+} Ions, *J. Lumin.* **87–89,** 999–1001 (2000).

Strek, W., Zych, E., and Hreniak, D., Size Effects on Optical Properties of $Lu_2O_3:Eu^{3+}$ Nanocrystallites, *J. Alloys Comp.* **344,** 332–336 (2002).

Vink, A. P., Dorenbos, P., and Von Eijk, C. W. E., Observation of the Photon Cascade Emission Process Under $4f^15d^1$ and Host Excitation in Several Pr^{3+}-Doped Materials, *J. Solid State Chem.* **171,** 308–312 (2003).

Weber, M. J., Science and Technology of Laser Glass, *J. Non-Cryst. Solid,* **123,** 208–222 (1990).

Wegh, R. T., Donker, H., Oskam, K. D., and Meijerink, A., Visible Quantum Cutting in Eu^{3+}-Doped Gadolinium Fluorides via Downconversion, *J. Lumin.* **82,** 93–104 (1999a).

Wegh, R. T., Donker, H., Oskam, K. D., and Meijerink, A., Visible Quantum Cutting in $LiGdF_4:Eu^{3+}$ Through Down Conversion, *Science* **283,** 663–666 (1999b).

Wegh, R. T., Van Loef, E. V. D., and Meijerink, A., Visible Quantum Cutting via Down Conversion in $LiGdF_4:Er^{3+}$, Tb^{3+} Upon Er^{3+} $4f^{11} \rightarrow 4f^{10}5d$ Excitation, *J. Lumin.* **90,** 111–122 (2000).

Wermuth, M., and Güdel, H. U., Photon Avalanche in $Cs_2ZrBr_6:Os^{4+}$, *J. Am. Chem. Soc.* **121,** 10102–10111 (1999).

Yang, H.-S., Hong, K. S., Feofilov, S. P., Tissue, B. M., Meltzer, R. S., and Dennis, W. M., Electron–Phonon Interaction in Rare Earth Doped Nanocrystals, *J. Lumin.* **83–84,** 139–145 (1999).

Yeh, D. C., Sibley W. A., Suscavage, M., and Drexhage, M. G., Multiphonon Relaxation and Infrared-to-Visible Conversion of Er^{3+} and Yb^{3+} Ions in Barium–Thorium Fluoride Glass, *J. Appl. Phys.* **62,** 266–275 (1987).

CHAPTER 7
Growth and Characterization of Nanomaterials

This chapter covers growth and characterization of nanomaterials. The chapter is divided into two main sections. Section 7.1 describes some principal methods for growth of nanomaterials. Section 7.2 discusses some commonly used techniques for characterization of nanomaterials. As new applications of nanomaterials emerge, the need for novel approaches to produce specifically tailored materials, which can be made reproducibly and economically, grows.

It is difficult to discuss all the methods currently employed in this introductory book. The principal methods of vapor-phase growth for inorganic semiconductors are molecular beam epitaxy (MBE), metal–organic chemical vapor deposition (MOCVD), and laser-assisted vapor deposition, all of which are described in this chapter. Another growth technique used for semiconductors is liquid-phase epitaxy (LPE).

More recently a chemical approach using wet chemistry, which we call nanochemistry, is emerging as a powerful method for growing nanostructures of metals, inorganic semiconductors, organic materials, and organic:inorganic hybrid systems. The advantage offered by nanochemistry is that surface functionalized nanoparticles and nanorods of metals or inorganic semiconductors, dispersible in a wide variety of media (e.g., water, polymer, biological fluids), can be readily prepared. Furthermore, nanochemistry also lends itself to a precise control of conditions to produce monodispersed nanostructures (all nanoparticles within a very narrow distribution of size). All of these growth methods are discussed in various subsections of Section 7.1. The advantages and disadvantages of these methods are also given.

An important aspect of the development of nanostructures is characterization of their composition, structure, and shape. Some growth methods, such as MBE, actually incorporate a number of these characterization techniques for *in situ* monitoring of growth. A very popular technique used in semiconductor growth analysis and incorporated in MBE machines is the reflection high-energy electron diffraction (RHEED) technique utilized to monitor layer-by-layer growth.

The utilized techniques focus on the composition and surface analysis as well as on the crystal structure and size determination. The techniques discussed in this chapter are microscopy, diffractometry, and spectroscopy. Transmission electron microscopy (TEM), scanning electron microscopy (SEM), and scanning probe microscopy [e.g., scanning tunneling microscopy (STM) and atomic force microscopy (AFM)] are used for direct imaging of the nanostructures. The X-ray diffraction technique provide information on the crystallography of the material. X-ray photoelectron spectroscopy (XPS) and energy-dispersive spectroscopy (EDS) provide information about the composition of the material. The principles and applications of these techniques are described in various subsections of Section 7.2 of this chapter. The highlights of the chapter are provided in Section 7.3.

For further details, the following readings are suggested:

Barnham and Vvedensky, eds. (2001): *Low-Dimensional Semiconductor Structures*
Kelly (1995): *Low-Dimensional Semiconductors*
Jones and O'Brien (1997): *CVD of Compound Semiconductors*
Hermann and Sitter (1996): *Molecular Beam Epitaxy*
Wang, ed. (2000): *Characterization of Nanophase Materials*
Bonnell, ed. (2001): *Scanning Probe Microscopy and Spectroscopy*

7.1 GROWTH METHODS FOR NANOMATERIALS

This section describes some of the principal growth methods used to produce nanomaterials. These methods utilize the starting materials, which contain the chemical constituents of the final nanomaterials in various physical forms such as solid, liquid, or gas. Based on this division, the growth methods can be classified as vapor-phase, liquid-phase, or solid-phase growth. Another consideration is growth of ordered nanostructures (in crystalline form) on a substrate, where the crystalline lattice of the substrate provides a template for the growing layer to orient itself. Here, lattice matching between the substrate and the growing nanostructure plays an important role. This type of growth is called *epitaxial growth* (or epitaxy), and the growth is considered to be pseudomorphic. For the growth of quantum-confined structures of semiconductors, epitaxy methods are widely used. Another scheme of classification, in the case of inorganic semiconductors, is based on the chemical nature of precursors. For molecular beam epitaxy, the source materials used are usually in elemental forms (homoatomic materials). If the precursors are in the form of complex molecules such as metal–organic compounds, the method is called metal–organic chemical vapor deposition (MOCVD).

An important and widely different approach, perhaps not so familiar to many working in the area of inorganic semiconductors, is that of nanochemistry. This approach utilizes a traditional chemical synthesis in the solution phase. To achieve the fabrication of nanostructures, nanochemistry is performed in a number of ways such as chemistry in nano-confined geometries (as in micelles and reverse micelles) or termination of reaction at a precise point of growth (chemical capping).

The various subsections of Section 7.1 provide descriptions of these principal methods of growth.

7.1.1 Epitaxial Growth

Molecular Beam Epitaxy. Molecular beam epitaxy (MBE) is a widely used method for epitaxial growth of quantum-confined nanostructures of both II–VI and III–V semiconductors as well as for Si and Ge (Vvedensky, 2001; Kelly, 1995). The growth is carried out in an ultrahigh vacuum (UHV $\approx 10^{-11}$ torr) environment of a stainless steel chamber, as shown schematically in Figure 7.1. It involves the evaporation of atoms (e.g., Ga) or homoatomic molecules [i.e., molecular form of an element (e.g., As_2)] that are the constituents of the growing semiconductor. These constituents are contained in heated cells, called *effusion cells* or *Knudsen cells*. The vapor, formed by heating of the cells, passes through a small orifice and is accelerated by the pressure differential on the two sides of the orifice. It forms a molecular beam within which the particles neither react nor collide with one another. Then the beam impinges on a substrate (e.g., GaAs) mounted on a holder that is controlled from the opposite side of the chamber. By controlling and monitoring the fluxes from different cells using shutters, together with the adjustment of the substrate temperature, the composition and the epitaxial growth rate on the substrate can be precisely controlled with a monolayer resolution. The substrate is rotated to ensure a uniform deposition across it.

Shutters in front of the orifices are used to open and close in less than 0.1 s to produce appropriate combinations of the constituent elements for various semiconductor alloys (e.g., $Al_xGa_{1-x}As$). Since the quality of epitaxial growth is very sensitive to the substrate temperature (e.g., 580°C is best for GaAs and 630°C is best for AlAs), a controlled heating of the substrate is necessary. The surrounding chamber is cooled to the liquid nitrogen temperature to ensure that any material in the beam

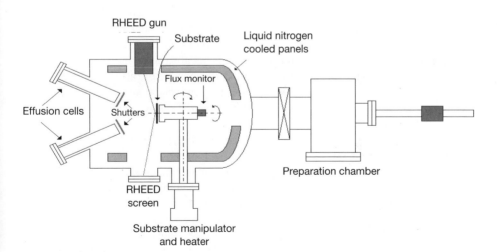

Figure 7.1. Schematic diagram of an MBE growth chamber. Courtesy of Hong Luo, University at Buffalo.

that misses the substrate on the first pass will stick to the cold surroundings and is not reflected to make a second pass.

The MBE chamber also incorporates a number of *in situ* characterization techniques for monitoring the growth. Examples are glancing-angle reflection high-energy electron diffraction (RHEED) which provides the ability to monitor the crystalline integrity of the growing surface and layer-by-layer deposition. Mass-spectrometric techniques yield information on the fluxes of the atomic and molecular species, in order to control growth rate, alloy composition, and doping levels. Other useful characterization techniques are X-ray diffraction and scanning probe microscopy. Many of these techniques are discussed in detail in Section 7.2.

The growth chamber shown in Figure 7.1 is preceded by an adjoining preparation chamber. Wafers (solid forms) of the constituents are introduced in the preparation chamber and subjected to heating, electron beam irradiation, and other forms of cleaning, prior to introduction into the ultrahigh vacuum systems of the growth chamber. The preparation chamber may contain additional diagnostic equipment that is not needed during the growth. Even though the method uses epitaxial growth, best suited for growth of a layer lattice-matched with the substrate, it is possible to grow thin layers of materials that are not well lattice-matched.

The major advantage offered by MBE is that the ultrahigh vacuum environment of the chamber permits the application of many *in situ* analytical techniques for characterizing the growth of the materials and its composition at various levels of spatial resolution. The MBE method is well suited for fabrication of quantum wells, quantum wires, and quantum dots. The quantum well fabrication is achieved by controlled layer-by-layer growth. A convenient method of fabrication of quantum wires, and quantum dots utilizes a substrate with a patterned surface. Another method of production of a quantum dot array utilizes the formation of coherent three-dimensional islands by growth on a substrate where there is an appreciable lattice mismatch between the deposited layer (epilayer) and the substrate. This type of growth is also called *Stranski–Krastanov (SK) growth* (Vvedensky, 2001). The growth in the form of islands results from the strain due to lattice mismatch. An example is the growth of InAs on GaAs (001 face) (Guha et al., 1990). However, the growth mode depends on lattice mismatch and is not always SK mode.

The MBE technique has revolutionized the semiconductor (also metals and oxides) science and technologies, exemplified by semiconductor diode lasers and quantum dot laser diodes that typically involve MBE-grown quantum well and quantum dot structures.

Figure 7.2 is a high-resolution TEM image (micrograph) of a ZnSe/ZnCdSe superlattice). An extreme case of application of this technique is the fabrication of what is referred to as *digital alloys*, in which different compounds are not randomly alloyed but are combined in ultrathin layer form, down to a fraction of an atomic layer. A recent example is the fabrication of III–V/Mn ferromagnetic digital alloys made possible by this technique, which have shown a range of interesting properties, including the high crystal quality and improved magnetic properties in some materials. Shown in Figure 7.3 is the cross-sectional transmission electron micrograph of a GaSb/Mn digital alloy, where the dark lines are half-monolayers of Mn

Figure 7.2. An illustrative example of TEM micrograph of an MBE grown film containing a ZnSe/ZnCdSe superlattice. The bright layers are ZnSe layers, and the dark layers are ZnCdSe layers. Courtesy of Hong Luo, University at Buffalo.

inserted in the GaSb host lattice (Chen et al., 2002). These structures have shown Curie temperatures (temperature defining the ferromagnetic–paramagnetic transition) higher than 400 K, which is a dramatic increase from their counterparts of conventional alloys. Intensive device research is currently being conducted to exploit these improved properties.

Metal–Organic Chemical Vapor Deposition (MOCVD). MOCVD is a chemical vapor deposition method in which the precursors for the elemental constituents of the semiconductor structure to be grown are in the metal–organic chemical form (Jones and O'Brien, 1997). Hence the precursors used here are different from the ones used in the case of MBE in which case the precursors are in the ele-

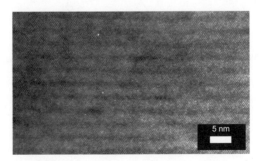

Figure 7.3. High-resolution TEM image of a GaMnSb digital-layer alloy. From Chen et al. (2002), reproduced with permission.

mental form. When the MOCVD method is used for epitaxial growth on a substrate, it is also called *metal–organic vapor-phase epitaxy* (MOVPE) and is an example of vapor-phase epitaxy. The method essentially consists of the following steps:

- Evaporation and transport of suitable precursors of the semiconductors to produce the semiconductor material
- Deposition and growth of the semiconductor on a substrate
- Removal of the remaining decomposition products from the chamber

Figure 7.4 shows the schematic diagram of an MOCVD growth chamber (Kelly, 1995). It consists of a glass reactor that contains a heated substrate situated at an angle to a laminar flow of gas. The precursors are carried over the substrate by a carrier gas, often hydrogen. For example, trimethyl gallium ($Ga(CH_3)_3$) is a precursor for Ga and arsine (AsH_3) is a precursor for As for growing the III–V semiconductor, GaAs. The chemical reaction produced by pyrolyses in this case is

$$Ga(CH_3)_3 + AsH_3 \rightarrow 3CH_4 + GaAs$$

By the same scheme, the precursors for AlAs are $Al(CH_3)_3$ and AsH_3. Similarly, in the case of a II–VI semiconductor CdS, the precursors are $Cd(CH_3)_2$ and H_2S. The chemical composition of the growth layer is determined by the relative ratio of the metal–organic precursors in the incoming gas mixture. For a multilayer growth, uniform premixing of the precursors is performed in a premixing chamber.

The MOCVD method provides the advantage of being a simpler growth technique, with a growth rate typically 10 times that of MBE. However, the precursors, particularly for III–V semiconductors, are quite toxic and demand extreme safe-

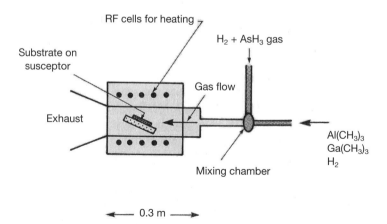

Figure 7.4. Schematic diagram of MOCVD reactor chamber. From Kelly (1995), reproduced with permission.

guard and care in handling. Furthermore, the hydrodynamic condition of gas flow does not permit the extensive *in situ* characterization offered by MBE.

Chemical Beam Epitaxy (CBE). This method is essentially a combination of MOCVD and MBE. It utilizes an ultrahigh vacuum chamber like in MBE, but the constituent elements are derived from metal–organic precursors as in MOCVD. The method thus offers advantages of both MOCVD and MBE. The growth of the semiconductor occurs in ultrahigh vacuum, but the metal–organic precursors are external to the system, which facilitates replenishment, thus decreasing the reactor "downtime." In the CBE approach, therefore, one can use the *in situ* diagnostic techniques, mentioned above under MBE.

Liquid-Phase Epitaxy (LPE). Liquid-phase epitaxy (LPE) is a film deposition method from the liquid phase, either solution or melt. A substrate is brought into contact with a saturated solution of the film material at an appropriate temperature. Typically, growth of epitaxial layers occurs in two different types of LPE setups: one for the *dipping* process and the other for the *tipping* process. Dipping refers to a process in which the substrate is dipped vertically in the melt. Tipping is a process whereby a horizontally placed graphite crucible contains both the melt and the substrate. By tilting the crucible, the melt flows onto the substrate. The substrate is then cooled at a suitable rate to lead to film growth.

LPE is a method of choice when a high deposition rate and crystalline perfection of the films are desired. As a robust and cost-saving method, it is a widespread production process in the semiconductor industry. Typically, compounds and alloys of III–V semiconductors (similar to MBE) are manufactured by this method.

The advantage offered by LPE is the lower cost and higher deposition rate, compared to those for MBE. LPE also provides a control of stoichiometry and yields low defect concentration. A disadvantage of LPE is that solubility considerations greatly restrict the number of materials for which this method is applicable. Furthermore, morphology (crystal orientation) control is difficult and surface quality is often poor.

7.1.2 Laser-Assisted Vapor Deposition (LAVD)

Laser-assisted processes have been utilized for deposition of films and nanoparticles. One type of laser-assisted deposition involves laser ablation of a solid target where the ablated material can be, by itself, deposited on a substrate. Alternatively, the ablated material can be mixed with a reactive gas to make the appropriate material, then carried by an inert gas through a nozzle into a vacuum chamber to produce a molecular beam, and then deposited on a substrate whose temperature is controlled. This method is also called *laser-assisted molecular beam deposition* (LAMBD) (Wijekoon et al., 1995). Nanoparticles of TiO_2 are examples produced by this method. This method has also been used to produce organic:inorganic nanocomposites (Wijekoon et al., 1995).

Another example of laser-assisted vapor deposition involves laser decomposition of precursor gaseous molecules to produce the desired materials. This approach has been utilized at our Institute for Lasers, Photonics, and Biophotonics, by Swihart and co-workers (Li et al., 2003) to produce silicon nanoparticles. The silicon nanoparticles were synthesized by laser-induced heating of silane to temperatures where it dissociates, in the reactor shown schematically in Figure 7.5. A continuous-wave CO_2 laser beam was focused to a diameter of about 2 mm just above the central reactant inlet. Silane weakly absorbs the laser energy at a wavelength of 10.6 μm, and is thereby heated. Sulfur hexafluoride (SF_6) may be added to the precursor stream as a photosensitizer. SF_6 has a large absorption cross section at the laser wavelength and can, therefore, dramatically increase the temperature achieved for a given laser power. Helium and hydrogen flows confine the reactant and photosensitizer (SF_6) to a region near the axis of the reactor and prevent them from accumulating in the arms of the six-way cross from which the reactor is constructed. Hydrogen also serves to increase the temperature at which particle nucleation occurs, and to decrease the particle growth rate, since it is a byproduct of the silane dissociation and particle formation. All gas flow rates to the reactor were controlled by mass flow controllers. The resulting particles were collected on cellulose nitrate

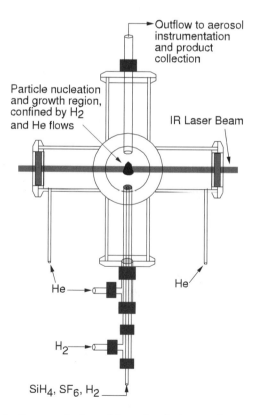

Figure 7.5. Schematic diagram of a laser-driven chemical vapor deposition. From Li et al. (2003), reproduced with permission.

membrane filters. The effluent was directed to a furnace where it was heated to 850°C to decompose any residual silane. This method can produce silicon nanoparticles at 20–200 mg/hr in the present configuration using a 60-W laser. With a commercially available multikilowatt laser focused into a thin sheet, one could readily scale this up by one to two orders of magnitude.

7.1.3 Nanochemistry

Nanochemistry is an active new field that deals with confinement of chemical reactions on nanometer length scale to produce chemical products that are of nanometer dimensions (generally in the range of 1–100 nm) (Murray et al., 2000). The challenge is to be able to use chemical approaches that would reproducibly provide a precise control of composition, size, and shape of the nanomaterial product formed. Nanochemistry has been used to make nanostructures of different composition, size and shapes. Nanoscale chemistry also provides an opportunity to design and fabricate hierarchically built multilayer nanostructures to incorporate multifunctionality at nanoscale.

Nanochemistry offers the following capabilities:

- Preparation of nanoparticles of a wide range of metals, semiconductors, glasses, and polymers
- Preparation of multilayer, core-shell-type nanoparticles
- Nanopatterning of surfaces, surface functionalization, and self-assembling of structures on this patterned template
- Organization of nanoparticles into periodic or aperiodic functional structures
- *In situ* fabrication of nanoscale probes, sensors, and devices

A number of approaches provide nanoscale control of chemical reactions, some of which are described here.

Reverse Micelle Synthesis. An example of a reaction in a confined geometry is the synthesis of nanoparticles such as CdS (quantum dots) in a reverse micelle cavity also known as a microemulsion nanoreactor. Figure 7.6 shows a schematic representation of the reverse micelle nanoreactor aqueous core dispersed in a bulk oil medium.

The reverse micellar system is generally composed of two immiscible liquids, water and oil, where the aqueous phase is dispersed as nanosize water droplets encapsulated by a monolayer film of surfactant molecules in a continuous nonpolar organic solvent such as a hydrocarbon oil. The continuous oil phase generally consists of isooctane or hexane. Sodium bis(2-ethylhexyl)sulfosuccinate (Aerosol-OT or AOT) serves as the surfactant. In addition to water, aqueous solutions containing a variety of dissolved salts, including cadmium acetate and sodium sulfide, can be solubilized within the reverse micelles (Petit et al., 1993). The size of the reverse micelle, and subsequently the volume of the aqueous pool contained with-

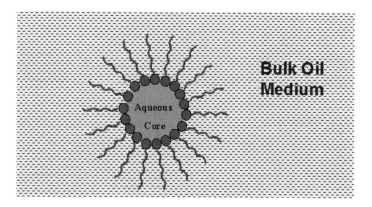

Figure 7.6. Schematic diagram of a reverse micelle nanoreactor aqueous core.

in the reverse micelle, is governed by the water-to-surfactant ratio, also termed W_0, where $W_0 = [H_2O]/[\text{surfactant}]$ (Petit et al., 1993). A continuous exchange of the reverse micellar contents through dynamic collisions enables the reaction to proceed. However, since the reaction is confined within the reactor cavity, growth of a crystal beyond the dimensions of the cavity is inhibited. In the final stage of this synthesis, a passivating reagent, such as p-thiocresol, is added to the continuous oil phase. This species is then able to enter the aqueous phase as an RS^- anion and bond to the surface of the contained nanocrystal, eventually rendering the surface of the nanocrystal hydrophobic and inducing precipitation of the capped CdS nanoparticles.

Metallic nanoparticles have been made using this approach. By appropriate choice of the surfactant or a mixture of surfactants, the shape of the cavity can be made cylindrical to produce nanorods. The reverse micelle chemistry also lends itself readily to a multistep synthesis of a multilayered nanoparticle such as a core-shell structure (Lal et al., 2000).

Colloidal Synthesis. This method involves growing nanoparticles or nanorods of inorganic materials (elements and compounds) by chemical reaction of their precursors in a carefully chosen solvent. This has, of late, been a very fruitful method of producing nanocrystalline materials with a reasonable uniformity in sizes.

Generally, when a solid is formed from its precursors through a chemical reaction, it starts with the fast formation of a multitude of nuclei. More and more of the solid product then add to the nuclei, and the sizes of the crystallites grow slowly. The idea is to truncate the growth process of the crystallites, before they grow infinitely big in sizes, thus yielding nanoparticles or nanocrystals (sometimes abbreviated as NC). This is achieved by suitably surface capping the crystallites by an appropriate surfactant, which is usually a long-chain organic molecule with a functional group. Whether the surfactant should be added during the reaction, or be generated *in situ*, or added post synthesis, of course, depends on the material under study. Also, the selection of surfactant depends on the nature of NC-forming materials. At our laboratory, we have used this method to produce Au nanoparticles, stabilized by long-chain thiols or amines, and oxide nanoparticles, such as Fe_3O_4, stabilized by carboxylic acids or amines. Figure 7.7 shows the transmission electron microscope (TEM) images of Au particles bound through thiol linkage to mercaptoundecanoic acid (MUDA), mercaptophenol (MP), mercaptohexanol (MH), and aminothiophenol (ATP). The TEM image of magnetic nanoparticles of iron ferrite bound to oleic acid is shown in Figure 7.8.

In almost all cases of colloidal synthesis, a systematic adjustment of the reaction conditions—time, temperature, concentrations of the precursors, and chemical nature of reagents and surfactants—can be used to control NC size and shapes of NC samples (Fendler and Meldrum, 1995; Murray et al., 2000). Bifunctional organic molecules can potentially act as linkers to bind two types of nanoparticles.

Figure 7.7. Transmission electron micrographs of Au particles. (a) Au-MUDA, (b) Au-MP, (c) Au-MH, (d) Au-ATP.

188 GROWTH AND CHARACTERIZATION OF NANOMATERIALS

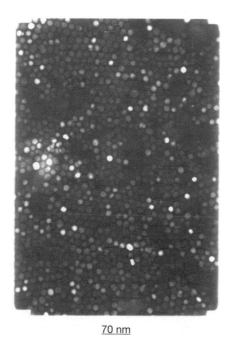

70 nm

Figure 7.8. TEM of iron ferrite in sizes of 10–12 nm.

Figure 7.9 shows TEM images of Au nanoparticles bound to the surface of polystyrene, produced in our laboratory, using aminoethanethiol.

An approach used in the synthesis of III–V semiconductor nanocrystals (i.e., InP, InAs, GaP, and GaAs) involves dissolution of the Group III halide (i.e., $InCl_3$ or $GaCl_3$) in a coordinating solvent, with added coordinating ligands or surfactants (Alivisatos et al., 2001). Then, the Group V precursors ($P(SiMe_3)_3$ or $As(SiMe_3)_3$) are added at an elevated temperature. This reaction is then allowed to proceed for a long period of time (3–6 days) at the elevated temperature (300–500°C). Then the nanocrystals are isolated and size-selectively precipitated. Size-selective precipitation is a process by which relatively monodispersed particle sizes are selectively precipitated in multisteps, from a sample containing large distribution of sizes.

New methods for the synthesis of III–V nanocrystals involve the dissolution of Group III salts [i.e., $In(acetate)_3$ or In_2O_3] in a noncoordinating solvent with the addition of coordinating ligands or surfactants (Battaglia and Peng, 2002). Then the reaction mixture is heated to a temperature where rapid injection of a Group V precursor [$P(SiMe_3)_3$ or $As(SiMe_3)_3$] results in a crop of III–V nuclei. The reaction is followed by removing aliquots. By varying the reaction concentration and reaction temperatures and doing successive injections of precursors, relatively monodisperse crops of nanocrystals are isolated.

A method for semiconductor nanocrystals synthesis, recently used in our laboratory, does not require the use of coordinating solvents, ligands, or surfactants

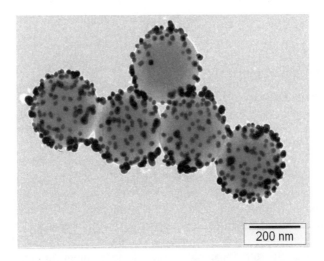

Figure 7.9. TEM of Au particles on polystyrene particle surface.

(Lucey et al., 2004). Specially synthesized Group II and Group III precursors that are soluble in noncoordinating solvents (i.e., octadecene, benzene, toluene, or hexane) are used. Upon rapid injection with Group VI or Group V precursors, these precursors produce a crop of II–VI or III–V seeds, respectively, as well as generate a surfactant that controls the size distribution. By varying the precursor concentrations and temperature of the reactions, highly monodispersed nanocrystals are obtained.

7.2 CHARACTERIZATION OF NANOMATERIALS

The properties and improved performance exhibited by nanomaterials are strongly dependent on their composition, size, surface structure, and interparticle interactions. Hence characterization of these properties is of immense importance in the development of nanomaterials and in understanding structure–function relationship. This requires the use of a range of microscopic and spectroscopic methods suitable for investigation of nanostructures (Wang, 2000). This part of Chapter 7 discusses some important microscopic and diffraction techniques widely used for characterization of nanomaterials. The three principal categories covered here are X-ray characterization, electron microscopy, and scanning probe microscopy. X-ray characterization includes X-ray diffraction (XRD) and X-ray photoelectron spectroscopy (XPS), which are discussed in Section 7.2.1. The electron microscopic techniques, transmission electron microscopy (TEM) and scanning electron microscopy (SEM), are described in Section 7.2.2. Other electron beam techniques, such as reflection high energy diffraction (RHEED) and energy dispersive spectroscopy (EDS), are covered in Section 7.2.3. Section 7.2.4 introduces scanning tunneling microscopy

(STM) and atomic force microscopy (AFM) as examples of scanning probe microscopy.

7.2.1 X-Ray Characterization

7.2.1.1 X-Ray Diffraction. X-ray diffraction, often abbreviated as XRD, is extensively used to characterize the crystalline form of nanoparticles and to estimate the crystalline sizes (see Figure 7.10). It often utilizes X-ray diffraction from nanoparticles in powder form, which is thus called *powder diffraction*. A comprehensive review of powder diffraction is provided by Langford and Louër (1996).

X-ray diffraction of a crystalline material is based on the principle of elastic scattering of X rays by a periodic lattice characterized by long-range order. It is a reciprocal space-based method (that is, it gives information about types of periodicity present in the material, rather than the real space distribution of individual atoms), and it gives an ensemble average information on the crystal structure and the particles size of a nanostructured material. When a beam of monochromatic X rays impinges on a sample, the ray penetrates the sample and gets diffracted by the periodic lattice of a crystalline material, according to the well-known Bragg's equation:

$$n\lambda = 2d \sin \theta \quad (7.1)$$

where n is an integer, λ is the X-ray wavelength, d is the spacing between crystallographic planes giving rise to a particular diffracted beam, and θ is the incidence angle.

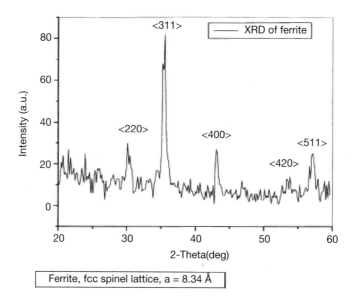

Figure 7.10. XRD of Fe_3O_4 nanoparticles.

The diffraction pattern generated by constructive interference of the scattered X rays provides crystallographic information on the materials. In the case of nanomaterials, the sample can be in form of powder or as a thin film that is exposed to a beam of X rays, where the angle of incidence is varied. For a sample consisting of polycrystalline (powder) materials, the X-ray diffractogram is a series of peaks conforming to the 2θ values, where Bragg's law of diffraction is satisfied. Figure 7.10 shows the powder X-ray diffraction pattern of Fe_3O_4 nanoparticles that can be indexed as a face centered cubic lattice.

Nanoparticles, less than approximately 100 nm in size, show appreciable broadening of their X-ray diffraction lines. The observed line broadening can be used to estimate the average size of the nanocrystals (or crystalline domains). In the simplest case of stress-free particles, the size can be estimated from a single diffraction peak. The particle/grain size, D, is related to the X-ray line broadening by Scherrer's formula (Cullity, 1978):

$$D = 0.9\lambda/\beta \cos\theta \qquad (7.2)$$

where λ is the wavelength, θ is the diffraction angle, and β is the full width (in radian) at half-maximum intensity. In those cases where stress may be present, a more rigorous method involving several diffraction peaks is required. However, Scherrer's formula should be used only as a rough guide for particle size. Particle size distribution (if there is polydispersity) and/or the presence of stress often provides results for crystallite sizes (average values) that are different from those obtained by TEM.

Figure 7.11 shows the XRD pattern of Er^{3+}-doped TiO_2 nanocrystals processed at different temperatures (Patra et al., 2003). The Er^{3+} concentration is 0.25 mole %. For nanoparticles processed at 500°C, the diffraction patterns are indicative of the antase crystalline form of TiO_2. For the nanoparticles processed at 1000°C, the diffraction patterns are mainly those of the rutile crystalline form of TiO_2. The sample

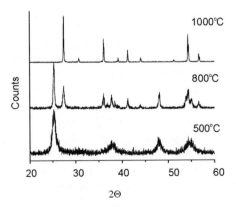

Figure 7.11. Powder X-ray diffraction patterns of 0.25 mole % Er-doped TiO_2 nanoparticles obtained after heating to three different temperatures.

processed at 800°C contains both crystalline forms. Another evident feature is the sharpening of the diffraction peaks with increase in the processing temperature, indicating an increase in the particle size. The nanoparticles processed at 500°C exhibit broad features indicating a relatively small nanosize. The average sizes of the nanocrystallites, obtained using Scherrer's formula, are ~ 14, 31, and > 100 nm for samples processed, respectively, at 500°C, 800°C and 1000°C. The corresponding values obtained by TEM are 15, 40, and > 100 nm.

A significant development in the XRD study of nanostructured materials is the use of synchrotron sources providing very bright and continuous spectrum of X-ray radiations, allowing one to study small quantities of nanoparticles.

7.2.1.2 X-Ray Photoelectron Spectroscopy.

X-ray photoelectron spectroscopy (XPS), also commonly known as electron spectroscopy for chemical analysis (ESCA), is used to study the composition and electronic states of the surface region of a sample. It makes use of the photoelectric effect in which a photon (X ray in this case) strikes the surface of a material to eject electrons that leave the surface with various energies. This technique is capable of providing information on the oxidation states, immediate chemical environment, and concentration of the constituent atoms. A good reference source for XPS is the three-volume-set handbook by Crist (2000).

The sample, in the form of a thin film of the nanomaterial under study, is irradiated with a monochromatic soft X-ray beam of energy in the range 200–2000 eV. These X-ray photons have a limited penetration depth (depending on the angle of incidence, photon energy, and the material).

The kinetic energy, with which the photoelectron leaves, is a function of the binding energy and, hence, the chemical environment of the particular atom. Only the electrons from the first few nanometers of the sample surface are generally probed in this method. The electrons leaving the sample are detected by an electron spectrometer according to their kinetic energy.

The XPS of a specimen is represented by a spectrum displaying the flux of emitted electrons, $N(E)$, as a function of the binding energy (determined from the kinetic energy of the ejected electron). Figures 7.12a–c show this spectrum for InP nanoparticles prepared by a novel procedure at our laboratory. (See Section 7.1.3.) Figure 7.12a shows a survey spectrum which gives information about different elements on the sample surface. The peaks due to O and C are derived from contaminants. A high-resolution scan, as shown in Figures 7.12b and 7.12c, is used to calculate atomic concentrations. The peak areas are used (with appropriate sensitivity factors) to determine the concentrations of the constituent elements. The shape of each peak and the binding energy can be slightly altered by the chemical state of the emitting atom. Figure 7.12b shows the characteristic peaks of indium $3d_{5/2}$ and $3d_{3/2}$ electrons and 7.12c shows the characteristic peaks of phosphorus $3p$ core level electrons from InP (130 eV) and oxidized phosphorus (~ 133 eV). Thus, XPS gives chemical bonding information of the elements in a nanomaterial compound. It may be noted that XPS is not suitable for light atoms such as hydrogen and helium.

7.2 CHARACTERIZATION OF NANOMATERIALS   **193**

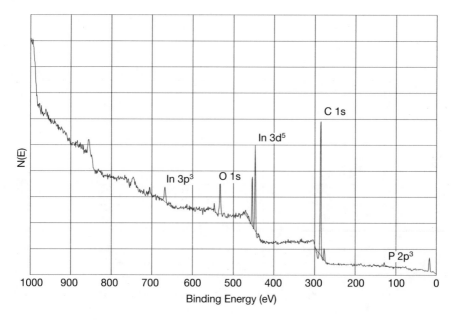

Figure 7.12a. XPS survey spectrum of InP nanoparticles.

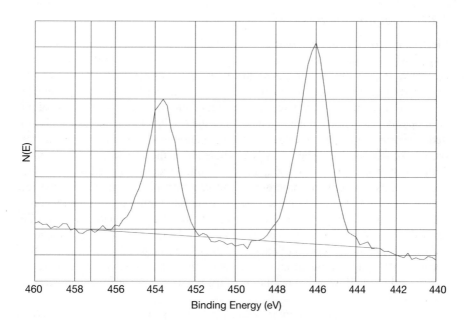

Figure 7.12b. Binding energy of indium $3d^5$ electrons.

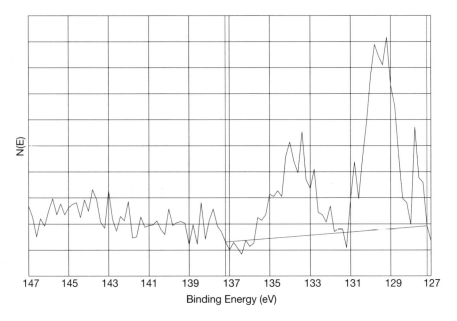

Figure 7.12c. Binding energy of phosphorus $2p^3$ electron in InP.

7.2.2. Electron Microscopy

Electron microscopy has emerged as a powerful tool for characterization of the size and shape of nanostructured materials (Heimendahl, 1980). Like optical microscopy, it provides an image of the material in direct space. However, the nanometer scale resolution provided by electron microscopes are highly suitable for imaging of nanomaterials, whereas the resolution provided by common optical microscopes is in hundreds of nanometers. Since the resolution of a microscope is limited to approximately the wavelength of the radiation used to image (the spatial resolution is usually taken to be $0.61\lambda/NA$, where NA is the numerical aperture of the optical system), even for UV light, an optical microscope is limited to the resolutions of ~200 nm. Electron microscopes utilize electrons of energies in the few thousand electron volt (eV) range, which is a thousand times greater than that of a visible photon of energy ~2 to 3 eV. Using the de Broglie equation $\lambda = h/p$ described in Chapter 2 for relation between the wavelength λ, and the momentum p, the wavelength of an electron of energy ~3600 eV is calculated to be 0.02 nm. But, because of the aberrations of an electron lens, the resolution actually achieved is significantly less. However, a resolution of 0.1 nm can be obtained with an electron microscope.

The two principal electron microscopic methods described here are transmission electron microscopy (TEM) and scanning electron microscopy (SEM). TEM is the electron microscopic analogue of transmission optical microscope in which a focused beam of electrons is used, instead of light, to see through (using transmitted electrons) the specimen. SEM involves scanning of a focused electron beam across the sample.

7.2.2.1 Transmission Electron Microscopy (TEM).
Transmission electron microscopy is a tool utilized to analyze the structures of very thin specimens through which electrons as probes are transmitted. A transmission electron microscope is constructed similar in principle to an optical microscope, where an electron beam, like light in a transmission microscope, travels through the sample and is affected by the structures in the specimen. However, the sample in the case of TEM is in a very high vacuum. The transmitted electron beam is projected onto a phosphor screen for imaging or digitally processed in a computer. The electromagnetic lenses in TEM consist of current carrying coils surrounded by iron. The electron beam is thus focused by these electromagnetic lenses, passed through the specimen under examination. The transmitted beam is intercepted by objective lenses and finally captured on a fluorescent screen.

A TEM consists of the following components:

- An electron gun that produces a stream of monochromatic electrons
- Electromagnetic condenser lenses that focus the electrons into a small beam
- A condenser aperture to restrict the beam by eliminating the high-angle electrons
- A sample stage on which the sample is placed
- An objective lens to focus the transmitted beam
- Optional objective and selected area metal apertures to enhance the contrast by blocking out high-angle diffraction, as well as to obtain electron diffraction
- Subsequent intermediate and projector lenses to enlarge the image, allowing optical recording of the image carried by the transmitted electron beams

For TEM, the sample should be thin and thus allow the high-energy electrons to pass through. As the electrons pass through the specimen, scattering occurs at the atoms because of Coulomb interactions. The degree of scattering (which can be elastic or inelastic) depends on the constituent atoms of the specimen. The heavy atoms (the ones with high atomic numbers) scatter the highest. The ray intensity transmitted through such areas, compared with that through the areas with light atoms, is lower. The intensity distribution of the electrons reaching the fluorescent screen is determined by the number of electrons transmitted. This results in a relative darkness in the specimen image area, which is rich in heavy atoms.

A TEM image provides reliable information about the size and shapes of nanoparticles, as illustrated by Figures 7.13a and 7.13b.

7.2.2.2 Scanning Electron Microscopy (SEM).
Scanning electron microscopy is a technique to obtain the image of a sample, by scanning of an electron

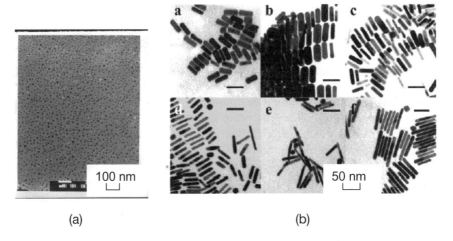

Figure 7.13. TEM of (a) GaP nanoparticles and (b) Au nanorods. Figure 7.13b is from Nikoobakht and El Sayed (2003), reproduced with permission.

beam across the surface of specially prepared specimens. SEM provides a greatly enlarged and highly resolved three-dimensional view of the specimen's exposed structure. For this purpose, an electron beam from an electron gun is focused onto the specimen surface by condenser lenses. It is possible to focus the electron beam to a small area, typically 10–20 nm in diameter. This spot size determines the resolution of the image, which is finally obtained by scanning. A set of scanning coils (made of wires carrying controlling current) below the condenser lens deflects the electron beam. The beam deflection allows scanning of the surface in a grid fashion (like in television), dwelling on each point for a period of time determined by the scan speed (usually in the microsecond range).

There are several types of signals which are produced when the focused electron beam impinges on a specimen surface, and which can be used to form an SEM image. They include back-scattered electrons, secondary electrons, cathodoluminescence, and low-energy characteristic X rays generated by the impinging electrons. The signals are obtained from specific emission volumes within the sample and are used to measure many characteristics of the sample (composition, surface topography, crystallography, magnetic or electric character, etc.).

The secondary and back-scattered electrons are captured by a detector and are primarily responsible for topographic images. The detector converts every electron into a light flash that, in turn, produces electrical pulses. These pulses are amplified by a signal amplifier and used to modulate the brightness of spots in a two-dimensional picture that can be displayed on a cathode ray tube (CRT) or processed digitally. Thus, in principle, every point spot on a specimen is transposed to a corresponding point on the CRT. The brightness of a spot in a SEM image is a measure of the intensity of secondary electrons, which critically depends on the local surface topography. The detector is asymmetrically positioned such that the elevated surface areas that directly face the detector appear bright whereas "holes" and crevices are imaged darker. Figure 7.14 shows an example of a SEM picture produced by multilayer, well-ordered, close-packed polystyrene spheres which form a photonic crystal structure (discussed in Chapter 9).

It should be noted that it is desirable to make the sample conductive, since in the case of insulating samples, surface charges build up and deflect the course of secondary electrons. Therefore, for nonconducting samples such as nonmetal surfaces, a thin layer of graphite or metal is usually deposited to make them conductive and prevent surface charging.

The advantage of SEM over TEM is that it provides a tremendous depth of focus. Thus, a three-dimensional image of the exposed surface of a sample, as illustrated in Figure 7.14, can be obtained. In contrast, TEM provides only a transmission contrast in a thin sample. But the resolution obtained by TEM is significantly higher than that for SEM. In the latter case, typical resolutions are about ~ 10 nm. Therefore, for information on size and shapes of nanoparticles less than 10 nm, TEM is the appropriate technique.

7.2.3 Other Electron Beam Techniques

Reflection High-Energy Electron Diffraction (RHEED). This technique is generally incorporated in an MBE growth chamber, because it is a standard method to monitor layer by layer growth. The method utilizes a high-energy (10–20 keV)

Figure 7.14. Cross-sectional SEM image of close-packed 200-nm polystyrene spheres.

beam of electrons from an electron gun that impinges on the growth surface at a glancing angle (~ 0.5°–3.0°). The electrons penetrate a few layers into the surface and the exiting electrons are recorded on a phosphorescence screen (Vvedensky, 2001). The images recorded are diffraction patterns (reciprocal space).

RHEED, by monitoring the diffraction patterns as a function of growth of the film on the substrate, provides information on the crystalline symmetry of the film, the extent of long-range order (from the sharpness of the pattern), and the growth pattern (whether three-dimensional or two-dimensional). However, the most practical application of RHEED is in monitoring the growth thickness, layer-by-layer, using the intensity of the specular beam for which the angles of incidence and reflection are the same. Figure 7.15 shows a typical example of the observed oscillation and decay of the intensity as a function of deposition time of the ZnSe layer. The oscillations result from the repeated formation of ZnSe layers, providing in situ monitoring of the layer-by-layer growth. The maximum corresponds to the addition of a full ZnSe bilayer. The decay of the envelope is derived from the imperfection in the growth, implying that, as the layer grows, a subsequent layer begins to form before the preceding layer is complete. The period of oscillations, in seconds, corresponds to the time required to form complete two successive layers (a bilayer). This type of intensity pattern can be used to precisely control the material deposition to within a fraction of a layer, by using a shutter in the molecular beam, as described in Section 7.1.1 under MBE. The use of various shutters together with RHEED monitoring can also serve to deposit the desired number of layers of one material (e.g., GaAs) followed by a selected number of layers of another type of composition.

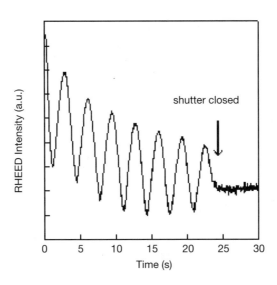

Figure 7.15. RHEED oscillations during the growth of ZnSe by MBE. Courtesy of Hong Luo, University at Buffalo.

7.2 CHARACTERIZATION OF NANOMATERIALS

Energy-Dispersive Spectroscopy (EDS). In tandem with the scanning electron microscope (SEM), it is possible to determine the chemical composition of a microscopic area of a solid sample using energy-dispersive spectroscopy (EDS). The principle is that the electron beam of the SEM, as described in Section 7.2.2, causes atoms in the field of the view to emit X rays, the energy of which is characteristic of the emitting element. By this method, it is possible to analyze a particular point on a surface. This technique allows one to obtain chemical composition information on a wide range of elements of samples, such as polished surfaces, fracture surfaces, powders, and surface films. Figure 7.16 illustrates the EDS of InP nanoparticles obtained in our laboratory using this technique.

7.2.4 Scanning Probe Microscopy (SPM)

Scanning probe microscopy (SPM) has emerged as a powerful technique to obtain three-dimensional real-space images of nanostructures and to perform nanoscopically localized measurements of physical properties such as local density of electronic states (Bonnell and Huey, 2001; Chi and Röthig, 2000). The images in SPM are produced utilizing local interactions between a small tip (radius in nanometers) and the sample surface. Depending on the nature of local interactions used (hence the nature of the tip), one can obtain images that provide spatial distributions of surface topography, electronic structure, magnetic structure, or any number of other local properties. A major advantage of the SPM technique is that it provides a rich variety of information (such as local physical properties or bonding), with resolution approaching the single-atom level and without destroying the sample. No special sample preparation is needed. In contrast, the electron microscopy discussed above requires special sample preparation and relatively high vacuum conditions.

All SPM techniques, regardless of the physical nature of interaction between the tip and the surface to be imaged, involve bringing the tip sufficiently close to the surface to measure the interaction. The tip is then raster scanned over the surface, with movement controlled on a length scale (nanometers) comparable to the tip

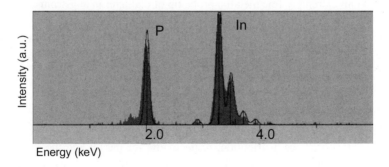

Figure 7. 16. EDS of InP nanoparticles.

size. For example, a topographic image of the surface can be obtained by making measurements at different locations, by scanning the probe at constant height.

The most common scanning probe microscopic techniques are scanning tunneling microscopy (STM) and atomic force microscopy (AFM). These two types of microscopic methods are described here. In STM, the interaction between the probe tip (metallic) and the surface is electrical, producing electron tunneling, a phenomenon already described in Chapter 2. It is the oldest of the scanning probe microscopic techniques, which measures the tunneling current flowing between the probe tip and the sample, as they are held a small distance apart and an electric field is applied between them. AFM is based on atomic force between the tip and the surface. Depending on the separation between the tip and the sample, various forces may play the dominant role. For example, when tip is extremely close to the sample surface, it is the short-range repulsive part of the van der Waals forces. When the tip is lifted above the surface, long-range attractive interactions play an important role. Other variations of AFM are: (i) magnetic force microscopy (MFM), which utilizes a magnetic tip to probe local magnetic interactions, (ii) electric force microscopy (EFM), which utilizes a metallic tip to probe local electrostatic interactions (force and not tunneling currents), and (iii) chemical force microscopy, which utilizes a probe coated with a chemical structure to probe specific chemical interactions. Near-field optical microscopy (NSOM), already discussed in Chapter 3, is another example of scanning probe microscopy which generally utilizes an optical fiber tip. However, the interaction in the case of NSOM is optical and, therefore, resolution achievable by NSOM is not as high as in AFM and STM, which are capable of providing atomic resolution. Described below in some detail are STM and AFM.

Scanning Tunneling Microscopy (STM). STM was originally introduced (Binnig et al., 1982) for analysis of conducting surfaces under ultrahigh vacuum conditions. [See Bai (2000) for early work and subsequent developments.] Since then the technique has been modified to be of a much wider utility, because it can be used in a variety of environmental conditions such as in air (Park and Quate, 1986), water (Sonnenfield and Hansma, 1986), oil, and electrolytes (Manne et al., 1991).

In STM an extremely fine metallic probe tip of diameter in angstroms is brought to a distance of 5–50 Å from the sample surface. Within this distance, an overlap of the electronic wavefunctions of the tip and the sample allows tunneling. Consequently, a flow of electrons, dependent on the applied voltage (called bias voltage), produces a tunneling current that is given as (Bonnell and Huey, 2001)

$$I = CU\rho_t\rho_s e^{-2kz} \quad . \tag{7.3}$$

In this equation, C is a constant; U is the applied bias voltage; ρ_t and ρ_s are the electronic densities of states distribution, respectively, for the tip and the sample surface; z is the sample-tip distance in the Å unit. The term k relates to the barrier height and is given (see Chapter 2)

$$k = \frac{\sqrt{2m_e(V-E)}}{\hbar} \tag{7.4}$$

where m_e is the mass of an electron, $\hbar$ is Planck's constant, E is the energy of electrons, and V is the potential barrier. As the current is exponentially dependent on z, a decrease of mere 1 Å in the sample-tip separation produces an increase of one order of magnitude in the tunneling current, providing an extremely high vertical resolution.

STM images are generally produced in two different modes as shown in Figure 7.17. The probe tip, which is in effect a metal wire with an etched sharp tip, is raster-scanned across the sample surface using a piezoelectric tube scanner. The application of a scanning voltage across the tube produces a lateral displacement, while application of a voltage between the inside and outside of the tube produces a vertical displacement. In the constant current mode, the tip is moved vertically to maintain a constant current, as the tip is scanned laterally across the surface, keeping the basis voltage constant. A voltage signal from a comparison electron circuit provides a feedback electrical signal to control scan in the z direction. The image, generated by the feedback signal as a function of (x, y) lateral position, provides the topographic image of the surface of the sample.

In the constant height mode, the distance between the tip and the surface is kept constant, while variation in the tunneling current at a fixed bias voltage is recorded as a function of the tip's lateral position. In this case the current variation provides image and can be related to the variation of the charge density. In general, the STM image relates to a convolution of the electronic structures of the sample and the tip

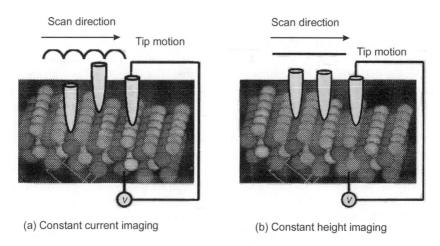

Figure 7.17. STM. A bias is applied between the sample and tip. As the tip scanned from left to right, either (a) the tip is moved vertically to keep current constant or (b) the vertical position is held constant and the current varies. From Bonnell and Huey (2001), reproduced with permission.

[see Eq. (7.3)]. However, often the assumption is made that the variation of the contrast in the image is attributed to the electronic properties of the sample.

Tunneling spectroscopy can also be performed in an STM arrangement to obtain information on sample density of states as a function of energy (applied bias voltage).

Though all the SPM techniques are generally used for the surface topography information, the STM technique can also provide information about three-dimensional topography: size, shape, and periodicity of features, surface roughness, electronic structure, and possible elemental identity. In STM technique, the dependence of the image on the electronic characteristics of the system can be used to extract information on the electronic structure of the materials being imaged (Feenstra, 1994).

Atomic Force Microscopy (AFM). Atomic force microscopy detects the overall forces between a probe tip and the sample surface (Meyer, 2003). In this case, the probe is attached to a cantilever-type spring. In contrast to STM, this imaging technique is not dependent on the electrical conductivity of the sample. The force exerted on the tip by the sample surface produces a bend of the cantilever. By using a cantilever with known spring constant C, the net force F can be directly obtained from the deflection (bend), Δz, of the cantilever according to equation $F = C\Delta z$. The two frequently used modes employed to produce AFM images are the contact mode and the tapping mode.

In essence, AFM operates by measuring attractive or repulsive forces between a tip and the sample. In the contact mode the tip operates in the repulsive regime (during scan, the tip comes in contact with the sample), while in the noncontact mode it operates in the attractive regime (the tip is very close to the sample, but not in contact with it). The contact mode can be used on samples in air and liquids, while the noncontact mode cannot be used in liquids.

The common contact mode AFM operates in constant force (like constant current in STM) or constant height mode. In the constant force mode, a feedback loop is used to keep the deflection of the cantilever constant. In the constant height mode, the deflection of the cantilever is measured. The scan involves a piezoelectric tube. Figure 7.18 shows a schematic diagram often used to measure cantilever deflection involving an optical method. In this case, a photodiode measures the position of a laser beam that has been reflected off the top of the cantilever. The feedback electronics controls the position of the tip. The sample is mounted on a piezoelectric tube scanner for *xyz* motion, as described earlier in the case of STM.

Another mode of operation of AFM is called the "tapping mode," in which the "cantilever" is made to vibrate at close to its natural frequency and the proximity of the surface is determined by the damping of this oscillation. In this case, the oscillation amplitude and the resonance frequency of the cantilever decrease with a decrease in the tip-sample distance (hence an increase in the force). The feedback loop here, as shown in Figure 7.18, is used to keep the amplitude or the frequency constant. This eliminates lateral shear forces on the tip and reduces the force normal to the tip and the surface, which could damage soft samples. Hence the tapping mode is commonly used in the case of soft samples, even though the resolution is slightly

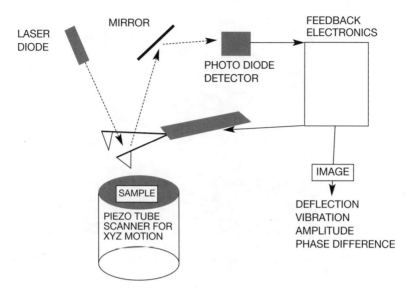

Figure 7.18. AFM surface scan by cantilever. From Bonnell and Huey (2001), reproduced with permission.

lower than in the contact mode. This method, sometimes referred to as ac-AFM or dynamic AFM, is sensitive to the force gradient rather than the force itself.

There are many other modes of AFM techniques, such as intermittant contact mode microscopy, lateral force microscopy (LFM), magnetic force microscopy (MFM), electrostatic force microscopy (EFM), and scanning thermal microscopy (SThM). For example, MFM is used to obtain the size and shape of magnetic features, as well as the strength and polarity of magnetic fields at different locations. EFM can be used to obtain electrostatic field gradients on the sample surface, due to dopant concentrations.

The tips used in AFM are commonly fabricated from silicon or silicon nitride and are left bare (for regular AFM) or coated with a special material (such as metal for EFM and rare-earth boride for MFM) to suit a particular application. Currently, commercially available AFM tips are microfabricated in three geometries, namely, (1) conical, (2) tetrahedral, and (3) pyramidal. Conical tips can be made sharp with a high aspect ratio (the ratio of tip length to tip diameter), making them especially useful for imaging features that are deep and narrow. Though the conical tips with diameters of 5 nm have been made, they are easily broken, and hence the commonly used tip diameters are in the range of 10–50 nm. But with the introduction of carbon nanotubes as AFM or STM tips, the tip diameter has been reduced to the 1- to 2-nm range (Wong et al., 1998a,b).

As an example of AFM imaging, Figure 7.19 provides a topographic image of close-packed polystyrene microspheres, obtained in our laboratory. It shows a high degree of order needed for a perfect photonic crystal, a topic discussed in Chapter 9.

Figure 7.19. AFM image of the surface of close-packed polystyrene sphere (diameter 200 nm) photonic crystal.

7.3. HIGHLIGHTS OF THE CHAPTER

- A variety of methods are available for the growth of nanomaterials. They include both vapor-phase and liquid-phase approaches.
- Cited examples of vapor-phase methods are molecular beam epitaxy (MBE), metal–organic chemical vapor deposition (MOCVD), and laser-assisted vapor deposition.
- Examples of liquid-phase methods are liquid-phase epitaxy (LPE) and nanochemistry, with the latter emerging as a powerful technique for producing nanoparticles of a wide variety and surface characteristics.
- Epitaxial growth refers to oriented growth of a nanostructure on a substrate whose crystalline lattice structure is matched (hence the term lattice-matched) with the crystalline lattice structure of the growing nanomaterials.
- Molecular beam epitaxy (MBE) involves the evaporation, in an ultrahigh vacuum chamber, of atoms (e.g., Ga or Al) or homoatomic molecules (e.g., As_2) that are constituents of the growing nanostructure. Their combination forms a molecular beam, with subsequent deposition on a lattice-matched substrate (e.g., GaAs).
- An MBE chamber also incorporates a number of *in situ* characterization techniques for monitoring growth.
- The MBE method has been widely utilized for fabrication of quantum wells, quantum wires, and quantum dots.
- Metal–organic chemical vapor deposition (MOCVD) is a chemical vapor deposition method in which the precursors for the elemental constituents of the semiconductor to be grown are metal–organic compounds.

- When the MOCVD method is used for epitaxial growth on a lattice-matched substrate, the technique is also called metal–organic vapor-phase epitaxy (MOVPE).
- MOCVD is a simpler growth technique that is ~ 10 times faster than MBE. However, the chemical precursors used in MOCVD are often toxic, which requires extreme safeguards and careful handling.
- Chemical beam epitaxy (CBE) combines MOCVD and MBE. Like MBE, it utilizes an ultrahigh vacuum chamber, but constituent elements are derived from metal–organic precursors as in MOCVD.
- Liquid-phase epitaxy (LPE) utilizes a lattice-matched substrate that is brought into contact with a saturated solution (or melt) of the film forming material at an appropriate temperature. It provides a high deposition rate and crystalline perfection in the deposited film.
- Laser-assisted vapor deposition (LAVD) utilizes a laser beam to generate materials in the vapor phase which then deposit on a substrate.
- One type of LAVD involves laser ablation of a solid material which can be directly deposited, or be mixed with a reactive gas to make the appropriate material, and then carried by an inert gas for controlled deposition.
- Another type LAVD involves laser-induced dissociation of precursor materials such as silane (SiH_4) to produce silicon nanocrystals.
- Nanochemistry involves solution phase chemistry confined to nanometer length. It has been used to produce metal, semiconductor, as well as organic:inorganic hybrid nanoparticles and core-shell structures of different sizes and shapes.
- Two examples of nanochemistry are: (i) reverse micelle synthesis, also known as microemulsion chemistry, in which chemical reaction is produced in a well controlled nanocavity formed by micellar-type molecules; (ii) colloidal synthesis, in which nanoparticles are grown by chemical reaction of their precursors in a carefully chosen solvent and then truncating the growth by surface capping.
- A wide variety of techniques are available for characterization of nanomaterials. Some principal methods are described in Section 7.2.
- X-ray characterization plays an important role in structural determination. Powder X-ray diffraction provides information on the crystalline form of the nanoparticles. The broadening of the diffraction peaks can be related to the size of the nanoparticles by Scherrer's formula, and thus can be used to estimate the size.
- X-ray photoelectron spectroscopy (XPS) utilizes photoelectric effect produced by an X-ray photon. The kinetic energy with which an electron leaves provides information on its binding energy, and hence the chemical environment of a particular constituent atom.
- Electron microscopy is a powerful method to image nanostructures and provide direct information on their size and shape.

- The two principal types of electron microscopy are (i) transmission electron microscopy (TEM), which involves imaging with electrons transmitted through a thin sample, and (ii) scanning electron microscopy (SEM), which utilizes electrons scattered when the incident electron beam is scanned across the surface of the sample.
- Reflection high-energy electron diffraction (RHEED) is another technique utilizing an electron beam. This technique, generally incorporated in an MBE growth chamber, is utilized to monitor layer-by-layer growth by using diffraction of a high-energy beam of electrons.
- Energy-dispersive spectroscopy (EDS) utilizes the analysis of characteristic X rays emitted by the constituent atoms when an electron beam, used in the SEM arrangement, is incident on the sample.
- Scanning probe microscopy (SPM) produces images with nanometer resolution by utilizing local interactions between a sample and a small tip which is scanned over the sample surface. The most common SPM are scanning tunneling microscopy (STM) and atomic force microscopy (AFM).
- In STM, image is generated by using tunneling current between a metallic tip and the sample surface under a bias voltage between them. The method can be used to obtain topographic information as well as information on electronic structure (density of state distribution) of the sample.
- In AFM, an image is generated by detecting the force between a probe tip and the sample. In this case, the probe is attached to a cantilever whose deflection is a measure of the force.

REFERENCES

Alivisatos, A. P., Peng, X., Manna, L., Process for Forming Shaped Group III–V Semiconductor Nanocrystals, and Product Formed Using Process, U.S. Patent 6,306,736, Oct. 23, 2001.

Bai, C., *Scanning Tunneling Microscopy and Its Applications,* Springer-Verlag Series in Surface Sciences, Vol. 32, Springer-Verlag, Berlin, 2000.

Barnham, K., and Vvedensky, D., eds., *Low-Dimentional Semiconductor Structures,* Cambridge University Press, Cambridge, U.K., 2001.

Battaglia, D., and Peng, X., Formation of High Quality InP and InAs Nanocrystals in a Noncoordinating Solvent, *Nano. Lett.* **2,** 1027–1030 (2002).

Binnig, G., Rohrer, H., Gerber, C., and Weibel, E., Surface Studies by Scanning Tunnelling Microscopy, *Phys. Rev. Lett.* **49,** 57–61 (1982).

Bonnell, D., and Huey, B. D., Basic Principles of Scanning Probe Microscopy, in *Scanning Probe Microscopy and Spectroscopy—Theory, Techniques, and Applications,* 2nd edition, D. Bonnell, ed., John Wiley & Sons, New York, 2001, pp. 7–42.

Bonnell, D., ed., *Scanning Probe Microscopy and Spectroscopy—Theory, Techniques, and Applications,* 2nd edition, John Wiley & Sons, New York, 2001.

Chen, X., Na, M., Cheon, M., Wang, S., Luo, H., McCombe, B., Sasaki, Y., Wojtowicz, T.,

Furdyna, J. K., Potashnik, S. J., and Schiffer, P., Above-Room-Temperature Ferromagnetism in GaSb/Mn Digital Alloys, *Appl. Phys. Lett.* **81,** 511–513 (2002).

Chi, L., and Röthig, C., Scanning Probe Microscopy of Nanoclusters, *Characterization of Nanophase Materials,* in Z. L. Wang, ed., John Wiley & Sons, New York, 2000, pp. 133–163.

Crist, B. V., *Handbook of Monochromatic XPS Spectra, 3-Volume Set,* John Wiley & Sons, New York, 2000.

Cullity, B. D., *Elements of X-Ray Diffraction,* Addison-Wesley, Menlo Park, CA, 1978.

Feenstra, R. M., Scanning Tunneling Spectroscopy, *Surf. Sci.,* **299–300,** 965–979 (1994).

Fendler, J. H., and Meldrum F. C., Colloid Chemical Approach to Nanostructured Materials, *Adv. Mater.* **7,** 607–632 (1995).

Guha, S., Madhukar, A., and Rajkumar, K. C., Onset of Incoherency and Defect Introduction in the Initial Stages of Molecular Beam Epitaxical Growth of Highly Strained $In_xGa_{1-x}As$ on GaAs(100), *Appl. Phys. Lett.* **57,** 2110–2112 (1990).

Heimendahl, M. V., *Electron Microscopy of Materials: An Introduction,* Academic Press, New York, 1980.

Herrman, M. A., and Sitter, H., *Molecular Beam Epitaxy,* Springer-Verlag, Berlin, 1996.

Jones, A. C., and O'Brien, P., *CVD of Compound Semiconductors,* VCH, Weinheim, Germany, 1997.

Kelly, M. J., *Low-Dimensional Semiconductors,* Clarendon Press, Oxford, U.K., 1995.

Lal, M., Levy, L., Kim, K. S., He, G. S., Wang, X., Min, Y. H., Pakatchi, S., and Prasad, P. N., Silica Nanobubbles Containing an Organic Dye in a Multilayered Organic/Inorganic Heterostructure with Enhanced Luminescence, *Chem. Mater.* **12,** 2632–2639 (2000).

Langford, J. I., and Louër, D. Powder Diffraction, *Rep. Prog. Phys.* **59,** 131–234 (1996).

Li, X., He, Y., Talukdar, S. S., and Swihart, M. T., Process for Preparing Macroscopic Quantities of Brightly Photoluminescent Silicon Nanoparticles with Emission Spanning the Visible Spectrum, *Langmuir* **19,** 8490–8496 (2003).

Lucey, D. W., MacRae, D. J., Furis, M., Sahoo, Y., Manciu, F., and Prasad, P. N., Synthesis of InP and GaP in a Noncoordinating Solvent Utilizing *in-situ* Surfactant Generation, in preparation (2004).

Manne, S., Massie, T., Elings, B., Hansma, P. K., and Gewirth, A. A., Electrochemistry on a Gold Surface Observed with the Atomic Force Microscope, *J. Vac. Sci. Technol.* **B9,** 950–954 (1991).

Meyer, E., *Atomic Force Microscopy: Fundamentals to Most Advanced Applications,* Springer-Verlag, New York, 2003.

Murray, C. B., Kagan, C. R., and Bawendi, M. G., Synthesis and Characterization of Monodisperse Nanocrystals and Close-Packed Nanocrystal Assembles, *Annu. Rev. Mater. Sci.,* **30,** 545–610 (2000).

Nikoobakht, B., and El-Sayed, M. A., Preparation and Growth Mechanism of Gold Nanorods (NRs) Using Seed-Mediated Growth Method, *Chem. Mater.* **15,** 1957–1962 (2003).

Park, S.-I., and Quate, C. F., Tunneling Microscopy of Graphite in Air, *Appl. Phys. Lett.* **48,** 112–114 (1986).

Patra, A., Friend, C. S., Kapoor, R., and Prasad, P. N., Fluorescence Upconversion Properties of Er^{3+}-Doped TiO_2 and $BaTiO_3$ Nanocrystallites, *Chem. Mater.* **15,** 3650–3655 (2003).

Petit, C., Lixon, P, and Pileni, M. P., *In-Situ* Synthesis of Silver Nanocluster in AOT Reverse Micelles, *J. Phys. Chem.* **97,** 12974–12983 (1993).

Sonnenfeld, R., and Hansma, P., Atomic-Resolution Microscopy in Water *Science,* **32,** 211–213 (1986).

Vvedensky, D., Epitaxial Growth of Semiconductors, in *Low-Dimentional Semiconductor Structures,* K, Barnham and C. Vvedensky, eds., Cambridge University Press, Cambridge, U.K., 2001.

Wang, Z. L., ed. *Characterization of Nanophase Materials,* Wiley-VCH, Weinheim, Germany, 2000.

Wijekoon, W. M. K. P., Liktey, M. Y. M., Prasad, P. N., and Garvey, J. F., Fabrication of Thin Film of an Inorganic–Organic Composite via Laser Assisted Molecular Beam Deposition, *Appl. Phys. Lett.* **67,** 1698–1699 (1995).

Wong, S. S., Joselevich, E., Woolley, A. T., Cheung, C. L., and Lieber, C. M., Covalently Functionalized Nanotubes as Nanometer-Sized Probes in Chemistry and Biology, *Nature* **394,** 52–55 (1998a).

Wong, S. S., Harper, J. D., Lansbury, P. T., and Lieber, C. M., Carbon Nanotube Tips: High-Resolution Probes for Imaging Biological Systems, *J. Am. Chem. Soc.* **120,** 603–604 (1998b).

CHAPTER 8

Nanostructured Molecular Architectures

This chapter describes the very rich field of molecular engineering to produce novel and multifunctional optical materials by nanostructure control of their architecture. This nanostructure control is derived from manipulation of both covalent (chemical bonds) and noncovalent interactions that can yield specific optical and electronic interaction topologies and produce molecular nano-objects of particular shapes. For the sake of readers not well versed in chemical methods, details of chemical synthetic approaches used are avoided, but appropriate references are given for those interested in these details. The various kinds of covalent bonding (σ and π) were already discussed in Chapter 4, Section 4.6. In this chapter, Section 8.1 describes some important noncovalent interactions that determine the three-dimensional structure of a molecular architecture.

Section 8.2 describes polymeric media for photonics. The structural flexibility of a polymeric structure at nanoscale provides the ease to readily use molecular engineering to tailor its properties. Some approaches used to derive this nanostructure control are discussed here together with illustrative examples of photonic applications.

Section 8.3 discusses molecular machines that represent two molecular units mechanically interlocked—that is, held without any chemical bond. These two molecular units can undergo relative displacement (rotation or translation) with respect to each other under the influence of an external stimulus, such as the action of light in the context of this book. Thus, the result is nanoscale displacement (mechanical motion) by the action of light and the systems exhibiting this behavior can also be called opto-mechanical nanomachines or molecular motors. Two examples of molecular structures working as molecular machines are described here: rotaxanes and catenanes. The basic molecular design is provided together with illustrative examples of photonic functions.

Section 8.4 presents dendrimers that are hyperbranched structures, which are often also termed as bio-inspired or nature-inspired. The bio-inspired materials, along with dendrimers in relation to them, are discussed in detail in Chapter 12. Here the basic concept of a dendritic structure, approaches used for dendrimer architecture, and their merits are discussed.

Section 8.5 covers the field of supramolecular structures. Here the molecular units are held together by noncovalent interactions. The various noncovalent interactions used to produce different types of supramolecular architecture are described. The flexibility of multifunctionality offered by a supramolecular structure is outlined. Specific examples include a supramolecular dendrimer that involves a combination of supramolecular chemistry and dendrimer chemistry. Another example provided is of a helical foldmer with a tunable interior cavity.

Section 8.6 describes monolayer and multilayer molecular assemblies. The two specific methods described are the Langmuir–Blodgett technique and the self-assembled monolayer deposition. Examples of multilayer films exhibiting enhanced nonlinear optical response are presented.

As the various topics discussed in this chapter present highly diverse groups of molecular structures, no single book or review can be expected to cover them all. Therefore, a general reading reference for each group is provided in the respective sections covering them. The highlights of this chapter are given in Section 8.7.

8.1 NONCOVALENT INTERACTIONS

A molecular architecture involves a three-dimensional structure that is most frequently determined by noncovalent interactions that do not involve any chemical bonding. These interactions make a long, flexible molecular chain (such as protein) fold, and two complementary chains (strands) of DNA pair up to form the double-helix structure. Some important noncovalent interactions that determine the shape and function of the various molecular architectures, presented in this chapter, are discussed here (Whitesides et al., 1991; Brunsveld et al., 2001; Gittins and Twyman, 2003).

Hydrogen Bond. This bond involves a weak electrostatic interaction between the hydrogen atom bonded to an electronegative atom (a highly polar bond) in one molecule and another electron-rich atom (containing a nonbonded electron pair) on a different molecule or another part of the same molecule (as in the case of intramolecular hydrogen bonds). An example is the hydrogen bond in H_2O, shown in Figure 8.1. Typical groups capable of forming the hydrogen bonds are —NH_2, —C=O, and —OH.

It is this hydrogen bonding between the base pairs of two chains of nucleic acids

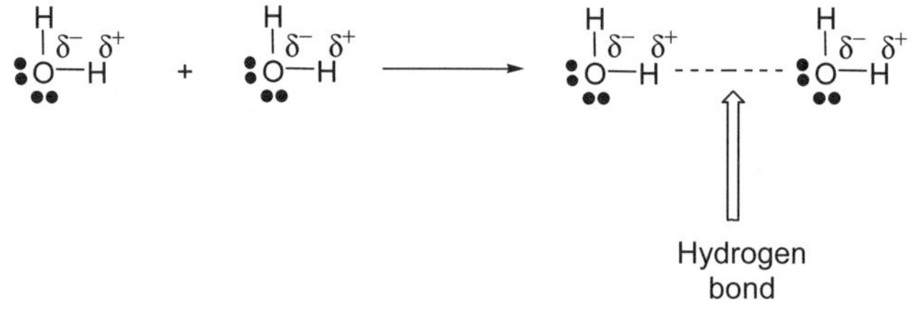

Figure 8.1. The hydrogen bond is represented by a dashed line.

that gives rise to the double-stranded Watson–Crick model of DNA (Watson and Crick, 1953). The base pairs, which can form hydrogen bonding, are quite specific. Often, multiple hydrogen bonds are the key factors in determining the architecture of large supramolecular or polymeric structures.

Metal Coordination. In this case the interaction utilized is between a metal ion and an organic ligand (an organic molecular unit containing an atom such as nitrogen with a lone pair of electrons that can be coordinated to the metal ion). Molecular segments can be then distributed around a central metal ion in various geometries. Often these ligands are characterized as monodentate, bidentate, and so on, depending on the number of binding sites in the ligand. Metal–ligand coordination is a stronger interaction than the hydrogen bond.

π–π (or Arene–Arene) Interactions. Arenes (e.g., a benzene ring) are planar conjugated structures containing delocalized π electrons around the ring (see Section 4.6). The π molecular orbitals are perpendicular to the plane of the ring. π–π interactions are generated when the π molecular orbitals of the arene rings overlap. This interaction stabilizes stacking of the aromatic rings to favor a discotic or columnar arrangement.

Electrostatic and Ionic Interactions. This interaction is manifested between molecular units or segments of opposite charges. It can also occur between an ionic group and another molecular unit with a permanent dipole moment. In such a case the interaction is in the form of ion-dipole.

Van der Waals Interactions. These are short-range, nonspecific interactions between two chemical species. They result from the momentary random fluctuations in the distribution of the electrons of any atom, giving rise to a transient, unequal electric dipole. The attraction of the unequal dipoles of two noncovalently bonded atoms gives rise to the van der Waals interaction. These forces are responsible for the cohesiveness of nonpolar liquids and solids that cannot form hydrogen bonds.

Hydrophobic Interactions. These interactions involve the nonpolar segments (segments containing only C C and C—H bonds) of a molecule. An example is the interaction between long alkyl chains which form the nonpolar tail parts of molecules that can assemble to form an outer-membrane-type bilayer structure. Another example is the clustering of nonpolar segments of a polymer, as discussed in Section 8.2.

The noncovalent interactions individually may be weaker than a covalent chemical bond. Collectively, they play a critical role in maintaining the three-dimensional structures of molecular systems, as well as of biological structures of DNA, proteins, and biomembranes. Noncovalent interactions of more than one type often determine how large molecular structures fold or unfold. The reason is that these noncovalent interactions are highly directional. They are also specific, thus providing recognition features to self-direct a molecular unit or segment for aligning at an appropriate binding (interacting) site. This feature is frequently referred to as self-assemblying. Thus, molecular structures that exhibit this feature through noncovalent interactions are called self-assemblying structures and can be used to automatically create molecular assemblies of a specific form or produce a pattern of them on a substrate.

8.2 NANOSTRUCTURED POLYMERIC MEDIA

Polymeric media are showing significant promise for various photonic applications. Figure 8.2 provides some areas of polymer photonics being pursued at our Institute and illustrates the various applications of polymer media for photonics. The two excellent articles presenting nanotechnologies based on polymeric media, particularly the class of block copolymers, are by Thomas and co-workers (Park et al., 2003) and by Bates and Fredrickson (1999).

Polymeric materials are molecular hierarchical systems in which the structure and functionalities can be controlled from the atomic level to the bulk level, hence providing a nanostructure control. This tremendous structural flexibility provides one with a true opportunity to introduce multifunctionality in a polymeric material. In essence, one can attempt to mimic nature. Like the structures of proteins, DNA, and other biological systems, one can control the primary, secondary, and tertiary structures by a sequential chemical synthesis approach. The schematics of a multifunctional polymer are shown in Figure 8.3. One can vary the main chain structure

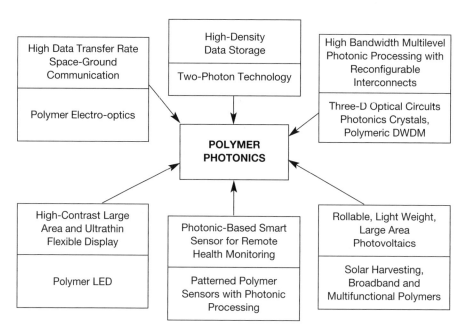

Figure 8.2. Chart illustrating various applications of polymer photonics.

by making alternating or block copolymers (e.g., block segments —B—B—B— and —C—C—C—); one can graft side chains of various functionalities. Also, by introducing flexible chain segments in the main chain or in the side chain, the polymer can be made soluble. By using a combination of these modifications, one can introduce different functionalities such as optical response, electronic response, mechanical strength, and improved processibility. Further improvements can be brought by making nanocomposites or blends of polymers. This feature is discussed in Chapter 10. Use can also be made of induced orientational alignment to produce or enhance a desired functional response.

A diblock copolymer with one segment of the polymer having one type of structure (e.g., hydrophobic character) and the other part having another structure (e.g., hydrophilic) can produce various morphologies by phase separation of the two blocks, depending on the length scale of these blocks (Bates and Fredrickson, 1999; Park et al., 2003). Thus, by adjusting the relative scale of the two components, one can make them pack in the bulk phase in different morphologies, as shown in Figure 8.4. In some cases, one can have them coiled up as a sphere. In another case, one can pack them like cylinders. The third case is a gyroid and the fourth one is a lamellar structure. Increasing volume fraction of the minority phase defined at the bottom of each curve of Figure 8.4 controls the morphology of the polymer (Bates and Fredrickson, 1999). This feature can be utilized to produce a self-assembled photonic crystal. This topic is discussed in Chapter 9.

A specific application of nanostructured polymers presented here is in the area of polymer light emitting diodes, which is highly active worldwide (Akcelrud, 2003). Polymer light-emitting diode (often abbreviated as PLED)-based luminescent displays are already in the marketplace. A block copolymer approach can be used to achieve color tunability (Kyllo et al., 2001; Yang et al., 1998). A block copolymer consists of blocks of repeat units of different kinds (such as —B—B—B— and —

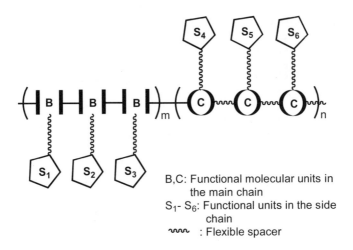

B,C: Functional molecular units in the main chain
S_1- S_6: Functional units in the side chain
∿∿ : Flexible spacer

Figure 8.3. Schematic representation of a multifunctional block copolymer.

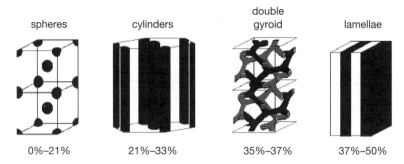

spheres
cylinders
double gyroid
lamellae
0%–21%
21%–33%
35%–37%
37%–50%

Figure 8.4. Various morphologies adopted by noncrystalline diblock copolymers. The percentage listed below each morphology represents the volume fraction of the minority phase (smaller segments) block, which is surrounded by the majority phase. Courtesy of E. L. Thomas, M.I.T.

C—C—C— in a di-block copolymer). By adjusting the block length (oligomer length) of the emitting block conjugated segment (say —B—B—B— in Figure 8.3), the color of emission can be varied. The behavior is similar to that exhibited by finite-size oligomers as discussed in Section 4.6 of Chapter 4. A strategy, often used to make the polymer soluble for solution casting, is to make the nonconjugated block (—C—C—C—) a flexible aliphatic chain. Alternatively, flexible side chains are introduced as depicted in Figure 8.3 to enhance solubility. An example of the type of block copolymer containing conjugate-nonconjugated blocks is provided in Figure 8.5. It consists of the conjugated segments of PPV (poly-*p*-phenylene vinylene) block in the form of a methoxy derivative, separated by nonconjugated aliphatic segments —OC_6H_{12}—O— (Kyllo et al., 2001). This type of block copolymer has shown that the color (wavelength) of emission is controlled by the conjugation length (length of the conjugated block) and not affected by the length of the nonconjugated block, sometimes called inert spacer. However, Karasz and co-workers found that the

H3CO
OCH3
OCH3
H3CO
H3CO
OC6H12O
OCH3
n
Conjugated block

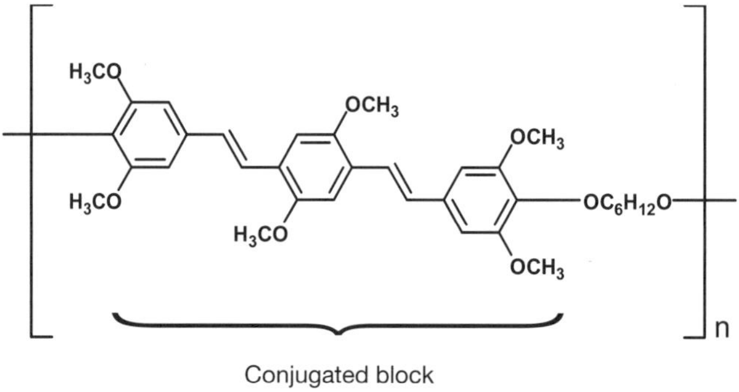

Figure 8.5. Electroluminescent polymer, with conjugation confinement, containing a dimethoxy PPV conjugation block and a nonconjugated aliphatic chain (Chung et al., 1997).

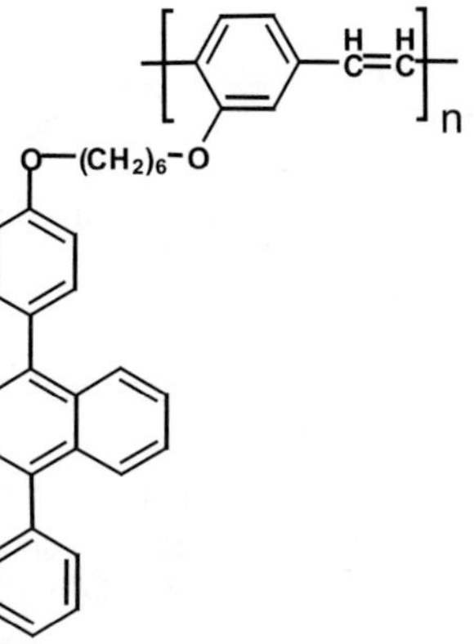

Figure 8.6. A side-chain polymer containing the PPV unit in the main chain and conjugated aromatic structure, 9,10-diphenylanthracene, in the side group.

electroluminescence efficiency is dependent on the length of the nonconjugated block, a longer inert spacer yielding a higher efficiency device (Yang et al., 1998).

Another design of a block copolymer, shown in Figure 8.6, contains conjugated segments both in the main chain and in the side chain, separated from the main chain by a flexible spacer —O—(CH$_2$)$_6$—O— (Chung et al., 1997). Here the light emission can originate from both conjugated units, but at different wavelengths. Such designs provide a strategy to produce white color LEDs (broadband emitters).

Park et al. (2003) describe numerous examples of the use of block copolymers as templates for nanoparticle synthesis, crystals, and high-density information storage media. They discuss that the application of block copolymers to nanotechnology stems from the scale of the microdomains and the convenient tunability of the size, shape, and periodicity afforded by changing their molecular parameters.

8.3 MOLECULAR MACHINES

Molecular machines or molecular motors, in the context of this book, refer to nanostructures that involve two or more separate molecular components that are not connected by any chemical bond. These components are intrinsically mechanically entangled so that they can only separate by cleavage of some chemical bond. The site of entanglement or interlock is determined by noncovalent interactions such as hydrogen bonds. Under an external stimulus such as light, electric field, or chemical change, the site of interlock opens and shifts to another location to lead to a mechanical displacement. Thus, these molecular machines can also be thought of as molecular level shuttles.

One method to produce such interlocked structures employs the use of specific noncovalent interactions such as hydrogen bonds or metal–ligand coordination to hold the molecular precursors of the interlocked units in correct orientation. Then a

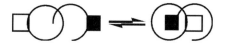

Figure 8.7. Rotational motion in a catenane.

chemical reaction forms the necessary chemical bond to produce the final interlocked structure.

The subject of these types of molecular machines is covered in many excellent reviews. A good earlier source of reference is a special issue of *Accounts of Chemical Research* (an American Chemical Society publication) on "Molecular Machines," Volume 34, pp. 409–522 (2001). Some other references are Balzani et al. (2000) and Leigh (2003a).

Two examples of this class of nanostructure presented here are (i) catenanes and (ii) rotaxanes. Catenanes consist of two or more interlocked cyclic structures as shown by a cartoon diagram in Figure 8.7. This figure represents a [2]catenane, where the number in the square brackets indicates the number of interlocked rings (in the case of Figure 8.7, it is 2). In the case where the two rings are identical, the molecule is sometimes referred to as a *homocircuit catenane*. If they are different, the molecule is called a *heterocircuit catenane*.

Figure 8.8 shows the cartoon representation of a rotaxane structure. Rotaxanes are formed by macrocyclic rings (a large molecular ring) trapped onto a linear molecular unit, often called a thread. Two bulky stopper groups on the two ends of the thread prevent any escape of the ring from the thread. As in the case of catenanes, a prefix number in square bracket represents the number of interlocked components. Thus the representation presented in Figure 8.8 is that of a [2]rotaxane.

The research groups of Sauvage (1998) and Stoddart (Balzani et al., 1998) developed synthetic routes to both catenane and rotaxane types of molecular architectures using assembly of π-electron-deficient cyclophanes and π-electron-rich aromatic polyether units and transition metal-coordinated ligands. Stoddart and co-workers reported a shuttle that can be switched via electric, photochemical, or chemical stimuli that created a proton transfer (Bissell et al., 1994). Sauvage and co-workers have shown that positional isomerism can be induced in the metal-based system by varying the oxidation state of the transition metal using electronic or photochemical redox (electron transfer) reaction (Jimenez et al., 2000).

Here a specific example is provided from the work of Brouwer et al. (2001) which shows the operation of a light-fueled translational molecular level motor.

Figure 8.8. Linear motion in a rotaxane (a molecular shuttle).

The structure of this hydrogen bond assembled rotaxane shuttle is shown in Figure 8.9. In this structure, translation of a macrocycle (large ring) between two stations takes place after photoexcitation with a laser pulse. This submolecular motion occurs on a microsecond time scale and is fully reversible. After the charge recombination, approximately ~100 μs later, the macrocycle shuttles back to its original site.

Another example is of writing a fluorescent pattern in a polymer film using submolecular motion (Leigh, 2003b). For this purpose, a rotaxane containing a fluorescent anthracene group on the thread and a quenching group on the macrocycle was synthesized. When the macrocycle is in a position close to the anthracene unit, no fluorescence from anthracene is observed due to the proximity of the quenching group. When the macrocycle moves to another position further away from the anthracene unit, under the influence of an external stimulus (such as light absorption or change of polarity), fluorescence is observed. This two-state molecular switch was incorporated into a polymer film to store visible images arising solely from controlled submolecular motion.

8.4 DENDRIMERS

The term *dendrimer* is derived from a combination of two Greek words: "dendron," meaning tree or branch, and "meros," meaning part. It refers to a vast class of large molecular architectures that contain structurally perfect, highly branched nanostructures. There has been an explosion of interest in dendrimers over the past decade, because dendritic nanostructures hold promise for many applications, including photonics. Naturally, this field is well covered by a number of books and excellent reviews. Some cited here are:

> Newkome, Moorefield, and Vögtle (1996): *Dendritic Molecules: Concepts, Syntheses, Perspectives*
> Grayson and Fréchet (2001): *Convergent Dendrons and Dendrimers: From Synthesis to Applications*
> Dykes (2001): *Dendrimers: A Review of Their Appeal and Applications*
> Vögtle et al. (2000): *Functional Dendrimers*

Hyperbranched polymers are produced by a noniterative (not in a series of sequence steps) polymerization process and hence exhibit an irregular molecular architecture with incompletely reacted branch points. Also, they are polydispersed systems with a broad distribution of molecular weight. In contrast, a dendrimer is a highly ordered, regularly branched globular macromolecule, produced by a stepwise iterative approach (Grayson and Fréchet, 2001). Thus dendrimers differ from hyperbranched polymers by their structural perfection, containing an exact number of concentric layers of branching points, referred to as *generations*. Figure 8.10 illustrates this feature of a dendrimer defining the different generations (Andronov

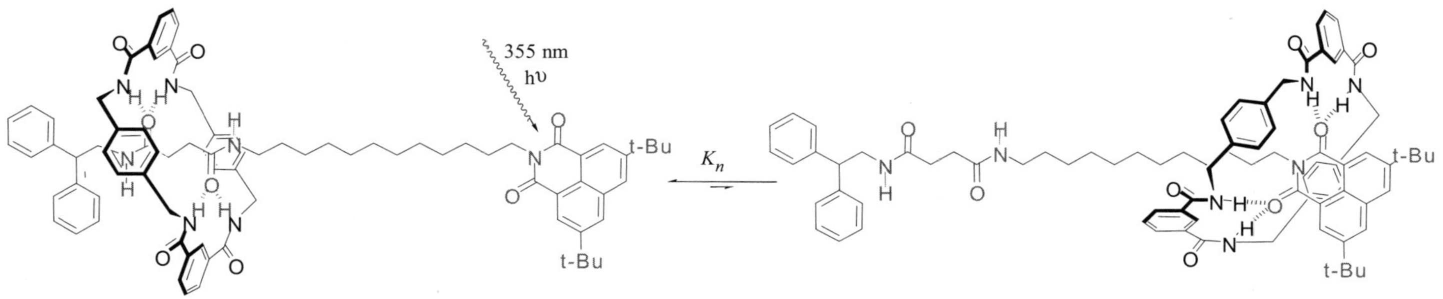

Figure 8.9. A light-fueled translational molecular level motor. From Brouwer et al. (2001), reproduced with permission.

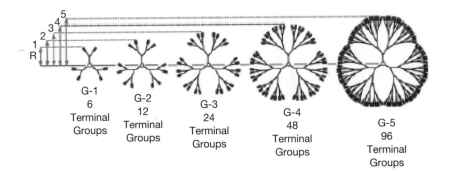

Figure. 8.10. Different generation dendrimers. As dendrimer generation increases, the number of terminal groups doubles, but the distance between the terminal groups and the core also increases. From Adronov and Fréchet (2000), reproduced with permission.

and Fréchet, 2000). The basic architecture of a dendrimer is divided into three principal regions (Grayson and Fréchet, 2001; Dykes, 2001):

- A core or focal moiety that forms the starting point (kind of nucleation site) for branching
- Layers of branched repeat units that emanate from this core and define the generation of a dendrimer
- The end groups on the outer layer of repeat units, defining the periphery (a multivalent surface) of a dendrimer

A dendrimer thus is essentially a monodispersed compound, well-defined by its generation, which is produced by a highly controlled growth process. Its macroscopic properties are largely determined by the nature of the end groups and their interaction with the external environment. This is not the case with a linear polymer (or a block copolymer) discussed in Section 8.2. In comparison to their linear polymeric analogues, dendrimers exhibit enhanced solubility.

Dendrimers can be produced by two general approaches that are described below.

Divergent Approach. This approach, developed by Tomalia et al. (1990) and Newkome et al. (1996), starts with initiation of growth from the core, outwards by repetition of a series of coupling and activation steps to produce a layer-by-layer buildup. The core is a monomer unit with a multiple coupling (branching) site which reacts with the complementary reactive group (unprotected) of other monomers. This is the coupling step. Other reactive groups on the reactive monomers are protected, until needed, to produce a controlled growth. This schematic is shown in Figure 8.11. The protected groups can be chemically activated by conversion to a reactive chemical group, for coupling to a second set of mole-

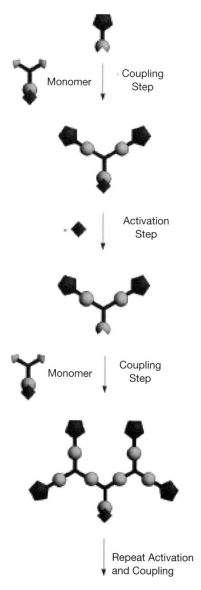

Figure. 8.11. Cartoon representation of dendron synthesis. From Grayson and Fréchet (2001), reproduced with permission.

cules to produce the next generation dendrimer. This step is the activation step. Thus a dendrimer of a well-defined generation can be produced. With an appropriate choice of coupling and activation steps, the divergent approach provides the advantage of large-scale preparation of dendrimers, because the quantity of dendrimer sample essentially doubles with each added generation. A major issue is the likelihood of incomplete functionalization or side reaction which grows exponentially with each generation. Thus, structural flaws are very likely to occur.

Convergent Approach. This approach, introduced by Fréchet and co-workers (Hawker and Fréchet, 1990; Grayson and Fréchet, 2001), starts with growth from what becomes the exterior of the dendrimer and progresses inward by coupling end groups to each branch of the monomer. The process is used to first produce a wedge-shaped dendritic fragment called *dendrons,* which are then activated and coupled to produce the globular dendritic molecule. The dendron unit and the resulting dendrimer are represented by their cartoon diagrams in Figure 8.12 (Adronov and Fréchet, 2000). In this diagram, the core is highlighted by a black circle, the rings of branching units are represented by gray circles, and the end groups are represented by rectangles. The advantage offered by the convergent approach is greater structural control, because each growth step requires a relatively low number of coupling reactions. Another advantage is the ability to precisely place functional groups throughout the structure. A disadvantage is that it is less readily scalable compared to the divergent approach.

The nanoscale architecture of a dendrimer offers a number of potential advantages. Some of these are:

1. *Multifunctionality.* Various photonic functions can be incorporated in the core, the branching units, and the peripheral groups. In addition, guest molecules

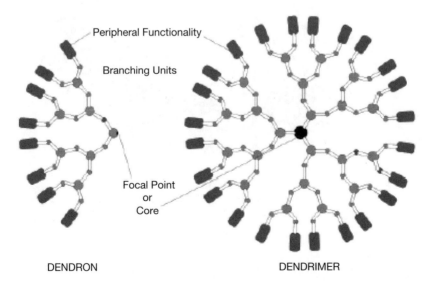

Figure. 8.12. Schematic diagram of the structure of a dendron and a dendrimer, highlighting the focal point or core (black) surrounded by rings of branching units (gray circles) and end groups (rectangles). From Adronov and Fréchet (2000), reproduced with permission.

can be physically trapped as inclusions inside the dendritic structure by chemically sealing the periphery. Consequently, the guest is trapped in a dendritic box. These dendritic boxes can be chemically manipulated to entrap even large-size molecules. An example is provided by Figure 8.13, which shows a poly(propylene imine) dendrimer entrapping the Rose Bengal dye, a large molecule, and p-nitrobenzoic acid, a small guest, within its structure (Jansen et al., 1994, 1995). The sterically bulky hy-

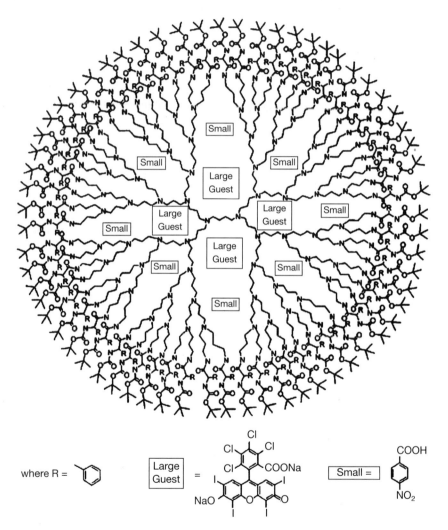

Figure 8.13. Poly(propylene imine) dendrimer, comprising a ^{t}BOC-L-phenylaniline surface, traps Rose Bengal (large guest) and p-nitrobenzoic acid (small guest) within its structure. The sterically bulky hydrogen-bonded shell makes this entrapment permanent. From Jansen et al. (1994), reproduced with permission.

drogen-bonded shell produces a permanent entrapment. An illustration of multifunctional dendritic structure is provided by dendrimer-based, light-emitting diodes. Dendrimer-based organic LEDs were reported by Moore and co-workers (Wang et al., 1996). They synthesized phenylacetylene dendrimers with peripheral triphenylamine groups for hole transport and the 9,10-bis(phenylethynyl)anthracene core as the light emitter. However, only modest electroluminescence efficiency was achieved.

2. *Site Isolation of the Core.* Dendritic structures provide isolation of the core from the surrounding environment, thereby reducing the chance of environmentally induced nonradiative quenching of emission of the emitter at the core. An example is provided by the work of Kawa and Fréchet (1998), who used dendrons with the carboxylate anion focal points as ligands to couple them to a lanthanide cation forming the core. This structure is shown in Figure 8.14. The luminescence properties of the lanthanide ion were dependent on the size of the dendritic shell. The dendritic shells isolated the lanthanide ions from each other, thus preventing concentration quenching derived from ion-pair interactions as discussed in Chapter 6. A similar strategy can be used to isolate an organic dye, particularly from the water

Figure 8.14. Lanthanide ions with poly(aryl ether) dendrimer ligands show improved luminescent properties. From Kawa and Fréchet (1998), reproduced with permission.

molecule in an aqueous medium which provides high-frequency phonons to enhance nonradiative processes. Thus a dendritic structure offers another approach to site isolation, a topic discussed in Section 6.7.

3. *Energy Funnels for Light Harvesting.* By appropriate choice of the peripheral groups, the branching units, and the core, photons can be absorbed by the peripheral groups and funneled (transferred) along the branches to the core to produce enhanced emission, amplification, or lasing from the core center. Thus the dendrimer branches can act as receiving antennas, similar to chlorophylls of the photosynthetic unit in nature which absorb solar energy and funnel it to the reaction center (Devadoss et al., 1996). The light-harvesting dendrimers are further discussed in Chapter 12 under bio-inspired materials.

4. *Modulation of Interactions.* By appropriate choice of nanostructure, a dendritic architecture can provide an opportunity to enhance or considerably reduce certain types of intermolecular interactions. An example provided here is from the application of second-order optically nonlinear organic molecules in electro-optic modulators. The electro-optic modulators utilize the linear electro-optic effect (also called the Pockels effect; see Prasad and Williams, 1991) which is a second-order nonlinear optical property, producing a linear change in refractive index with the applied electric field. Organic molecular dipolar structures, containing an electron donor group and an electron acceptor group separated by a conjugated segment, have been produced which exhibit large second-order nonlinear optical effects at the microscopic level (Dalton, 2002). In the bulk level, these dipoles have to be aligned to produce a noncentrosymmetric structure required for net second-order nonlinearity and thus for electro-optic effect. A method commonly used is to incorporate the electro-optically active molecule in a polymer matrix, either as an inclusion (guest–host system) or a chemically bonded unit (e.g., in the side chain), and align the dipoles by application of a high electric field at an elevated temperature. These polymer electro-optic modulators are drawing considerable attention, because work by Dalton and Steier (Dalton et al., 2003) has demonstrated very low V_π voltage (biasing voltage required to produce a change of refractive index that provides a 180° phase shift along the device length) and very high speed modulation in excess of 150 GHz. A major problem in the guest–host polymeric system is that at high concentration the dipolar molecules tend to pair up by dipole–dipole interactions (Dalton et al., 2003). This pairing produces centrosymmetry and reduces the effective second-order nonlinearity. Jen, Dalton, and co-workers have shown that by incorporating the electro-optically active dipolar molecule in the interior of a dendritic globular structure, this dipole-dipole interaction can significantly be reduced to achieve higher loading without pairing (Dalton et al., 2003). Jen and co-workers (Luo et al., 2003) have also demonstrated that by combining the site isolation effect and the self-assembly effect on the dendronized polymers, it greatly reduces the intermolecular electrostatic interaction between the dipolar chromophores and increases the nanoscale order of the resulting polymers. As a result, it significantly improves the poling efficiency of the nonlinear optical polymer (a factor of 2.5–3) to obtain extremely high electro-optic coefficients (>100 pm/V).

Another example is provided by choice of the peripheral groups to enhance hydrophilic or hydrophobic interactions, thereby facilitating their dispersion in a particular matrix. Hawker et al. (1993) used this strategy to solubilize a hydrophobic molecule in water by placing hydrophilic carboxylate groups on the pheriphery.

5. *Dendritic Fluorescence Sensors.* Placing many fluorescence sensing dyes on the periphery in a dendritic structure can serve to amplify the sensitivity of a sensor. Balzani et al. (2000) demonstrated it by preparing a poly(propylene amine) dendrimer containing 32 dansyl units at its periphery. The dansyl molecules serve as fluorescence sensors for metal ions such as Co^{2+} by quenching of fluorescence. They showed that the sensor was able to respond to very low concentrations of Co^{2+}.

8.5 SUPRAMOLECULAR STRUCTURES

As described in the introduction section of this chapter, a supramolecular structure represents association of two or more chemical species through noncovalent intermolecular interactions of the types described in Section 8.1. An example is a supramolecular polymer which, contrary to a regular polymer involving a covalent bond between the monomeric units, is defined as an array of monomers held together by reversible and highly directional noncovalent interactions. These highly directional interactions provide the ability to tailor their three-dimensional structure and produce nano-objects of various shapes.

Since the pioneering work of Lehn in supramolecular chemistry, this field has evolved into a multidisciplinary branch of supramolecular science. Consequently, a vast amount of literature, which is continuously expanding, exists. Some general references, suggested for further reading, are:

Lehn (1995): *Supramolecular Chemistry*
Brunsveld et al. (2001): *Supramolecular Polymers*
Gale (2002): *Supramolecular Chemistry*

The supramolecular assemblies, produced by noncovalent association, may exhibit excited states and properties different from their constituents, allowing new photoinduced energy and electron transfer processes. Some of the important features achieved by a supramolecular architecture are:

- *Nano-objects* of different shapes, such as spheres, cylinders, and discs.
- *Mesoporous foldmers* containing a folded molecular assembly with a central hollow tube.
- *Discotic structures* exhibiting strong chirality that yield a difference in optical interactions between left and right circularly polarized light.

The supramolecular chemistry involves spontaneous organization (self-assemblying) of molecular constituents, directed by noncovalent interactions which produce self recognition of a molecular site with a complementary (binding) site. Thus, hydrogen-bonded supramolecular assemblies can be formed from molecular components that have a hydrogen-bond donor site, such as an amide hydrogen, and a complementary acceptor site, such as a carbonyl oxygen. Furthermore, molecular pair-specific hydrogen bonding, such as in base pairing in DNA, can also be invoked. These types of hydrogen bonding can be utilized to produce folded helical structures that are described below in detail.

Metal–ligand coordination has also been used extensively to produce helicates, grids, and cylindrical cages. These self-assembled architectures are reversible. The helicates are molecular strands that wrap themselves around a row of metal ions forming double-helical metal complexes. Discotic molecules, which are structures with a disc-shaped core generally formed by planar aromatic rings and a periphery made of flexible side chains, self-organize by strong π–π interaction (attraction) to stack and form a rod-like or worm-like supramolecular polymeric structure. In the discotic supramolecular materials, the interdisc π–π interaction is several orders of magnitude stronger than the van der Waals interaction between the columns.

Two specific examples are presented here as illustrations of the supramolecular approach for producing complex optoelectronic and photonic materials.

Percec et al. (2002) showed that the self-assembly of fluorinated tapered dendrons can lead to the formation of supramolecular liquid crystals. By attaching conducting organic donor or acceptor groups to the apex of the dendrons, they produced supramolecular nanometer-scale columns that contained in their cores π-stacks of donors, acceptors, or donor–acceptor complexes by π–π interactions. A schematic representation of this liquid crystal assembly process is shown in Figure 8.15. These supramolecular liquid crystals exhibit high carrier mobilities, thus offering a distinct advantage over polymeric optoelectronic devices (e.g., light-emitting diodes, photovoltaics, and photorefractivity, with the latter being discussed in Chapter 10) where low charge carrier mobility limits the efficiency.

Percec et al. (2002) also found that when these functionalized dendrons are mixed with amorphous polymers containing compatible side groups, they co-assemble to incorporate the polymer in the center of the column through electrostatic donor–acceptor interactions. This process is also illustrated in Figure 8.15. These mixed supramolecular assemblies also exhibited enhanced charge carrier mobilities.

Another example is that of oligomers that adopt folded, helical conformations. These folding oligomers, or foldmers, involve an oligomeric unit with a rigidified backbone that involves at least one additional set of interactions (e.g., hydrogen bonds) to minimize conformational freedom (Gong, 2001; Sanford and Gong, 2003). This type of oligomer can be viewed as a tape. A short oligomer of such a conformation has a crescent shape that is similar to a macrocycle, as shown in Figure 8.16a. As the rigidified backbone reaches a certain length, the oligomer can be forced to adopt a helical conformation in which one end of the molecule must lie

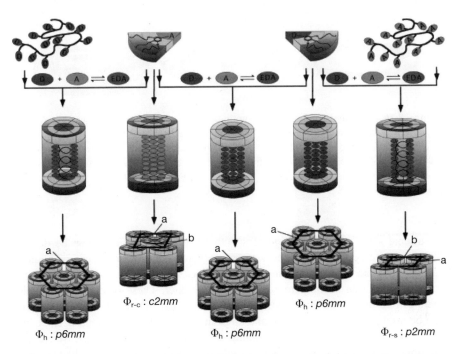

Figure 8.15. Schematic illustration of the liquid crystal assembly processes. Shown are the self-assembly, co-assembly, and self-organization of dendrons containing donor (D) and acceptor (A) groups with each other and with disordered amorphous polymers containing D and A side groups. The different systems form hexagonal columnar (Φh), centered rectangular columnar (Φr-c), and simple rectangular columnar (Φr-s) liquid crystals. a and b are lattice dimensions; space groups are shown. From Percec et al. (2002), reproduced with permission.

above the other because of crowding up (Figure 8.16b). The all-aromatic helicene oligomers whose backbones are rigidified by localized hydrogen-bonds belong to this category (Berl et al., 2000). Eventually, polymers and oligomers with long enough backbones can fold into hollow helical foldmers (Figure 8.16c) (Zhu et al., 2000; Gong et al., 2002).

Changing the curvature of the backbone can lead to the adjustment of the diameter of the interior cavity. Gong and co-workers (Gong, 2001; Sanford and Gong, 2003) produced foldmers with tunable interior cavities based on oligoamides and polyamides with different backbone curvatures and thus different, nanosized (10–60 μm) interior cavities that are featured by amide oxygen atoms and are thus electrostatically negatively charged. The cavity, therefore, is hydrophilic and can easily include polar nanoparticles such as inorganic quantum dots for photonic applications. A representative structure of oligoarylamide is shown in Figure 8.17. The curved backbone is produced by *meta*-disubstituted benzene rings, and the intramolecular hydrogen bonds rigidify each of the amide linkages in the molecule.

Figure 8.16. Oligomers and polymers with rigidified, crescent backbones. (a) A short oligomer can be viewed as a macrocycle. (b) When the backbone reaches a certain length, a helical conformation is adopted. (c) A helix of multiple turns can result from a long oligomer or polymer chain. Changing the curvature of the backbone leads to the adjustment of the diameter of the interior cavities in each case. From Gong (2001), reproduced with permission.

Additional stabilization for oligomers with six or seven residues can be derived from π–π interaction between benzene rings. Tuning of the interior cavity diameter can be achieved by incorporating building blocks with two amide linkers on the same benzene ring in the para position to each other. These types of helical foldmers can be useful nanomaterials for photonics. First the nanosize interior cavity can be used as a template to bind and align inclusions such as nanoparticles, quantum dots, or quantum rods. This binding can readily be accomplished by creating binding sites on the interior surface of foldmers. Second, their helical shape imparts chirality, which can also be useful to produce chiral optical and nonlinear optical response for photonics.

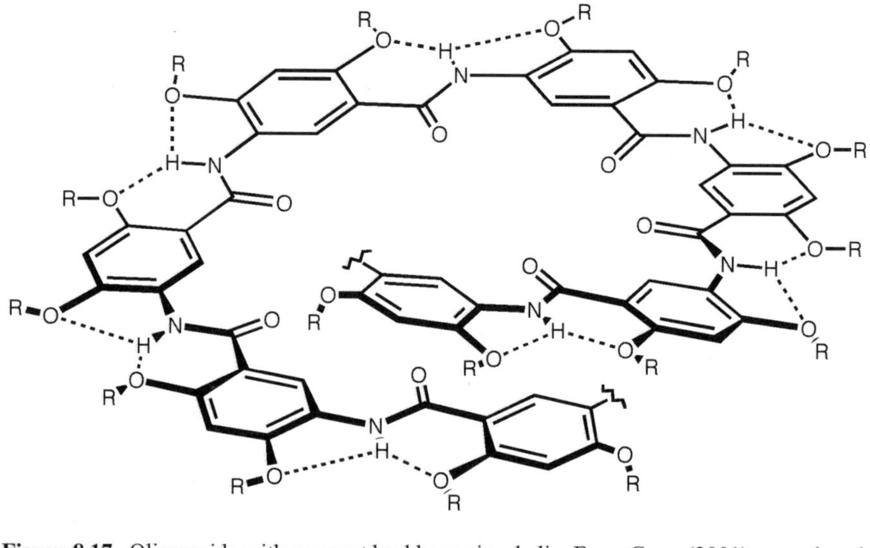

Figure 8.17. Oligoamide with crescent backbones in a helix. From Gong (2001), reproduced with permission.

8.6 MONOLAYER AND MULTILAYER MOLECULAR ASSEMBLIES

Molecular assemblies can be formed in a monolayer form with nanometer thickness. These monolayers can be successively deposited with the nanoscopic resolution of one layer at a time to form a multilayer film. Using alternating deposition, one can also build superlattices. The two principal methods described here are the Langmuir–Blodgett (abbreviated as LB) technique and the self-assembled monolayers (abbreviated as SAM) (Ulman, 1991). Using a patterned surface, the techniques can be utilized to build nanostructures of molecular assemblies for a number of applications such as nano LED. These techniques also provide the flexibility to align the molecular axes in a specific direction to produce well-oriented molecular assemblies. Furthermore, the multilayer single component or alternate layer superlattice structures can be fabricated in a noncentrosymmetric form to produce a medium with high second order optical nonlinearity, which exhibits second-harmonic generation and linear electro-optic effect. Therefore, no electric field poling, as required for a polymeric medium (see Section 8.4), is needed.

Langmuir–Blodgett Films. Langmuir–Blodgett (LB) films consist of assemblies of molecules with polar (hydrophilic) head groups and long aliphatic (hydrophobic) tails deposited onto a substrate from the surface of water. A key feature of these films is that forces at the water surface and lateral surface pressure are used to condense a randomized set of such molecules from a gas-like phase to one that is highly organized and stabilized by van der Waals forces between molecules. These forces are sufficiently cohesive to allow the films to be transferred to a substrate as a coherent film. Here we briefly review the preparation and properties of these layers. Some examples of photonic applications are discussed.

The deposition behavior for a typical amphiphilic molecule, such as stearic acid $C_{17}H_{35}COOH$ used in LB film formation, is shown in Figure 8.18. In the figure, the part of the molecule represented by a dark cricle is the polar head group and the tail represents the aliphatic section. The first withdrawal of a polar (hydrophilic) substrate such as glass results in typical deposition behavior illustrated in the figure. Reimmersion results in a favorable interaction between aliphatic tails so that a tail-to-tail film is formed. The next withdrawal step results in head-to-head deposition. This process can be repeated numerous times, and films with thicknesses in the range of ~1 μm can be obtained. This type of deposition is commonly referred to as Y type. It should be apparent that deposition of this type leads to a film with overall centrosymmetric structure. Occasionally, a chemical structure is encountered where deposition occurs only on immersion or withdrawal. These deposition types are referred to as X and Z, respectively, for historical reasons. They both lead to noncentrosymmetric structures but are often quite unstable, and rearrangement can occur, leading to centrosymmetric structures. A review of LB films and guide to the literature is given by Ulman (1991).

A number of chemical approaches have been devised for obtaining stable noncentrosymmetric LB film structures for second harmonic generation or electro-optic effect, both being second-order nonlinear optical processes. In one of the first ap-

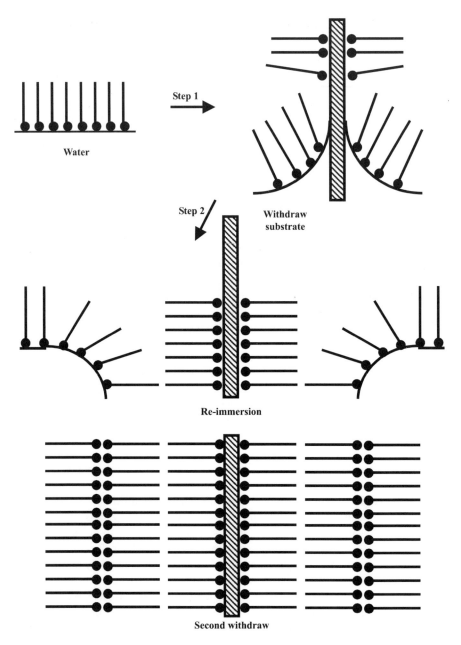

Figure 8.18. The Langmuir–Blodgett Y-type deposition process. In the first step a hydrophilic substrate is withdrawn from the surface of water and the hydrophilic heads of the compressed surface film adhere to the surface. In the second step the substrate and film are reimmersed and tail-to-tail deposition occurs. The process is repeated to build thick film.

proaches for constructing multilayers for second-order nonlinear optics, films consisting of alternating layers of a merocyanine dye and ω-tricosenoic acid were constructed by Girling et al. (1987). The arrangement of molecules is illustrated schematically in Figure 8.19. From inspection it is apparent that the film illustrated in Figure 8.19 has a polar axis perpendicular to the plane of the film. In general, the molecular axes exhibit a tilt with respect to the Z axis defined in the figure. The distribution of tilt angles α may be sharply peaked about some average value α_0 or could exhibit a broad distribution, depending on the chemical nature of the system. A tilt angle α implies an azimuthal distribution ϕ as well. Experiments on LB films have generally supported a uniform distribution of ϕ, at least on the scale of the physical dimensions of the laser beams employed in the experiment.

Wijekoon et al. reported second harmonic generation in the blue spectral region (400 nm) using multilayer LB films fabricated with two blue transparent nonlinear optical polymers (Wijekoon et al., 1996). The structures of these polymers are shown in Figure 8.20 where they are referred to as normal polymer (N) and reverse polymer (R). For microscopic enhancement of second-order optical nonlinearity, the molecular design utilizes an electron donor group separated from an electron acceptor group by a conjugated segment. In the N polymer, the electron acceptor group —SO_2— is attached to the hydrophobic tail $C_{10}F_{21}$. On the other hand, the R polymer has the electron donating group —N— attached to the hydrophobic $C_{18}H_{37}$ tail. Hence the nonlinear molecules are oppositely oriented with respect to the tail. Thus an alternate deposition of monolayers of these two nonlinear optical polymers produces second-order nonlinearities in the same direction, because their dipoles point in the same direction. The net result is an additive enhancement of the nonlinearities of the composite bilayer (one layer of N and one layer of R). The intensity of the frequency-doubled light generated in these multilayers was found to increase quadratically with the increasing number of bilayers, indicating that the optical non-

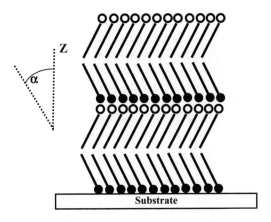

Figure 8.19. A Langmuir–Blodgett Y-type film made from two different amphophiles with an average tilt angle α relative to the surface normal.

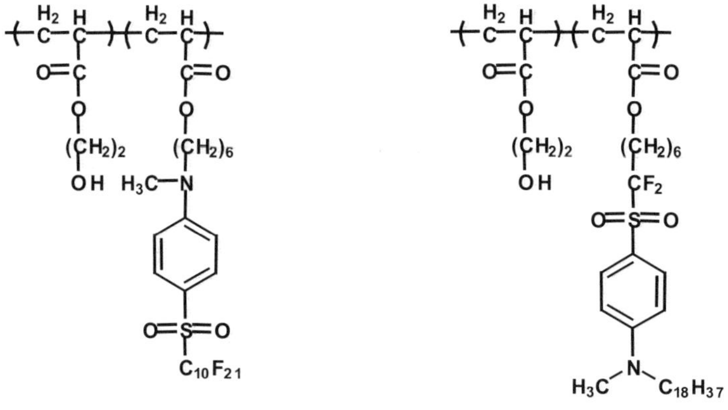

Figure 8.20. Chemical structure of blue transparent organic polymers for alternating layer LB film (Wijekoon et al., 1996).

Normal Polymer (N)

Reverse Polymer (R)

linearity of the bilayer is a linear sum of the optical nonlinearities of the individual monolayer.

Self-Assembled Monolayers (SAM). A method related to the LB films, but distinctly different, is that of self-assembled monolayers (SAM). This has potential advantages (Ulman, 1991). Molecules used to form SAMs generally consist of a head group, a long alkyl chain, and a reactive tail. A typical head group is the trichlorosilyl group, which reacts with hydroxylated surfaces such as Si to form a covalently bonded polymerized monolayer. The long alkyl group provides physical stability for the layer via van der Waals interactions. A typical tail group is an ester, which can be reduced with $LiAlH_4$ to form a new hydroxylated surface that can be subsequently used to deposit a new layer. Studies of the formation of self-assembled monolayers were conducted by Moaz and Sagiv (1984). High-quality multilayer films can also be formed using the SAM approach (Ulman, 1991).

Marks and co-workers have produced the self-assembled superlattices (alternate layer growth) with very large electro-optic coefficient (r_{33} coefficients, a measure of electro-optic strength, in excess of ~65 pm/V at 1064 nm) (Zhu et al., 2002; van der Boom et al., 2002). Marks and co-workers (Zhu et al., 2001) utilized an improved "all"-wet-chemical approach that involves iterative combination of only two steps: (i) self-limiting polar chemisorption of a protected highly nonlinear azobenzene chromophore monolayer and (ii) *in situ* trialkylsilyl group removal and self-limiting capping of each chromophore monolayer using octachlorotrisiloxane. This scheme is shown in Figure 8.21. The second step deposits a polysiloxane layer that is robust and thus stabilizes the polar microstructure via interchromophore crosslinking. It also regenerates a reactive hydrophilic surface. They demonstrated deposition of up to 80 alternating chromophore and capping layers that are struc-

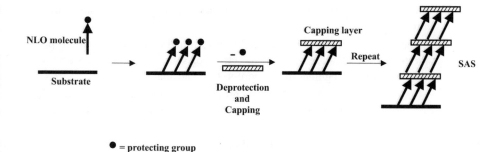

Figure 8.21. Schematic representation of two-step layer-by-layer self-assembly of self-assembled films.

turally regular and polar. In comparison to the LB films involving relatively weak interactions between the chromophore layers, their self-assembled superlattices (SAS) involved strong Si—O covalent bonds. These self-assembled superlattices exhibit the following advantages:

- Much higher electro-optic coefficients and lower dielectric constants compared to the standard electro-optic material: $LiNbO_3$
- A closely packed and robust film in which the chromophore molecules are covalently bonded to the substrate
- Molecular orientations intrinsically noncentrosymmetric, thus not requiring any post-deposition poling step
- Precise control of film refractive index at the subwavelength level (layer-by-layer deposition)
- Large area films readily prepared
- Compatibility with silicon or a variety of related substrates, allowing ready device integration.

8.7 HIGHLIGHTS OF THE CHAPTER

- A molecular architecture provides a flexible platform to produce a rich variety of multifunctional nanomaterials by manipulation of both covalent and noncovalent interactions.
- A molecular architecture involves a three-dimensional structure which is frequently determined by noncovalent interactions that do not involve any chemical bond. These three-dimensional structures determine the shape and function of the molecular architecture.
- Examples of the various types of noncovalent interactions are (I) hydrogen bond, (ii) metal coordinations, (iii) π–π interactions, (iv) electrostatic and

ionic interactions, (v) van der Waals interactions, and (vi) hydrophobic interactions.
- Noncovalent interactions can be highly directional and regiospecific to provide recognition features for self-directing (or self-assembling) of a specific molecular unit or segment to an appropriate molecular site.
- Polymeric materials are emerging as promising photonic media in which the structures and functionalities can be controlled from the atomic level to the bulk level, hence providing nanocontrol of strucutres.
- Multifunctionality can be introduced in a polymeric structure by introducing block segments in the main chains (such polymers being called block copolymers) or by attaching side chains of various functionalities—for example, pendant to the main chain (such polymers being called side-chain polymers).
- Orientational alignment, such as electric-field poling of polymers, can be used to produce or enhance a desired functional response. An example is the electro-optic effect in electrically poled polymers.
- Di-block copolymers containing two different segments of different optical and mechanical features and having noncovalent interactions offer tremendous opportunities to produce nanoobjects of different morphologies and optical functions.
- An illustrative example is a di-block polymer containing a π-conjugated segment whose length can be varied to provide electroluminescence of different colors.
- Molecular machines are another example of a nanostructured molecular architecture in which two molecular units are mechanically interlocked without involving a chemical bond. By the action of light, the two molecular units can undergo relative displacement with respect to each other.
- Dendrimers represent a vast class of large molecular architectures that contain structurally ordered, highly branched nanostructures. They contain an exact number of concentric layers of branching points, referred to as *generations*.
- The two approaches used for dendrimer synthesis are (i) the divergent approach, which starts with initiation of growth from the core, outwards by repetition of a series of coupling and activation steps to produce a layer-by-layer buildup, and (ii) the convergent approach, which starts with growth from what becomes the exterior and progresses inward by coupling end groups to each branch of the monomer.
- Some photonic examples provided are (i) site isolation of luminescent core to reduce environmentally induced nonradiative quenching, (ii) use of peripheral groups as photon-harvesting antennas that absorb light and funnel it by energy transfer to produce enhanced emission or photovoltic generation at the core, (iii) control of intramolecular interaction (such as electrostatic repulsion) to enhance a photonic effect (such as electric field poled structures for enhanced electro-optic effect), and (iv) use of peripheral groups to amplify the sensitivity for sensing applications.

- A supramolecular structure is another large class of nanostructured molecular structures that represent an association of two or more chemical species through noncovalent intermolecular interactions.
- An example is a supramolecular polymer in which an array of monomers are held together by reversible and highly directional noncovalent interactions that also can produce nano-objects of various shapes.
- Another example is a mesoporous foldmer containing a folded molecular assembly with a central hollow tube that can act as a cavity or a template to incorporate molecular units or nanoparticles of various photonic functions.
- Yet another type of supramolecular system involves a discotic structure that can produce strong chirality or enhanced interdisc $\pi-\pi$ interaction.
- Monolayer and multilayer molecular assemblies represent a class of layered molecular architecture with nanometer-thickness scale control.
- The two principal methods to produce mono- or multilayer assemblies are (i) the Langmuir–Blodgett techniques and (ii) the self-assembled monolayer method.
- The Langmuir–Blodgett method involves transfer of assemblies of molecules with polar (hydrophilic) head groups and long aliphatic (hydrophobic) tails from a water surface, on which they are spread as a film.
- Multilayer Langmuir–Blodgett films that are produced by alternative transfer of monolayer films of two different types, often as a superlattice, have been utilized for second-order nonlinear optical effects such as second harmonic generation.
- The self-assembled monolayer method utilizes covalent bonding between the head group of the depositing monolayer and the substrate. Chemical bonding between successive layers can be used to build multilayers.
- The self-assembled monlayer method produces films that are more robust than those formed by the Langmuir–Blodgett techniques.
- The self-assembled approach has been utilized to produce superlattices exhibiting a very large electro-optic coefficient, without any need for electric field poling.

REFERENCES

Adronov, A., and Frèchet, J. M. J., Light-Harvesting Dendrimers, *Chem. Commun.,* 1701–1710 (2000).

Akcelrud, L., Electroluminescent Polymers, *Prog. Polym. Sci.* **28**, 875–962 (2003).

Balzani, V., Gomez-Lopez, M., and Stoddart, J. F., Molecular Machines, *Acc. Chem. Res.* **31**, 405–414 (1998).

Balzani, V., Ceroni, P., Gestermann, S., Kauffmann, C., Gorka, M., and Vögtle, F., Dendrimers as Fluorescent Sensors with Signal Amplification, *Chem. Commun.,* 853–854 (2000).

Bates, F. S. and Fredrickson, G. H., Block Copolymers—Designer Soft Materials, *Physics Today* **52**, 32–38 (1999).

Berl, V., Hue, I., Khoury, R. G., Krische, R. G., and Lehn, J. M., Interconversion of Single and Double Helices Formed from Synthetic Molecular Strands, *Nature* **407**, 720–723 (2000).

Bissell, R. A., Cordova, E., Kaifer, A. E., Stoddart, J. F., and Tolley, M. S., A Chemically and Electrochemically Switchable Molecular Shuttle, *Nature* **369**, 133–137 (1994).

Brouwer, A. M., Frochot, C., Gatti, F. G., Leigh, D. A., Mottier, L., Paolucci, F., Roffia, S., and Wurpel, G. W. H., Photoinduction of Fast, Reversible Translational Motion in a Hydrogen-Bonded Molecular Shuttle, *Science* **291**, 2124–2128 (2001).

Brunsveld, L., Folmer, B. J. B., Meijer, E. W., and Sijbesma, R. P., Supramolecular Polymers, *Chem. Rev.* **101**, 4071–4097 (2001).

Chung, S. J., Jin, J., and Kim, K. K., Novel PPV Derivatives Emitting Light over a Broad Wavelength Range, *Adv. Mater.* **9**, 551–554 (1997).

Dalton, L., Nonlinear Optical Polymeric Materials: From Chromophore Design to Commercial Applications, in *Advances in Polymer Science: Polymers for Photonics Applications I*, Vol. 158, K. S. Lee, ed., Springer-Verlag, Berlin, 2002, pp. 1–86.

Dalton, L., Robinson, B. H., Jen, A. K.-Y., Steier, W. H., and Nielsen, R., Systematic Development of High Bandwidth, Low Drive Voltage Organic Electro-optic Devices and Their Applications, *Opt. Mater.* **21**, 19–28 (2003).

Devadoss, C., Bharathi, P., and Moore, J. S., Energy Transfer in Dendritic Macromolecules: Molecular Size Effects and the Role of an Energy Gradient, *J. Am. Chem. Soc.* **118**, 9635–9644 (1996).

Dykes, G. M., Dendrimers: A Review of Their Appeal and Applications, *J. Chem. Technol. Biotechnol.* **76**, 903–918 (2001).

Gale, P. A., Supramolecular Chemistry, *Annu. Rep. Prog. Chem. Sect. B* **98**, 581–605 (2002).

Girling, I. R., Gade, N. A., Kolinsky, P. V., Jones, R. J., Peterson, I. R., Ahmad, M. M., Neal, D. B., Petty, M. C., Roberts, G. G., and Feast, W. J., Second-Harmonic Generation in Mixed Hemicyanine: Fatty-Acid Langmuir–Blodgett Monolayers, *J. Opt. Soc. Am. B*, **4**, 950–955 (1987).

Gittins, P. J., and Twyman, L. J., Dendrimers and Supramolecular Chemistry, *Supramol. Chem.* **15**, 5–23 (2003).

Gong, B., Crescent Oligoamides: From Acyclic "Macrocycles" to Folding Nanotubes, *Chem. Eur. J.* **7**, 4337–4342 (2001).

Gong, B., Zeng, H. Q., Zhu, J., Yuan, L. H., Han, Y. H., Cheng, S. Z., Furukawa, M., Parra, R. D., Kovalevsky, A. Y., Mills, J. L., Skrzypczak-Jankun, E., Martinovic, S., Smith, R. D., Zheng, C., Szyperski, T., and Zeng, X. C., Creating Nanocavities of Tunable Sizes: Hollow Helices, *Proc. Natl. Acad. Sci. U.S.A.* **99**, 11583–11588 (2002).

Grayson, S. M., and Fréchet, J. M. J., Convergent Dendrons and Dendrimers: From Synthesis to Applications, *Chem. Rev.* **101**, 3819–3867 (2001).

Hawker, C. J., and Fréchet, J. M. J., Preparation of Polymers with Controlled Molecular Architecture. A New Convergent Approach to Dendritic Macromolecules, *J. Am. Chem. Soc.* **112**, 7638–7647 (1990).

Hawker, C. J., Wooley, K. L., and Fréchet, J. M. J., Unimolecular Micelles and Globular Amphiphiles—Dendritic Macromolecules as Novel Recyclable Solubilizing Agents, *J. Chem. Soc.-Perkin Trans.* **1**, 1287–1297 (1993).

Jansen, J., de Brabander van den Berg, E. M. M., and Meijer E. W., Encapsulation of Guest Molecules into a Dendritic Box, *Science* **266,** 1226–1229 (1994).

Jansen, J., Meijer, E. W., and de Brabander van den Berg, E. M. M., The Dendritic Box-Shape-Sensitive Liberation of Encapsulated Guests, *J. Am. Chem. Soc.* **117,** 4417–4418 (1995).

Jimenez, M. C., Dietrich-Buchecker, C., and Sauvage, J. P., Towards Synthetic Molecular Muscles: Contraction and Stretching of A Linear Rotaxane Dimer, *Angew. Chem. Int. Ed.* **39,** 3284–3287 (2000).

Kawa, M., and Fréchet, J. M. J., Self-Assembled Lanthanide-Cored Dendrimer Complexes: Enhancement of the Luminescence Properties of Lanthanide Ions Through Site-Isolation and Antenna Effects, *Chem. Mater.* **10,** 286–296 (1998).

Kyllo, E. M., Gustafson, T. L., Wang, D. K., Sun, R. G., and Epstein, A. J., Photophysics of Segmented Block Copolymer Derivatives, *Synth. Met.* **116,** 189–192 (2001).

Lehn, J.-M., *Supramolecular Chemistry,* VCH, Weinheim, Germany, 1995.

Leigh, D. A., Molecules in Motion: Towards Hydrogen Bond-Assembled Molecular Machines, in F. Charra, V. M. Agranovich, and F. Kajzar, *Organic Nanophotonics,* NATO Science Series II. Mathematics, Physics and Chemistry, Vol. 100, Kluwer Academic Publishers, The Netherlands, 2003a, pp. 47–56.

Leigh, D. A., private communication, 2003b.

Luo, J., Haller, M., Li, H., Kim, T.-D., Jen, A. K.-Y., Highly Efficient and Thermally Stable Electro-optic Polymer from a Smartly Controlled Crosslinking Process, *Adv. Mater.* **15,** 1635–1638 (2003).

Moaz, R., and Sagiv. J., On the Formation and Structure of Selfassembling Monolayers. I. A Comparative ATR-Wettability Study of Langmuir–Blodgett and Adsorbed Films on Flat Substrates and Glass Microbeads, *J. Colloid Interface Sci.* **100,** 465– 496 (1984).

Newkome, G. R., Moorefield, C. N., and Vögtle, F., *Dendritic Molecules: Concepts, Syntheses, Perspectives,* VCH, Weinheim, Germany, 1996.

Park, C., Yoon, J., and Thomas, E. L., Enabling Nanotechnology with Self-Assembled Block Copolymer Patterns, *Polymer,* **44,** 6725–6760 (2003).

Percec, V., Glodde, M., Bera, T. K., Miura, Y., Shiyanovskaya, I., Singer, K. D., Balagurusamy, V. S. K., Heiney, P. A., Schnell, I., Rapp, A., Spiess, H.-W., Hudson, S. D., and Duan, H., Self-Organization of Supramolecular Helical Dendrimers into Complex Electronic Materials, *Nature* **419,** 384–387 (2002).

Prasad, P. N., and William, D. J., *Introduction to Nonlinear Optical Effects in Molecules and Polymers,* John Wiley & Sons, New York, 1991.

Sanford, A. R., and Gong, B., Evolution of Helical Foldmers, *Curr. Org. Chem.* **7,** 1–11 (2003).

Sauvage, J. P., Transition Metal-Containing Rotaxanes and Catenanes in Motion: Toward Molecular Machines and Motors, *Acc. Chem. Res.* **31,** 611–619 (1998).

Tomalia, D. A., Naylor, A. M., and Goddard, W. A., III, Startburst Dendrimers: Molecular Level Control of Size, Shape, Surface Chemistry, Topology, and Flexibility from Atoms to Macroscopic Matter, *Angew. Chem. Int. Ed. Engl.* **29,** 138–175 (1990).

Ulman, A., *An Introduction to Ultra Thin Organic Films from Langmuir–Blodgett to Self-Assemblies,* Academic Press, San Diego, 1991.

van der Boom, M. E., Zhu, P., Evmenenko, G., Malinsky, J. E., Lin, W., Dutta, P., and Marks, T. J., Nanoscale Consecutive Self-Assembly of Thin-Film Molecular Materials for Electro-optic Switching, Chemical Streamlining and Ultrahigh Response Chromophore, *Langmuir* **18,** 3704–3707 (2002).

Vögtle, F., Gestermann, S., Hesse, R., Schwierz, H., and Windisch, B., Functional Dendrimers, *Prog. Polym. Sci.* **25,** 987–1041 (2000).

Wang, P.-W., Liu, Y.-J., Devadoss, C., Bharathi, P., and Moore, J. S., Electroluminescent Diodes from a Single-Component Emitting Layer of Dendritic Macromolecules, *Adv. Mater.* **8,** 237–241 (1996).

Watson, J. D., and Crick, F. H., Molecular Structure of Nucleic Acid. A Structure of Deoxyribose Nucleic Acid, *Nature* **171,** 737–738 (1953).

Whitesides, G. M., Mathias, J. P., and Seto, C. T., Molecular Self-Assembly and nanochemistry: A Chemical Strategy for the Synthesis of Nanostructure, *Science* **254,** 1312–1319 (1991).

Wijekoon, W. M. K. P., Wijaya, S. K., Bhawalkar, J. D., Prasad, P. N., Penner, T. L., Armstrong, N. J., Ezenyilimba, M. C., and Williams, D. J., Second Harmonic Generation in Multilayer Langmuir–Blodgett Films of Blue Transparent Organic Polymers, *J. Am. Chem. Soc.* **118,** 4480–4483 (1996).

Yang, Z., Hu, B., and Karasz, F. E., Contributions of Nonconjugated Spacers to Properties of Electroluminescent Block Copolymers, *J. Macromol. Sci. Pure Appl. Chem.* **A35,** 233–247 (1998).

Zhu, J., Parra, R. D., Zeng, H., Skrzypczak-Jankun, E., Zeng, X. C., and Gong, B., A New Class of Folding Oligomers: Crescent Oligoamides, *J. Am. Chem. Soc.* **122,** 4219–4220 (2000).

Zhu, P., van der Boom, M. E., Kang, H., Evmenenko, G., Dutta, P., and Marks, T. J., Efficient Consecutive assembly of large-response Thin-film Molecular Electro-optic Materials, *Polym. Prep.* **42,** 579–580 (2001).

Zhu, P., van der Boom, M. E., Kang, H., Evmenenko, G., Dutta, P., and Marks, T. J., Realization of Expeditious Layer-by-Layer Siloxane-Based Self-Assembly as an Efficient Route to Structurally Regular Acentric Superlattices with Large Electro-Optic Responses, *Chem. Mater.* **14,** 4983–4989 (2002).

CHAPTER 9

Photonic Crystals

This chapter provides a more detailed coverage on photonic crystals, a subject already introduced in Chapter 2. Photonic crystals represent a class of nanomaterials in which alternating domains of higher and lower refracting indices produce an ordered structure with periodicity on the order of wavelength of light. Photonic crystals have emerged as a major thrust area of nanophotonics and have witnessed a remarkable growth of research activities worldwide. This growth has been fueled both by a quest for fundamental understanding of optical processes in photonic crystals and by their promise for many technological applications. Both of these aspects are described in this chapter.

Section 9.1 provides some basic concepts pertaining to photonic crystals. Methods for calculating the band structure of a photonic crystal and its optical properties are presented in Section 9.2. Readers less theoretically inclined may skip this section.

Photonic crystals exhibit a number of unique optical and nonlinear optical properties as well as enhancement of some photonic functions. Section 9.3 lists some important features of photonic crystals using conceptual rather than mathematical descriptions. This is an important section in order for a reader to fully appreciate the field of photonic crystals.

A wide variety of methods have been used to fabricate photonic crystals of varying dimensions (1D, 2D, and 3D). These methods are described in Section 9.4. This section should be of value to chemists and material scientists in getting some insights into producing novel photonic crystals. Section 9.5 describes methods to create defects of different sizes and shapes in a photonic crystal. These defects can produce a microcavity effect to enhance emission for producing low-threshold lasing, or they can be used as channels for guiding light in a very controlled way.

Section 9.6 deals with the topic of nonlinear photonic crystals—in other words, photonic crystals exhibiting strong nonlinear optical effects near its band edge. The two specific examples presented are harmonic generation and two-photon excited fluorescence. Section 9.7 describes the features of a specific type of photonic crystals, the photonic crystal fibers. They are two-dimensional photonic crystals consisting of a strand of glass with an array of microscopic air channels running along its length.

Section 9.8 describes the application of photonic crystals in optical communications in providing both active and passive functions. Section 9.9 introduces another general application of photonic crystals, such as chemical, environmental, and biological sensors. Highlights of this chapter are given in Section 9.10.

For supplemental reading, suggested references are:

Joannopoulos et al. (1995): *Photonic Crystals*

Kraus and De La Rue (1999): *Photonic Crystals in the Optical Regime—Past, Present and Future*

Slusher and Eggleton, eds. (2003): *Nonlinear Photonic Crystals*

9.1 BASIC CONCEPTS

As introduced in Chapter 2, photonic crystals are ordered nanostructures in which two media with different dielectric constants or refractive indices are arranged in a periodic form (John, 1987; Yablonovitch, 1987). Thus they form two interpenetrating domains, one domain of higher refractive index and another of lower refractive index. It is a periodic domain with periodicity on the order of wavelength of light. The analogy between an electronic crystal of a semiconductor with a periodic lattice and a photonic crystal with a periodic dielectric has already been presented in Chapter 2. In a photonic crystal, for a certain range of energy of photons and certain wave vectors (that is, direction of propagation), light is not allowed to propagate through this medium. If we generate light inside, it cannot propagate in this direction. If we send light from outside, it is reflected. Figure 9.1 shows typical transmission and reflection spectra of a photonic crystal produced by close-packing of polystyrene spheres.

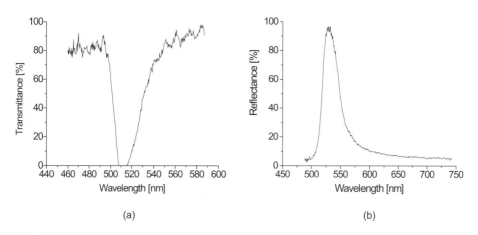

Figure 9.1. Typical transmission (a) and reflection (b) spectra of a photonic crystal produced by close-packing of polystyrene spheres. The diameters of the polystyrene spheres are 220 nm for transmittance study and 230 nm for reflection measurement.

9.1 BASIC CONCEPTS

In the case of a large refractive index contrast (defined by the ratio n_1/n_2) photonic crystal with a proper shape of building blocks (domains) and proper crystal symmetry, a complete bandgap develops. In this case, the bandgap is not dependent on the direction of wavevector, which defines the light propagation; also, the density of photon states goes to zero in the bandgap region. These materials with a complete gap are often called photonic bandgap materials.

Figure 9.2 shows the effect of dimensionality on the properties of photonic crystals. In the 3D arrangement, shown on the top left, spheres are packed together with a face-centered cubic arrangement. These spheres of submicron sizes can be made of glass or of plastic such as polystyrene. They are prepared by colloidal chemistry and, for this reason, were originally called colloidal crystals by Asher and co-workers (Carlson and Asher, 1984; Holtz and Asher, 1997).

A two-dimensional photonic crystal can be produced by a stack of cylinders (or parallellopipes) of given refractive index. The gap between them may simply be air or another medium of a different refractive index. This arrangement is two-dimensional, because the variation of refractive index occurs in a plane. An example is the case of a photonic crystal fiber, one possible geometry of it being a number of fibers stacked together in a manner to make a bundle of fibers.

The last case is a one-dimensional photonic crystal, which is a layer structure where alternating layers of different refractive indices are stacked. So the stacking and the refractive index variation are along only one of the axes (the vertical axis in the picture). A simple example is a Bragg grating. The reciprocal space shows in this case the bandgap region by the shadowed space in the figure.

Structures in which the periodic domains (e.g., spherical particles) of high refractive index close-pack with the media (such as air) of lower refractive index fill-

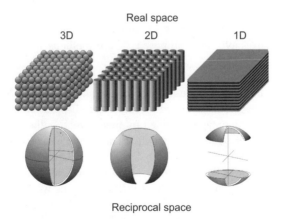

Figure 9.2. Real-space representation of a photonic crystal and reciprocal space regions for which propagation is not allowed. From http://www.icmm.csic.es/cefe/pbgs.htm, reproduced with permission.

ing the void are called regular structure photonic crystals. In the case of packing of colloidal particles, such a structure is also called an opal structure. In this case the bandgap is not really a full gap, because the density of photon states does not go to zero in the bandgap. One needs an inverse structure with voids of high refractive index and packing units (as spheres) of low refractive index to develop a full bandgap. One approach is to produce an inverse opal structure in which the packed spheres are hollow (void space, hence air of refractive index 1) and the void (interstitial space) is filled with a material of high refractive index (>2.9).

One way to produce such structures is to use polystyrene sphere packing and then fill the interstices with a high refractive index material (GaP in our work). Then polystyrene is burned off to leave holes (air). GaP has a refractive index of ~3.5 which, compared to air, creates a large dielectric contrast. If silica spheres are used to close-pack, they can be etched with HF to produce an inverse opal structure.

A detailed understanding of the optical properties of a photonic crystal is provided by the band structure, which yields the details of the allowed photon frequencies for different wavevectors (both magnitude and direction). The basic eigenvalue equation used for this purpose was already discussed in Chapter 2. The different methods used for such calculations are discussed in Section 9.2.

9.2 THEORETICAL MODELING OF PHOTONIC CRYSTALS

The band structure of a given photonic crystal defines its optical properties, such as transmission, reflection, and angular dependence. Therefore, the procedure for determining this structure has become a subject of extensive research. An exciting aspect of the photonic bandgap structure is that it can be calculated from first principles and that these results are consistent with experimental conclusions. Moreover, photonic crystals with desired properties can be precisely tailored (calculated) for specific applications.

Many methods for photonic band structure determination—that is, theoretical as well as numerical—have been proposed. These methods fall in two broad categories:

- Frequency-domain techniques where the photon eigenvalue equation, described in Chapter 2, is solved to obtain the allowed photon states and their energies. The advantage provided by this method is that it directly provides the band structure. Examples are the Plane Wave Expansion Method, PWEM (Ho et al., 1990), and the Transfer Matrix Methods, TMM (Pendry and MacKinnon, 1992; Pendry, 1996). Other methods are a combination of (a) finite-element discretization of Maxwell's equations with a fast Fourier transform preconditioner and a preconditioned subspace iteration algorithm for finding eigenvalues (Dobson et al., 2000) and (b) the Korriga–Kohn–Rostoker (KKR) method, which has been applied to the calculation of the frequency band structure of electromagnetic and elastic fields in relation to photonic crystals (Wang et al., 1993).

- Time-domain techniques, which calculate the temporal evolution of the input electromagnetic field propagating through the crystal. Then, the band structure is calculated by the Fourier transform of the time-dependent field to the frequency domain. A widely used time-domain method is Finite-Difference Time Domain (FDTD) (Arriaga et al., 1999).

The FDTD method calculates time evolution of the electromagnetic waves by a direct discretization of Maxwell's equations. In this method, the differentials in Maxwell's equations are replaced by finite differences to connect the electromagnetic fields in one time interval to the ones in the next interval. The advantage of this method is that results for a large frequency range can be obtained in a single run. This method is also suitable for wave packet propagation simulations and transmission and reflection coefficient calculations. Depending on the crystal structure to be analyzed, one method may have an advantage over the others.

Here, only the Plane Wave Expansion Method and the Transfer Matrix Method are given as the examples. The first one depends on the expansion of the electromagnetic field with a plane wave basis set.

$$\mathbf{H}(\mathbf{r}) = \sum_{\mathbf{G}} \sum_{\lambda} \mathbf{h}_{\mathbf{G},\lambda} \mathbf{e}_\lambda \exp[i(\mathbf{k} + \mathbf{G}) * \mathbf{r}] \qquad (9.1)$$

Here, $\mathbf{k}$ is a wavevector in the Brillouin zone, $\mathbf{G}$ is a reciprocal lattice vector, and $\mathbf{e}_\lambda$ are unit vectors perpendicular to $\mathbf{k} + \mathbf{G}$. These descriptions for the reciprocal lattice utilize the terminologies of solid-state physics (Kittel, 2003). The solution of the eigenvalue equation (9.2), also described in Chapter 2 and which is the *master equation* of fully vectorial electromagnetic waves

$$\nabla \times \left[\frac{1}{\varepsilon(\mathbf{r})} \nabla \times \mathbf{H} \right] = \frac{\omega^2}{c^2} \mathbf{H} \qquad (9.2)$$

provides the band structure of the crystal by numerical means. This approach is very useful for structures with no defects. Johnson (http://ab-initio.mit.edu/mpb/) of MIT has developed excellent software, which utilizes the PWEM. The package is publicly available under the GNU license. This method is recommended, especially for 3D bandgap structure calculations. As an example, we calculated, using this software, the band diagram for opals formed in close-packed fcc structure from polystyrene beads, which is shown in Figure 9.3.

The refractive index of $n = 1.59$ for polystyrene is used in our calculations. The directional gap at the <111> L point can be seen. The x axis contains the important symmetry lines in the Brillouin zone. The frequency is normalized through $f_n = (f*a)/c$, where a is the lattice constant and c is the speed of light.

The second method illustrated here is the Transfer Matrix Method, which calculates transmission and reflection coefficients and band structure for electromagnetic waves by using the transfer matrix of the system. In this system, the space is divided into small discrete cells with coupling to neighboring cells. The transfer matrix is

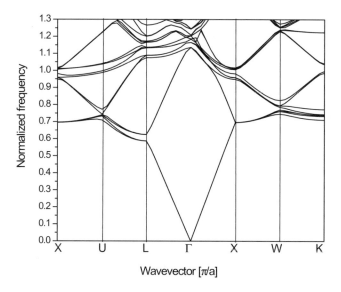

Figure 9.3. Band diagram for a face-centered close-packed polystyrene sphere structure, calculated using the plane wave expansion method. The L point corresponds to <111> direction.

the matrix that relates the field on one side of a unit cell of a periodic array to those on the other side.

Band structure can be calculated by diagonalizing the transfer matrix. The eigenvalues of the transfer matrix give the band diagram. Transmission and reflection coefficients can be calculated by utilizing multiple scattering formula. This method gives faster calculations of transmission and reflection coefficients and band diagrams than does PWEM. This method is not efficient for high frequencies. TMM is especially useful when one wants to calculate the propagation of electromagnetic modes in a finite photonic crystal.

As an example of calculations performed by this method, we determined reflection spectrum and band structure diagram for two kinds of photonic crystals by using the Translight package. This package was developed by Reynolds, based on TMM, and is also publicly available: http://www.elec.gla.ac.uk/groups/opto/photoniccrystal/Software/SoftwareMain.htm. The code for this method is very effective in the calculations of band structures of 1D and 2D photonic crystals as well as for calculations of transmission and reflection coefficients of 1D, 2D, and 3D photonic crystals.

Figure 9.4 shows the reflection spectra of an opal structure, formed in close-packed face-centered cubic form, in the <111> crystallographic direction. A refractive index of $n = 1.59$ (polystyrene) was used for the opal. Frequency is normalized through $f_n = (f*a)/c$, where c is the speed of light, a is the lattice constant and f is the normal frequency in hertz.

Figure 9.5 shows the calculated photonic band structure for a 1D photonic crystal, the so-called Bragg stack. The crystal consists of 10 periodic layers of dielectric

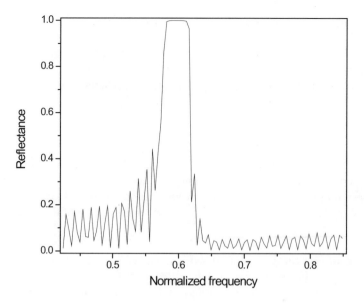

Figure 9.4. Reflection spectra of a face-centered cubic opal structure in the <111> crystallographic direction, calculated using the transfer matrix method with the Translight software.

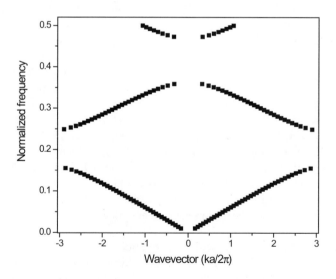

Figure 9.5. Band structure for a 1D photonic crystal, the Bragg stack, calculated using the transfer matrix method.

surfaces with dielectric constants $\varepsilon_1 = 1.96$ and $\varepsilon_2 = 11.56$. The incidence of electromagnetic waves is assumed to be perpendicular to the surfaces.

In summary, both methods are commonly used and they yield theoretical calculations that are consistent with experimental results. The Plane Wave Expansion Method is very useful for structures with no defects, whereas the Transfer Matrix Method is especially suitable for finite-size photonic crystals.

9.3 FEATURES OF PHOTONIC CRYSTALS

Photonic crystals exhibit a number of unique optical and nonlinear optical properties. These properties can find important applications in optical communications, low-threshold lasing, frequency conversion, and sensing. Some of these features are described here.

Presence of Bandgap. As discussed in Sections 9.1 and 9.2, photonic crystals exhibit bandgaps or pseudo-gaps (often called stop-gaps) in their band structure. The existence of these bandgaps can make them suitable for high-quality narrow-band filters, the wavelength of which can be tuned by changing the periodicity of the domains (lattice constant). The effective periodicity can also be changed in response to adsorption (or inclusion) of an analyte, making photonic crystal media suitable for chemical and biological sensing. This topic is discussed in detail in Section 9.9.

The position of stop-gap can be changed in a nonlinear crystal by using Kerr nonlinearity. The optical Kerr effect is a third-order nonlinear optical effect that produces a contribution Δn to the refractive index, where $\Delta n = n_2 I$ is linearly dependent on the intensity, with n_2 being the nonlinear index coefficient (Prasad and Williams, 1991). Thus by increasing the intensity, the refractive index of a nonlinear periodic domain can be changed, resulting in a change of the bandgap position. This manifestation can be utilized to produce optical switching for light of frequency at the band edge. By shifting the bandgap (and thus the band edge) with the change in intensity, light transmission can be switched from a low to a high value. This shifting of the bandgap frequency can be used for dynamic control of WDM channel routing. In birefringent photonic crystals, the reflective index has two different components for the two orthogonal polarizations. In such crystals, photonic bandgaps are different for two orthogonal polarizations, which can be used for polarization cleanup or for polarization-dependent optical switching.

Local Field Enhancement. Spatial distribution of an electromagnetic field can be manipulated in a photonic crystal to produce local field enhancement in one dielectric or the other. This field enhancement in a nonlinear photonic crystal can be utilized to enhance nonlinear optical effects that are strongly dependent on the local field.

Near the photonic bandgap, the low-frequency modes concentrate their energy in the high-refractive-index regions, and the high-frequency modes concentrate theirs in the low-refractive-index material (Joannopoulos et al., 1995). This behavior is shown in Figure 9.6. Therefore, if we illuminate a periodic structure with a strong fundamental light with a wavelength close to the low-frequency photonic bandgap edge, the field energy will be concentrated mostly in the high-refractive-index material, which may have a large value of nonlinear susceptibility. This strong field localization significantly increases nonlinear interactions of the fundamental field with the photonic crystals.

Anomalous Group Velocity Dispersion. The photonic band structure (discussed in Section 9.2), which gives the dispersion of the allowed propagation frequencies as a function of the propagation vector **k**, determines the group velocity with which an optical wave packet (such as a short pulse of light) propagates inside a medium. This frequency (energy) dispersion for photons also determines the effective refractive index and effects related to it (such as refraction, collimation,

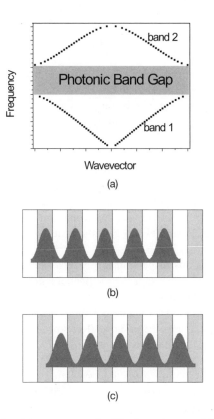

Figure 9.6. Photonic bandgap structure of a 1D photonic crystal (a); schematic illustration of the electric field distribution of the modes associated with band 1 (b) and band 2 (c). In parts b and c the dark region is the layer of high refractive index material.

etc.). For example, the group velocity, v_g, can be calculated from the photonic band structure by using the relation:

$$v_g = d\omega/dk \tag{9.3}$$

This group velocity is strongly modified in a photonic crystal because of the presence of a highly anisotropic and complicated band structure. It can show a wide variation from zero to values significantly slower than the vacuum speed of light, being in general dependent both on the frequency of light in relation to the bandgap and on the direction of its propagation. For example, the group velocity for light of frequency near the bandgap will be very low, being zero in the bandgap region (Imhof et al., 1999). Thus a photonic crystal medium can be used to manipulate the speed of propagation of an optical wave packet (and its energy). This feature can be used to enhance optical interactions that take longer time to manifest.

Another important feature is the group velocity dispersion, β_2, which can be defined as

$$\beta_2 = \frac{d^2k}{d\omega^2} \tag{9.4}$$

Thus β_2 is given by the curvature of the band structure. In the electronic band structure the curvature d^2E/dk^2 relates to the effective mass of the electron. Thus β_2, as defined by Eq. (9.4), can be seen to be inversely proportional to an effective mass of photon. This dispersion relates to the dispersion (variation) of the group velocity as a function of frequency of light. Thus it provides information on how a pulse of polychromatic light (containing many frequency components) will spread (broaden), because different frequency components will travel with different group velocities. Consequently, the group velocity dispersion quantifies the relative phase shift of the frequencies in a wave packet. An anomalous dispersion can be encountered near the bandgap, where the effective photon mass ($1/\beta_2$) is negative below and positive above the gap. Inside the stop gap, the group velocity dispersion shows a rapid decrease to zero.

Because of the anomalous dispersion, many new phenomena are predicted and have been verified. One is the *superprism* phenomenon, which is extraordinary angle-sensitive light propagation and dispersion due to refraction (Kosaka et al., 1998). Figure 9.7 illustrates the refraction behavior in a superprism photonic crystal by comparing it to that occurring in a normal prism. Hence a superprism exhibits the two dramatically enhanced properties of a prism, which are (i) superdispersion separating the various wavelength components over a much wider angle as demonstrated in Figure 9.7 and (ii) ultra-refraction producing a large increase in the angle of refraction (outgoing beam) produced only by a small change in the incidence angle. Thus Kosaka et al. (1998) reported a swing of the propagation beam from −70° to +70°, with a slight change in the incidence angle within ±7° in a 3D photonic crystal fabricated on a Si substrate (Figure 9.8). This effect is thus two orders of magnitude stronger than that in a conventional prism and can be used for beam

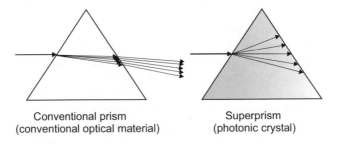

Figure 9.7. Diffraction in a superprism compared to that in a conventional prism.

steering over a considerably wide angle. Another observation is that this light path can show a negative bending, thus exhibiting negative refraction (Luo et al., 2002), even though the medium does not possess a negative refractive index.

Another observed feature derived from the anomalous group velocity dispersion behavior in a photonic crystal is the *self-collimating* phenomenon which shows collimated light propagation, insensitive to the divergence of the incident beam (Kosaka et al., 1998). In the same photonic crystal, both self-collimating and lens-like divergent propagation can be realized, depending on the band structure dispersion at the chosen wavelength and the direction of propagation. This self-collimating phenomenon is independent of light intensity (hence not derived from intensity dependent self-focusing) and is of significance to optical circuitry where beam propagation without divergence is useful.

Anomalous Refractive Index Dispersion. In nonabsorbing spectral regions, dielectric medium exhibits a steady decrease of refractive index with increasing wavelength (decreasing frequency). This is called normal refractive index dispersion. Near an optical absorption, the refractive index shows a dispersive behavior (derivative type curve) which is called anomalous dispersion of refractive index.

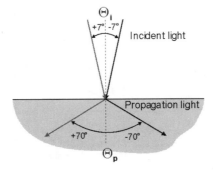

Figure 9.8. Ultra-diffraction in a photonic crystal, where a change of ±7° in the incidence angle produces a large change of ±70° in propagation of the refracted beam.

Photonic crystals exhibit anomalous dispersion of the effective refractive index near their high-frequency band edge. This feature is also related to the anomalous group velocity dispersion discussed above. Figure 9.9 (Markowicz et al., 2004) shows for a 3D photonic crystal the calculation of the effective index versus the normalized frequency, $f_n = a/\lambda$, where a is the lattice constant. This anomalous dispersion is not associated with absorption and thus does not involve any loss by absorption. It will be discussed in Section 9.6 on nonlinear photonic crystals that such anomalous dispersion can be utilized to produce phase-matching for efficient generation of second or third harmonic of light from the incident fundamental wavelength. Phase-matching requirement dictates that the phase velocities (c/n) for the fundamental and the harmonic waves are the same, so that power is continuously transferred from the fundamental wave to the harmonic wave, as they travel together in phase. This aspect is discussed in detail in that section.

Microcavity Effect in Photonic Crystals. A photonic crystal also provides the prospect of designing optical micro- and nanocavities embedded into it by creating defects (e.g., dislocations, holes) into it. Furthermore, by tailoring the size and the shape of these defect sites, microcavities of tunable sizes and dimensions can be created. These defect sites create defect associated photon states (defect modes) in the bandgap region, as shown in Figure 9.10. These defect modes are analogous to the impurity (dopant) states between the conduction and the valence bands of a semiconductor. For the purpose of illustration, let us assume a defect that is spherical in shape (for example, packing spheres missing in a self-assembled colloidal crystal described in Section 9.4). This spherical microcavity can support optical modes (allowed photonic states) for which the diameter, d, is given as

$$d = n\lambda/2 \tag{9.5}$$

The lowest mode correponds to $n = 1$. It can be seen that as the dimension of the cavity gets smaller (i.e., volume gets smaller), the number of cavity modes become

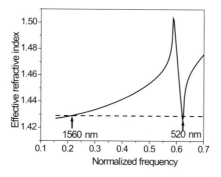

Figure 9.9. The plot of effective refractive index versus normalized frequency, showing anomalous dispersion at ~ 520 nm.

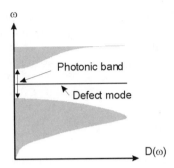

Figure 9.10. Schematic representation of defect modes in bandgap region. $D(\omega)$ is photon density of state at frequency ω.

significantly smaller. If an emitter is embedded in this microcavity, its emission can be cavity enhanced only if the emission is narrower than the cavity resonance peak and, at the same time, its frequency (wavelength) is matched to a cavity mode.

The cavity optical response is often described in terms of a quality factor, Q, defined as $Q = \omega/\Delta\omega$, where $\Delta\omega$ is the spectral width of a cavity mode. A cavity with desirable large enhancement of narrow emission requires a very high Q. Photonic crystals allow the achievement of a high-quality factor (Q) and formation of an extremely low mode volume nanocavity to provide the benefit of high selectivity of wavelengths and a large enhancement of the resonant electromagnetic field (cavity mode) within the cavity. This effect has already been used to demonstrate resonant cavity-enhanced photodetectors, light-emitting diodes, and low-threshold lasers (Temelkuran et al., 1998; Painter et al., 1999; Loncar et al., 2002). The convenient use of a photonic crystal to create high-Q microcavities has provided a strong impetus to study cavity quantum dynamics that produce novel optical phenomena (Haroche and Kleppner, 1989).

Two specific microcavity effects presented here are the *Purcell effect and photon recycling*. The Purcell effect is simply the enhancement of spontaneous emission under the following conditions: (i) The spontaneous emission spectrum is narrower than the cavity resonance peak, and (ii) the emission frequency matches one of the cavity mode frequencies (Purcell, 1946). The cavity enhancement factor, often referred to as Purcell factor, f_P, can be shown to be proportional to Q/V, where Q is the cavity quality factor and V is the cavity volume. Hence, a great deal of effort has been placed in enhancing Q and reducing V. Q values in excess of 10^4 with $V \approx$ 0.03 μm^3 in InGaAsP semiconductor have been achieved (Okamoto et al., 2003).

Photon recycling refers to the process of reabsorption of an emitted photon, followed by re-emission. This process can lead to an apparent increase in internal efficiency. A good illustration is provided by the work of Schnitzer and Yablonovitch (Schnitzer et al., 1993), where each photon was re-emitted for an average of 25 times, producing a total external efficiency of 72%. Photon recycling increases the photon lifetime and consequently slows down the response time of the device.

The example provided here is of a low-threshold photonic crystal laser (Loncar et al., 2002). Loncar et al. (2002) fabricated a photonic crystal microcavity laser by incorporating fractional edge dislocations in InGaAsP quantum well active materials. Very low threshold pumping at powers below 220 μW produced lasing action in the photonic crystal.

9.4 METHODS OF FABRICATION

A wide variety of methods have been used to fabricate photonic crystals. Some of them are more suitable for the fabrication of 1D and 2D photonic crystals, whereas others are useful when one needs to localize photons in three dimensions. Some of these methods are discussed here.

Self-Assembly Methods. Colloidal self-assembly seems to be the most efficient method for fabrication of 3D photonic crystals. In this method, predesigned building blocks (usually monodispersed silica or polystyrene nanospheres) spontaneously organize themselves into a stable structure. Although the currently available colloidal assemblies do not have a full photonic bandgap in the optical wavelengths because of their low index contrast, they do provide a template that can be infiltrated with material of higher refractive index (Vlasov et al., 2001).

A number of techniques are available for colloidal assembly fabrication. A widely used technique for creating colloidal crystals is gravity sedimentation. Sedimentation is a process whereby particles, suspended in a solution, settle to the bottom of the container, as the solvent evaporates. The critical point here is finding the proper conditions for liquid evaporation so that the particles form a periodic lattice.

Sedimentation under gravity is a slow process, taking as long as four weeks to get good crystals. If the process is accelerated to a few days, then crystallization produces a polycrystalline structure with many defects, as shown by the AFM image in Figure 9.11. The size of crystallites varies and depends on the quality of building blocks, time of crystallization, temperature, humidity, and so on.

Another self-assembly technique, introduced by Xia and co-workers (Park and Xia 1999; Gates et al., 1999), is called the cell method. The fabrication method is represented in Figure 9.12. An aqueous dispersion of spherical particles is injected into a cell formed by two glass substrates and a frame of photoresist or Mylar film, placed on the surface of the bottom substrate. One side of the frame has channels that can retain the particles, while allowing the solvent to flow through. The particles settle down in the cell to form an ordered structure (usually the fcc structure).

This technique is particularly useful for fabrication of thin polystyrene photonic crystals in water. Their thickness usually does not exceed 20 μm, and lateral dimensions are of the order of 1 cm. The method can be used when one wants to investigate emission from dye-incorporated photonic crystals. A disadvantage of this method is that when a crystal dries, defects are formed inside its structure. This is probably because particles are highly charged, and not all particles that form a crys-

9.4 METHODS OF FABRICATION

Figure 9.11. The AFM image of 450-nm silica beads produced by sedimentation method, using crystal growth for 5 days.

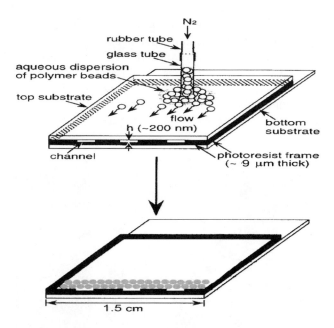

Figure 9.12. The cell method for self-assembling of colloidal crystals. From Gates et al. (1999), reproduced with permission.

tal are in physical contact with each other. When the water evaporates, the crystals form large domains, with cracks forming between these domains.

When one needs a larger area of dry crystalline structure (e.g., polystyrene/air), the vertical deposition method provides better results (Vlasov et al., 2001). This method produces a very long range order. Strong capillary forces at the meniscus region induce crystallization of colloidal spheres in a 3D ordered structure as shown in Figure 9.13. When the meniscus is swept across a vertically placed substrate, a colloidal crystal is formed. This movement of the meniscus can be realized, for example by solvent evaporation. In addition, as suggested by Zhu et al. (1997), one can apply gradient of temperature for large spheres ($d > 400$ nm). Figure 9.14 shows the scanning electron microscope (SEM) image of a sample, produced by this method. A highly ordered three-dimensional photonic crystal can readily be seen.

Another variation of vertical deposition is the convective self-assembly method used by Jiang et al. (1999). Wostyn et al. (2003b) have used this method to produce highly ordered face-centered close-packed arrays of polystyrene spheres.

As discussed in Section 9.1, a large refractive index contrast is required to create a full bandgap, where the density of photon states goes to zero at the gap. Thus, an area of major activity world wide is to use different fabrication and processing methods to increase the refractive index contrast. Commonly used methods involve infiltration of the void space (hollow domains) with high refractive index materials. Both dry (gas phase) and wet (chemical) infiltration techniques have been utilized for this purpose. An example is the infiltration of an opal structure (3D colloidal assembly) of silica with a high refractive material, GaP, at our Institute. The infiltration utilizes an MOCVD approach using metal organic precursors as discussed in Chapter 7. The silica spheres can then be etched out, using HF acid, to produce an inverse opal structure consisting of GaP and air as the alternating dielectric.

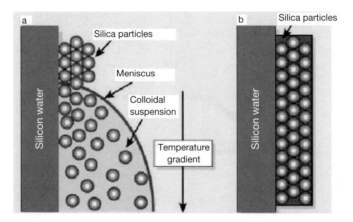

Figure 9.13. Vertical deposition method for crystallization of colloidal spheres. (Reproduced with permission from Joannopoulos, J. D., *Nature* **414**, 257–258 (2001).)

Figure 9.14. Scanning electron microscope (SEM) image of the polystyrene/air photonic crystal, prepared by the vertical deposition method.

In the wet method, chemical reaction through solution-phase infiltration of the reactants in the void space is used to make materials with high refractive index. This method has been used to prepare nanocrystals of semiconductors or high refractive index rigid polymers in the void. An example of liquid-phase infiltration is the formation of CdS in the void space of the closed-packed silica (Romanov et al., 1997).

Two-Photon Lithography. Two-photon lithography, discussed in Chapter 11, has been used for 3D photonic crystal fabrication. This technique utilizes the fact that certain materials, such as polymers, are sensitive enough to two-photon excitation to trigger chemical or physical changes in the material structure, with nanoscale resolution in three dimensions (Cumpston et al., 1999).

In two-photon excitation, discussed in Chapter 2, the simultaneous absorption of two identical photons in the infrared region (e.g., 800 nm) can be used for photofabrication. This topic is discussed in more detail in Chapter 11. When tightly focused in a photosensitive material, femtosecond laser pulses initiate two-photon polymerization and produce structures with a resolution down to 200 nm. Since the two-photon absorption is confined to a tiny volume, scanning the focus within the material can produce three-dimensional microscale patterns. Among the structures that have already been fabricated by two-photon lithography, the logpile-type photonic crystal is the most common one (see Figure 9.15).

E-Beam Lithography. Electron beam lithography is a method that enables one to create various photonic crystals with extremely high resolution. It is a relatively complicated method, because it includes many variables. Its main disadvantage is its high cost. In this method, the sample (wafer) is covered with an electron-sensi-

Figure 9.15. SEM image of a photonic crystal fabricated by two-photon lithography. From Cumpston et al. (1999), reproduced with permission.

tive material called *resist*. The material, used as resist, undergoes a substantial change in its chemical or physical properties, when it is exposed to an electron beam (Knight et al., 1996, 1997). The beam position and intensity are computer-controlled, and electrons are delivered only to certain areas to get the desired pattern. After exposition, a part of the resist is dissolved away and the sample can be further processed with etching procedures to get the final crystalline structure. Electron beam lithography is mostly used for fabrication of 2D photonic crystals. An example of a 2D photonic crystal fabricated by this technique is shown in Figure 9.16.

Etching Methods. These methods are more suitable for the fabrication of two-dimensional photonic crystals and have been used for semiconductors. These methods utilize marking of a planar pattern of unwanted areas on the surface of a semi-

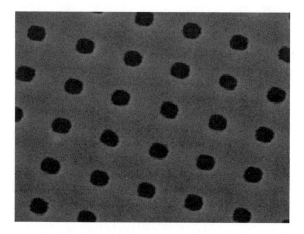

Figure 9.16. SEM image of a photonic crystal fabricated by E-beam lithography. From Beetz et al. (2002), reproduced with permission.

conductor, using a lithographic technique such as E-beam lithography. These marked areas are then etched to create holes. The two methods used are (Mizeikis et al., 2001):

(i) *Dry Etching.* An example is reactive-ion etching (RIE), which utilizes reactive ions generated by plasma discharge in a chlorine-based ($SiCl_4$ and Cl_2) or fluorine-based (CHF_3, CF_4, C_2F_6, and SF_6) reactive gas. These ions are accelerated toward the sample surface under an electric field. This dry etching provides a good control over the hole size, but has a limited maximum etching depth. The method has been used for many semiconductors, such as GaAs, AlGaAs, and Si.

(ii) *Wet Etching.* An example is electrochemical etching that has also been used for many semiconductors. Electrochemical etching of Si to produce microporous silicon photonic crystal is an example. In this case, a pre-pattern with etch pits was first created on the front face of a silicon wafer by using lithographic patterning and subsequent alkaline chemical etching using KOH solution. The wafer was then mounted in an electrochemical cell and electrochemically etched using an HF solution. The pre-etched pits form nucleation centers for electrochemical etching. The advantage provided by an electrochemical etching method is that deep holes can easily be produced.

Another approach to produce a 2D photonic crystal using an electrochemical method is provided by anodic oxidation of aluminum in acidic solutions, which is known to produce highly ordered porous honeycomb structure in resulting alumina (Al_2O_3) that consists of a closely packed array of columnar hexagonal cells (Keller et al., 1953; Almawlawi et al., 2000). The pore size and density of the pores in alumina can be precisely controlled by selecting the anodization condition (choice of acid, applied voltage) and pre-texturing of the aluminum surface with an array of nanoindentation using a SiC mold (i.e., creating pits) (Masuda and Fukuda, 1995; Masuda et al., 2000).

This porous Al_2O_3 can be used as a template to form other photonic media that can be grown in the pores, and Al_2O_3 can subsequently be etched out. Filling of a polymer in the pores can produce a negative replica that can subsequently be used for growth of other periodic structures such as those of semiconductors. For example, Hoyer et al. (1995) used electrochemical method to grow CdS.

Holographic Methods. Holographic methods, which utilize interference between two or more coherent light waves to produce a periodic intensity pattern, have been used to produce a periodic photoproduced photonic structure in a resin (photoresist). Here, the initial laser beam is split into several beams and allowed to overlap in the resin at angles predetermined by the desired periodicity. The simplest is the fabrication of a one-dimensional periodic structure (a 1D photonic crystal or a Bragg grating) produced by overlap of two beams, where the angle between the two beams determines the periodicity. This method has been in practice for a long time. Recent developments are the use of photopolymerizable (or photocrosslinkable) medium containing inorganic nanoparticles (as TiO_2, metallic nanoparticles) or liquid crystal nanodroplets (described in Chapter 10), in which the nanoparticles prefer domains that are not photomodified. In this case, the bright spot in the intensity pattern not only produces a photomodified region, but also aligns the nanoparticles in the non-photomodified region to enhance the refractive index contrast. This method has been utilized at our Institute, in collaboration with the Air Force Research Laboratory at Dayton, to produce 1D photonic crystals as shown in Figure 9.17. Interference of more than two beams have been utilized to produce 2D and 3D photonic crystals.

A 2D hexagonal photonic lattice was fabricated in a thin layer of photoresist by interference of three beams of $\lambda = 325$ nm (Berger et al., 1997). The resist layer with the pattern was used as a mask for reactive-ion etching to produce a 2D photonic crystal in GaAs with the depth of 3 μm. Shoji and Kawata (2000) used five overlapping beams from a continuous-wave HeCd laser at the wavelength of 442 nm to form a 2D triangular structure. Interference of three beams formed a 2D triangular structure, while the overlap of two additional beams with this structure produced modulation along the third dimension. The resulting 3D hexagonal light intensity pattern was transferred to a photopolymerizable resin to produce a 3D photonic crystal. The holographic method, utilizing interference of multiple beams, offers the advantage that highly periodic structures can be fabricated in a single laser exposure, thus avoiding step-by-step fabrication. However, a disadvantage of this approach may be that the refractive index contrast achieved is not large.

The holographic method can also be used to fabricate an electrically switchable polymer-dispersed liquid crystal photonic bandgap material (Jakubiak et al., 2003). Jakubiak et al. (2003) used polymer-dispersed liquid crystal photonic bandgap ma-

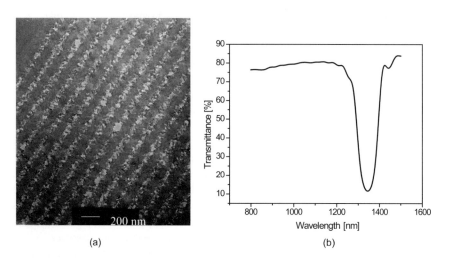

Figure 9.17. Holographic method used to produce one-dimensional photonic crystal using polymer-dispersed liquid crystal. (a) TEM picture, (b) transmission spectrum.

terials to demonstrate distributed feedback lasing action from an organic chromophore.

9.5 PHOTONIC CRYSTAL OPTICAL CIRCUITRY

In recent years there has been a growing number of research activities in creating point and extended defects in photonic crystals to produce microcavities and to introduce waveguiding structures, in order to build photonic crystal-based optical circuitry. Point defects have been created in a colloidal crystal by doping with impurity spheres of different volumes (Pradham et al., 1996). Conventional lithographic techniques have also been used to introduce defects into inverted opal structures (Jiang et al., 1999).

Braun and co-workers (Lee et al., 2002) used two- and three-photon polymerization to introduce waveguide structures into self-assembled photonic crystals. This multiphoton polymerization, involving the same type of approach described under two-photon lithography in Section 9.4, was used to fabricate high-resolution 3D patterns in the interior of colloidal crystals. The three-photon polymerization was found to produce smaller features.

Recently, our research laboratory has utilized two-photon polymerization to produce one- and two-dimensional photonic crystal-based structures. In Figure 9.18 we present the results of our studies on the fabrication of 1X3 photonic crystal splitter type waveguides, and photonic crystal grating structures. Both patterns were created in the interior of a polystyrene colloidal photonic crystal. The colloidal assemblies were prepared by using the vertical deposition technique of ~200-nm polystyrene particles. The resulting crystals were infiltrated with a methanol solution of

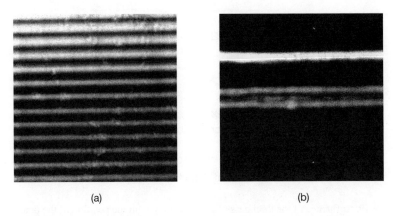

(a) (b)

Figure 9.18. Defect structures created intentionally inside 3D polystyrene photonic crystals. (a) Two-photon fluorescence (using 800 nm excitation) image of the fabricated grating structure. (b) One-photon confocal fluorescence image of the fabricated 1X3 beam splitter in the branching region.

ORMOCER, a hybrid inorganic/organic material, containing about 2% of a photoinitiator. The two-photon polymerizations were performed using ~ 100-fs pulses from a mode-locked Ti:sapphire femtosecond laser. After polymerization, the unpolymerized part of the photosensitive material was removed (washed out) with methanol, and single- as well as two-photon fluorescence techniques were used to perform imaging on the modified colloidal photonic crystals. Both structures presented in Figure 9.18 were imaged about 2 μm inside the crystals.

Wostyn et al. (2003a) introduced a two-dimensional planar defect into a self-assembled colloidal crystal by a combination of convective self-assembly, another variation of the vertical deposition method described by Jiang et al. (1999), and the Langmuir–Blodgett (LB) technique, described in Chapter 8. They used a multilayer colloidal crystal, created by vertical deposition, then used the LB techniques to deposit a single layer of nanospheres of a different size, which was followed by deposition, again, of multilayer colloidal crystal by vertical deposition. The LB deposited layer acted as a planar microcavity that induced the appearance of a localized state, a defect mode, into the forbidden bandgap. The energy of this localized state, between the valence band and the conduction band, depended on the thickness of the defect layer that was controlled by the size of the nanosphere deposited by the LB technique. This dependence is shown in Figure 9.19.

9.6 NONLINEAR PHOTONIC CRYSTALS

This section provides a detailed discussion of some features of nonlinear photonic crystals (Slusher and Eggleton, 2003). A standard application of nonlinear optics is frequency conversion of optical radiation and, in particular, the generation of vari-

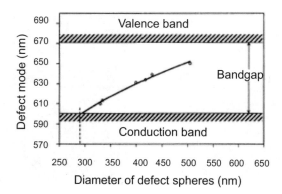

Figure 9.19. Influence of the thickness of the defect layer on the position of the defect mode in the photonic band gap. The colloidal crystal consists of 290 nm large spheres. Note that the Y axis represents wavelength instead of energy, so that the conduction band is below the valence band. On the energy axis, the conduction band is above the valence band. From Wostyn et al. (2003), reproduced with permission.

ous harmonics (Prasad and Williams, 1991). We illustrate the usefulness of a photonic crystal in third harmonic generation, from our work (Markowicz et al., 2004). Third harmonic generation (THG) is a very useful technique that can convert the coherent output of infrared lasers to shorter wavelengths in the visible and near ultraviolet. However, the smallness of the third-order optical susceptibility and the strong natural dispersion of the refractive index of most materials have prevented the practical utilization of one-step third harmonic generation. Therefore, in practice, one employs a cascaded two-step process to produce the third harmonic with high conversion efficiency (Shen, 1984, Boyd, 1992). However, it would be more convenient, as well as conceptually simpler, to find the right conditions for the implementation of direct one-step third harmonic generation. We found a dramatic enhancement in the efficiency of third harmonic generation using a 3D polystyrene photonic crystal medium.

Having perfect phase-matching of the pump and generated signal is essential for efficient nonlinear frequency conversion. One way to achieve this condition for third harmonic generation is to use an anomalous dispersion region where the refractive index of the medium decreases with the increase in the optical frequency. In bulk media, anomalous dispersion is typically accompanied by extremely high absorption, which prevents the useful implementation of this idea. However, in a photonic crystal, anomalous dispersion created by the periodic structure is not accompanied by loss, and strong third harmonic generation can be obtained. The crystals used in our experiments are fabricated by close-packing of colloidal polystyrene microspheres. Even though polystyrene possesses only a weak third-order nonlinear susceptibility, we observe a strong third harmonic generated beam.

Two photonic crystals consisting of polystyrene spheres, formed into a face-centered cubic (FCC) crystal structure, were used in our experiments. One crystal, which we call the blue crystal, was formed of 198-nm-diameter spheres; the other crystal (the green crystal) was formed of 228-nm-diameter spheres.

The wavelength of the pump beam was tunable from 1.1 μm to 1.6 μm by using a femtosecond amplified pump optical parametric generator. When the blue and the green crystals were illuminated by focused laser beams in the [111] crystallographic direction at wavelengths of 1.36 μm and 1.56 μm, respectively, a bright beam at the corresponding third harmonic wavelength was observed in both transmission and reflection. Figure 9.20 shows the wavelength dependence of the intensity of the third harmonic light generated in the blue and green polystyrene photonic crystals. The linear transmittance of the crystals in the [111] crystallographic direction is also shown in Figure 9.20. For both crystals, the maximum generation occurs when the third harmonic wavelength is tuned to the short-wavelength edge of the bandgap.

At low frequency, the effective refractive index depends only on the material dispersion, because the wavelength is much longer than the spatial periodicity of the structure. As the frequency is tuned toward the bandgap, the effective index deviates strongly from the natural refractive index of polystyrene, leading to poor phase-matching. However, phase-matching is restored at some point in the anomalous dispersion region of the bandgap, so that when light at this frequency propa-

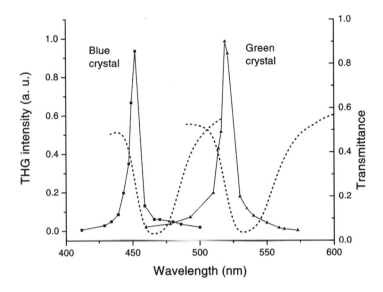

Figure 9.20. Intensity of the generated third harmonic signal plotted as a function of wavelength for the blue (solid square) and green (triangles) crystals. The dotted lines represent linear transmittance of the photonic crystal samples in the vicinity of their bandgaps.

gates in the forward direction in the crystal, its wavevector is phase-matched with that of the low-frequency pump. Thus a strong third harmonic generation could be observed due to phase-matching derived from the anomalous dispersion near the high-energy band edge where the third harmonic is produced. The calculated effective index dispersion in the region of the fundamental and third harmonic has been shown in Figure 9.9. In the same figure the dashed line shows the phase-matching.

Another type of nonlinear optical effect studied by our group is a two-photon excited emission. Even though two-photon absorption is like third harmonic generation, a third-order nonlinear optical process, two-photon emission is incoherent. In other words, this type of upconversion does not require phase-matching. We reported modification of two-photon excited, upconverted emission spectra from a highly efficient dye incorporated into photonic crystals that exhibit stop-gaps ranging in size from 1% to 3% of the gap center frequency (Markowicz et al., 2002). The photonic crystals were fabricated from polystyrene spheres with a diameter of 200 nm, and then infiltrated with dye. Coumarin 503 dye was chosen for the infiltration because the emission spectrum of this dye fits into the stop-gap of the transmission spectrum of the polystyrene photonic crystal and because its high fluorescence quantum efficiency (0.84) leads to mostly radiative decay.

We obtained two-photon excited fluorescence spectra of the dye in a photonic crystal by exciting the dye with a mode-locked Ti:sapphire laser at a wavelength of 800 nm (80-fs pulse width). The direction of the laser light was perpendicular to the surface of the sample. We collected spectra in the reflection mode.

In the two-photon excited emission spectra of coumarin 503, besides filter effects similar to those reported for transmission of external plane waves (Megens et al., 1999, Yamasaki and Tsutsui, 1998), a sharp maximum also appears (Figure 9.21). Although the concentration of solvated dye molecules was relatively high, we did not observe any changes in spectral shape, attributable to the dipole–dipole transfer of excitation. The emission spectrum of the dye solution, without a photonic crystal, overlapped well with the rescaled spectrum taken for the solution with a 100 times lower dye concentration. In both spectra, the sharp maximum did not appear. The maximum appears to be a result of amplification of light inside the photonic crystal. The position of the sharp maximum depends on the position of the stop-gap. Outside the stop-gap, the shapes of the emission spectra from coumarin 503 were similar for the polystyrene–methanol and polystyrene–DMSO structures. Both spectra also overlapped with the rescaled reference spectra outside the stop-gap region. The reference spectrum was measured for coumarin 503 placed in a cell with the same thickness as the polystyrene structure (8 μm).

We observed that the attenuation and amplification of light in emission spectra exist simultaneously. To check the position of the maximum with respect to the stop-gap more precisely, we performed the following experiment: Before the two-photon excited upconverted emission spectrum of coumarin 503 was measured, the transmission spectrum of exactly the same region of the sample was recorded. The results of this experiment show the amplification in the two-photon emission spectra appearing at the edge of the attenuation in the transmission spectra of external

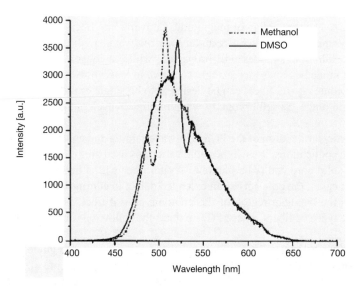

Figure 9.21. Two-photon excited upconverted emission spectra of coumarin 503 in polystyrene–methanol and polystyrene–DMSO photonic structures.

plane waves. This means that the amplification of two-photon excited upconverted emission took place at the edge of the (1,1,1) stop-gap.

9.7 PHOTONIC CRYSTAL FIBERS (PCF)

Photonic Crystal Fiber. A novel kind of photonic crystal, which is also an optical fiber, is a Photonic Crystal Fiber (PCF). This fiber features a periodic modulation of refractive index in its clad (Knight et al., 1997; Cregan et al., 1999). Typically, a PCF is a strand of glass with an array of microscopic air channels running along its length. These are periodic tubular structures made by melting and pulling a close stack of bare fibers to the size whereby the periodicity is matched with the wavelength of the stop-gap. The PCF, called holey fibers, utilize hollow core fibers. Hence, the two-dimensional structure produced consists of periodic array of air holes in glass. Fabrication of PCFs is a complex process, which includes preparation of a preform with precisely stacked tubes and rods. Their shape and distribution determine the final structure of the fiber. The preform is processed and stretched under careful conditions to fabricate the PCF. The balance of pressure within the hole against viscous forces of the material should be precisely maintained.

Light is guided in PCF at structural defects—points where the periodicity is interrupted—and its mechanism is based on the photonic bandgap effect. PCFs provide greatly enhanced control of dispersion, birefringence, and modal shape. Perhaps the most remarkable example is a hollow core fiber in which light is guided in a hollow tube by a photonic bandgap.

PCFs can be engineered to display a number of remarkable properties such as single-mode guidance at all wavelengths, novel dispersion properties, and mode size tailoring over three orders of magnitude. Conventional fibers can only operate usefully in a single mode, over a relatively narrow wavelength range. For shorter wavelengths, additional modes will propagate, while at longer wavelengths, the fundamental mode becomes increasingly sensitive to losses at bends in the fiber. In contrast, photonic crystal fibers can be designed to be single mode over an unlimited wavelength range, a useful property when multiple wavelengths must be carried in the fiber.

Most practical utilization of the PCF is in waveguiding and enhancing nonlinearity. For these applications, the two possible variations used are (i) a photonic crystal fiber with a solid core and (ii) a photonic crystal fiber with a hollow (air) core. In both of these cases, the core acts as an extended defect to confine and guide light of frequency in the bandgap region of the photonic crystal fiber. Figure 9.22 shows both of these types of fibers. These PCFs exhibit the following useful features.

Efficient Waveguiding. PCF of both solid and hollow cores can guide light efficiently as single-mode fibers, independent of the wavelength or core size. This is not the case with a conventional fiber. In the case of a solid-core PCF, the waveguiding is produced by light trapping that can involve total internal reflection due to high effective refractive index of the core as well as the existence of a photonic

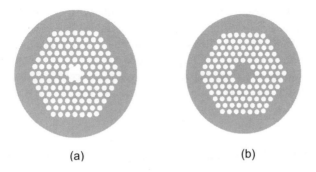

Figure 9.22. Photonic crystal fibers with a solid core (a) and with a hollow core (b).

bandgap in the surrounding PCF medium (acting as a cladding region). In the case of PCF with an air (hollow) core, guiding in the air core solely relies on trapping of light at the frequency of photonic bandgap of the PCF. The presence of this bandgap allows light to be reflected from the surface of the PCF surrounding the airhole. This leads to the exciting prospect of guiding light through air holes, with minimum loss and group velocity dispersion, the latter broadening a propagating pulse in an ordinary optical fiber (Cregan et al., 1999). This feature cannot be realized in a conventional fiber in which the guiding core must be a material of higher refractive index than that of the cladding region. In the case of a hollow (air) core guiding light, it is just the opposite. Thus, the air core PCF has the potential to outperform the conventional optical fiber by orders of magnitude in optical loss reduction. However, such realization has not yet been achieved, possibly due to defects in PCF structure. Another advantage offered is large bend angles permitted by photonic crystal guiding. In conventional fibers, a large bend angle leads to optical loss from the bend region, because of a relatively small refractive index contrast between the core and the cladding.

Zero Group Velocity Dispersion. As discussed in Section 9.3, group velocity dispersion shows a wide variation range in a photonic crystal, due to a highly complicated band structure. The group velocity dispersion causes a pulse of light to broaden in transmission through a conventional fiber. Specialty fibers have been produced with zero group velocity dispersion near 1550 nm, providing the major impetus to prefer this wavelength region for optical communications. PCF provides the flexibility to manipulate the band structure and adjust the core size, so that it is possible to form a fiber with zero group velocity dispersion at any selected wavelengths between 550 nm and 1.7 μm.

Enhanced Nonlinearity in Solid-Core PCF. In a solid-core PCF, the optical nonlinearity of the core is significantly enhanced by enhancement of the field. The enhanced Kerr effect (intensity-dependent refractive index, discussed in Section 9.3) means that a nonlinear phase shift more than an order of magnitude larger can be built up in a PCF compared to a conventional fiber. An example of the application of this nonlinearity is efficient broadband continuum generation in PCF using a train of pulses from a mode-locked laser (Ranka et al., 2000). The continuum generation occurs from a rapidly oscillating nonlinear phase shift due to the rapid intensity changes in the mode-locked pulses (producing picosecond pulses). Wavelengths covering the range from 400 nm to 1600 nm can be generated in just a few tens of centimeters of PCF.

9.8 PHOTONIC CRYSTALS AND OPTICAL COMMUNICATIONS

An important area of potential application for photonic crystals is optical communications. Here photonic crystals can be used to produce a light source consisting of low-threshold lasers and photodetectors with enhanced sensitvity. This topic has already been described in Section 9.3.

As described in Sections 9.3 and 9.5, a photonic crystal with a well-defined defect channel can be used to confine light in the defect region and guide it. Guiding through well-controlled defect channels allows sharp bending of light without significant optical loss. Let us examine this application utilizing a three-dimensional photonic crystal when light falls in the forbidden wavelength range. It cannot propagate directly through this crystal. But what if an extended defect is made in this photonic crystal. Then light will shy away from the region where it cannot propagate, and it will transmit along the extended defect. The extended defect region can have sharp bends. The advantage of this is that, as opposed to a regular waveguide, one can bend light by 90°. In other words, one can achieve a very

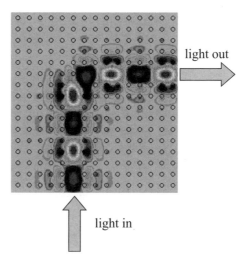

Figure 9.23. Theoretical modeling of field distribution around a defect path way, producing a sharp bend in light propagation. From Mekis et al. (1996), reproduced with permission.

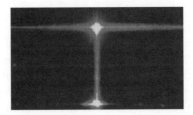

Figure 9.24. A helium–neon laser glows as it enters the waveguide at the bottom and hits the photonic crystal at the top, where it undergoes a sharp bend to propagate in the left and right directions. From Parker and Charlton (2000), reproduced with permission.

sharp bend which is not possible with a waveguide. If one wants to make an ordinary channel waveguide to produce a sharp bend, the light will leak with a considerable optical loss.

The field distribution around the defect can be theoretically modeled as shown in Figure 9.23. It shows the direction of light entering and going along the defect path and coming out with a sharp bend. Figure 9.24 shows another illustration of light propagating from the bottom and going in two different directions at a junction along the defect. One can make these defects in different topologies to guide light and create optical circuits involving sharp bends that provides opportunity for light manipulation.

Using a photonic crystal fiber, one can achieve zero group velocity dispersion over a broad range of wavelengths as described in Section 9.7. Hence, the carrier frequency for optical communication does not have to be limited to 1.3- and 1.55-μm regions.

The narrow-band filtering property of a photonic crystal, together with the superprism effect, is useful for wavelength division multiplexing (WDM), whereby many optical channels of closely spaced optical frequencies are separated over a wide angle range. Enhanced nonlinear phase shifting of central filter frequency can allow dynamic control (optical switching) for WDM channel routing.

Finally, a photonic crystal platform provides an opportunity for dense integration of emitter, receiver, amplifier, transmitter, and routers on the same chip. Thus both active and passive functions can be integrated to produce a true photonic chip.

9.9 PHOTONIC CRYSTAL SENSORS

Asher and co-workers in their pioneering work have used colloidal arrays to demonstrate a novel chemical and biosensing scheme (Holtz and Asher, 1997; Holtz et al., 1998; Lee and Asher, 2000; Reese et al., 2001). They used a three-dimensional periodic structure of colloidal crystal arrays (CCA) of highly charged polystyrene spheres of 100-nm diameter. Electrostatic interactions between these charged spheres lead them to self-assemble into a body-centered or a face-centered cubic structure.

For chemical and biochemical sensing, Asher's group fabricated a CCA of polystyrene spheres (of diameters ~100 nm) polymerized within a hydrogel that swells and shrinks reversibly in the presence of certain analytes such as metal ions or glucose. For this purpose, the hydrogel contains a molecular-recognition group that either binds or reacts selectively with an analyte. The result of the recognition process is a swelling of the gel owing to an increase in the osmotic pressure which, in turn, leads to a change in the periodicity (separaction between the spheres). As a result, the diffraction and scattering conditions changes to a longer wavelength. Results by Asher's group suggest that a mere change of 0.5% in the hydrogel volume shifts the diffracted wavelength by ~1 nm.

For the detection of metal ions such as Pb^{2+}, Ba^{2+}, and K^+, Holtz and Asher (1997) copolymerized 4-acryloaminobenzo-18-crown-6 (AAB18C6) into polymerized crystalline colloidal array (PCCA). This crown ether was chosen because of its selective binding ability with Pb^{2+}, Ba^{2+}, and K^+. In these cases, the gel swelling mainly results from an increase in the osmotic pressure within the gel.

For glucose sensing, Asher and co-workers (Holtz and Asher, 1997; Holtz et al., 1998) attached the enzyme, glucose oxidase (GO_x), to a polymerized crystalline colloidal array (PCCA) of polystyrene. For this purpose, the PCCA was hydrolyzed and biotinylated. This PCCA was polymerized from a solution containing ~ 7 wt% polystyrene colloidal spheres, 4/6 wt% acrylamide (AMD), and 0.4 wt% N,N'-methylenebisacrylamide (bisAMD), with water constituting the remaining fraction. This hydrogel was hydrolyzed in a solution of NaOH and was then biotinylated with biotinamidopentylamine, which was attached using a water-soluble carbodiimide coupling agent. Avidinated glucose oxidase was then directly added to the

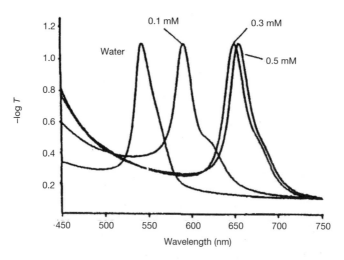

Figure 9.25. Transmission spectra of a polymerized crystalline colloidal array photonic crystal glucose sensor. The originate is given as $-\log T$, where T is the transmittance. From Holtz and Asher (1997), reproduced with permission.

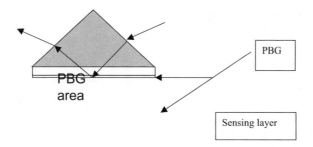

Figure 9.26. A sensor based on total internal reflection polarization in a PBG. From Nelson and Haus (2003), reproduced with permission.

PCCA. A glucose solution prepared in the air causes PCCA to swell and produces a red shift in the diffraction wavelength as shown in Figure 9.25.

Nelson and Haus (2003) have developed a chemical/biological sensor using a dielectric stack or metallo-dielectric stack. The new device shown in Figure 9.26 utilizes a photonic crystal to strongly localize light at the total internal reflection inteface in a Kretschmann-like geometry. This geometry has been discussed in Chapter 2. As the location of the photon localization is the last layer exposed to air, a sensing material may be placed there which acts to change the optical properties of the cavity in response to the presence of an analyte to be detected. Figure 9.27 displays the electric field as the layer thicknesses of two dielectrics are adjusted. The optimal design is chosen to maximize the local-field enhancement at the interface. Either the P- or S-polarization are affected based on the design parameters chosen.

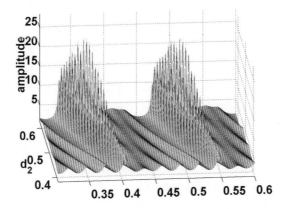

Figure 9.27. The field at the last interface as the thickness of each layer is adjusted. The resonances display the amplitude of the electric field. From Nelson and Haus (2003), reproduced with permission.

9.10 HIGHLIGHTS OF THE CHAPTER

- Photonic crystals are ordered nanostructures in which two dielectric media with different dielectric constants or refractive indices are arranged in a periodic form.
- When the periodicity matches the Bragg diffraction condition for certain range of wavelength of light, light in this wavelength range is reflected if incident externally, or trapped if generated internally.
- Diffraction leads to photon localization; that is, photons in this wavelength range cannot propagate through the photonic crystal.
- If the refractive index contrast (ratio of the refractive indices of the two dielectric media) is very large, complete photon localization occurs. This type of photonic crystal is often called photonic band gap material.
- If the refractive index contrast is not large, only pseudo-localization occurs, in which case the bandgap is often termed as pseudo-gap or stop-gap.
- Photonic crystals can be produced in different topologies, with one-, two- or three-dimensional periodicity.
- The band structure of a photonic crystal, defined by the plot of the energies (frequencies) of photons versus their wavevectors, is dependent on the refractive index contrast, the periodicity, and topology of the photonic crystal (the spatial periodic arrangement of the regions of high and low refractive indices).
- Optical properties, such as transmission, reflection, and angular dependence in a photonic crystal, are determined by the band structure.
- Band structure of a photonic crystal can be calculated by using known techniques and available computers programs, either in frequency domain or in time domain.
- Examples of the frequency-domain techniques presented in this chapter are: (i) the plane-wave expansion method, which uses expansion of the electromagnetic field with a plane-wave basis set, and (ii) transfer matrix method in which the space is divided into small discrete cells with coupling to neighboring cells and a transfer matrix relates the field on one side of the unit cell to those on the other side.
- Photonic crystals exhibit local field enhancement near a bandgap that results in low-frequency modes concentrating their energy in the high refractive index regions, while the high-frequency modes concentrate theirs in the low-refractive-index regions.
- The group velocity of photon wave packet is strongly modified in a photonic crystal and shows an anomalous dispersion of the group velocity near the bandgap where it rapidly changes value from a maximum to a minimum (derivative type curve), as the photon frequency is changed, exhibiting a strong dispersion.
- The anomalous dispersion leads to new optical phenomena such as (i) superprism, which produces superdispersion separating the various wavelength

components over a much wider angle, and (ii) ultra-refraction, which only by a small change in the incidence angle produces a large increase in the angle of refraction of the outgoing beam.
- Another manifestation of the anomalous group velocity dispersion is self-collimation whereby collimated light propagation is achieved that is insensitive to the divergence of the incident beam.
- Near their high-frequency band edge, photonic crystals exhibit anomalous dispersion of the effective refractive index whereby the variation of the refractive index versus photon frequency goes from a maximum to minimum as the photon frequency is varied.
- A photonic crystal also provides the prospect of designing optical micro- and nanocavities of different dimensionalities. This would be accomplished, by creating defects (e.g., dislocations, holes) in it.
- The two important microcavity effects are: (i) the Purcell effect, which is the enhancement of spontaneous emission, and (ii) photon recycling, which refers to the process of reabsorption of an emitted photon, followed by re-emission.
- A wide variety of methods have been used to fabricate photonic crystals of different dimensionalities.
- Colloidal self-assembly produces 3D photonic crystal by close-packing of monodispersed silica or polystyrene spheres of appropriate dimensions. This type of crystal is also called *colloidal crystal* and is often referred to as an *opal structure*. The structure produced by removing the packing polystyrene spheres by burning them out or the glass sphere produced by etching them out, is consequently called an *inverse opal structure*.
- Two-photon lithography using two-photon photopolymerization is another method of fabrication of photonic crystals. Another lithographic technique used to fabricate photonic crystal involves E-beam technology.
- Etching methods involving dry-type reactive-ion etching and wet-type electrochemical etching have been used to fabricate photonic crystals.
- Holographic methods involving the interference between two or more coherent light waves have been used to produce a periodic photoproduced photonic structure in a resin (photoresist).
- Fabrication of defect structures in photonic crystals, in order to introduce microcavity function and/or complex optical circuitry, is another area of activities.
- Various techniques such as lithography, multiphoton polymerization, and colloidal self-assembling coupled with the Langmuir–Blodgett technique have been used for fabrication of controlled defects of different shapes and sizes.
- Photonic crystals are suitable media to produce enhancement of nonlinear optical effects.
- The anomalous refractive index dispersion near the high-frequency edge of a stop-gap (or a bandgap) can be used for phase-matching between the fundamental and the harmonic. This produces a significantly enhanced phase-

matched harmonic generation by nonlinear optical interactions. Experimental observation of phase-matched third harmonic generation is presented.
- Another nonlinear optical effect presented here is a two-photon excited up-converted emission, which is an incoherent process that requires no phase-matching. In this case, simultaneous observation of attenuation of emission at one wavelength and amplification at another wavelength is observed as a result of the filtering effect due to the bandgap.
- A photonic crystal fiber is a two-dimensional photonic crystal with an array of microscopic air channels running along its length. An example is a holey fiber drawn from the stacks of hollow tubes, by melting and pulling.
- The presence of a bandgap in a photon crystal fiber, consisting of a hole in the core, produces reflection of light from the surface of the photonic crystal fiber, and this traps light in the core hole. This provides the unique prospect of guiding light through the air hole, with minimum loss.
- Photonic crystal fibers can be designed to produce zero group velocity dispersion for optical communications.
- A photonic crystal fiber, with a solid core made of nonlinear optical material, can produce enhanced optical nonlinearity in the core.
- Optical communications are an important potential application for photonic crystals. A photonic crystal medium can be used for active functions, such as low-threshold lasing and photodetection with enhanced sensitivity. It can also be used for passive functions, such as low loss optical wave guiding (through holes), without any complication of group velocity dispersion and with a sharp bend in guiding without significant loss.
- Narrow-band filtering by a photonic crystal, together with the superprism effect, is useful for dense wavelength division multiplexing (DWDM) in optical communications.
- Photonic crystals can also be used for chemical or biosensing, where the detection of an analyte is based on changes in the bandgap characteristics (e.g., shift of the gap position) affected by the presence of the analyte in the surrounding medium.

REFERENCES

Almawlawi, D., Bosnick, K. A., Osika, A., and Moskovits, M., Fabrication of Nanometer-Scale Patterns by Ion-Milling with Porous Anodic Alumina Masks, *Adv. Mater.* **12**, 1252–1257 (2000).

Arriaga, J., Ward, A. J., and Pendry, J. B., Order-N Photonic Band Structures for Metals and Other Dispersive Materials, *Phys. Rev. B* **59**, 1874–1877 (1999).

Beetz, C., Xu, H., Catchmark, J. M., Lavallee, G. P., and Rogosky, M., SiGe Detectors with Integrated Photonic Crystal Filters, *NNUN Abstracts 2002 / Optics & Opto-electronics,* p. 76.

Berger, V., Gauthier-Lafaye, O., and Costard, E., Photonic Bandgaps and Holography, *J. Appl. Phys.* **82**, 60–64 (1997).

Boyd, R. W., *Nonlinear Optics,* Academic Press, New York, 1992.

Carlson, R. J., and Asher, S. A., Characterization of Optical Diffraction and Crystal Structure in Monodisperse Polystyrene Colloids, *Appl. Spectrosc.* **38,** 297–304 (1984).

Cregan, R. F., Mangan, B. J., Knight, J. C., Birks, T. A., Russell, P. St. J., Roberts, P. J., and Allan, D. C., Single-Mode Photonic Bandgap Guidance of Light in Air, *Science* **285,** 1537–1539 (1999).

Cumpston, B. H., Ananthavel, S. P., Barlow, S., Dyer, D. L., Ehrlich, J. E., Erskine, L. L., Heikal, A. A., Kuebler, S. M., Lee, I.-Y. S., McCord-Maughon, D., Qin, J., Rockel, H., Rumi, M., Wu, X.-L., Marder, S. R., and Perry, J. W., Two-Photon Polymerization Initiators for Three-Dimensional Optical Data Storage and Microfabrication, *Nature* **398,** 51–54, (1999).

Dobson, D. C., Gopalakrishnan, J., and Pasciak, J. E., An Efficient Method for Band Structure Calculations in 3D Photonic Crystals, *J. Comp. Phys.* **161,** 668–679 (2000).

Gates, B., Qin, D., and Xia, Y., Assembly of Nanoparticles into Opaline Structures over Large Areas, *Adv. Mater.* **11,** 466–469 (1999).

Haroche, S., and Kleppner, D., Cavity Quantum Electrodynamics, *Phys. Today* **42,** 24–30 (1989).

Ho, K. M, Chan, C. T., and Soukoulis, C. M., Existence of a Photonic Gap in Periodic Dielectric Structures, *Phys. Rev. Lett.* **65,** 3152–3155 (1990).

Holtz, J. H., and Asher, S. A., Polymerized Colloidal Crystal Hydrogel Films as Intelligent Chemical Sensing Materials, *Nature* **389,** 829–832 (1997).

Holtz, J. H., Holtz, J. S. W., Munro, C. H., and Asher, S. A., Intelligent Polymerized Crystalline Colloidal Arrays: Novel Chemical Sensor Materials, *Anal. Chem.* **70,** 780–791 (1998).

Hoyer, P., Baba, N., and Masuda, H., Small Quantum-Sized CdS Particles Assembled to Form a Regularly Nanostructured Porous Film, *Appl. Phys. Lett.* **66,** 2700–2702 (1995).

Imhof, A., Vos, W. L., Sprik, R., and Lagendijk, A., Large Dispersive Effects near the Band Edges of Photonic Crystals, *Phys. Rev. Lett.* **83,** 2942–2945 (1999).

Jakubiak, R., Bunning, T. J., Vaia, R. A., Natarajan, L. V., and Tondiglia, V. P., Electrically Switchable, One-dimensional Polymeric Resonators from Holographic Photopolymerization: A New Approach for Active Photonic Bandgap Materials, *Adv. Mater.* **15,** 241–243 (2003).

Jiang, P., Bertone, J. F., Hwang, K. S., and Colvin, V. L., Single-Crystal Colloidal Multilayers of Controlled Thickness, *Chem. Mater.* **11,** 2132–2140 (1999).

Joannopoulos, J. D., Meade, R. D., and Winn, J. N., *Photonic Crystals,* Princeton University Press, Princeton, NJ, 1995.

John. S., Strong Localization of Photons in Certain Disordered Dielectric Superlattices, *Phys. Rev. Lett.* **58,** 2486–2489 (1987).

Johnson, S. G., The MIT Photonic-Bands package, http://ab-initio.mit.edu/mpb/.

Keller, F., Hunter, M. S., and Robinson, D. L., Structural Features of Oxide Coatings on Aluminum, *J. Electrochem. Soc.* **100,** 411–419 (1953).

Kittel, C., *Introduction to Solid State Physics,* 7th edition, John Wiley & Sons, New York, 2003.

Knight, J. C., Birks, T. A., Russell, P. St. J., and Atkin, D. M., All-Silica Single-Mode Fiber with Photonic Crystal Cladding, *Opt. Lett.* **21,** 1547–1549 (1996); Errata, *Opt. Lett.* **22,** 484–485 (1997).

Kosaka, H., Kawashima, T., Tomita, A., Notomi, M., Tamamura, T., Sato, T., and Kawakami, S., Superprism Phenomena in Photonic Crystals, *Phys. Rev. B.* **58**, 10096–10099 (1998).

Kraus, T. F. and De La Rue, R. M., Photonic Crystals in the Optical Regime—Past, Present and Future, *Prog. Quant. Electron.* **23**, 51–96 (1999).

Lee, K., and Asher, S. A., Photonic Crystal Chemical Sensors: pH and Ionic Strength, *J. Am. Chem. Soc.* **122**, 9534–9537 (2000).

Lee, W., Pruzinsky, S. A., and Braun, P. V., Multi-photon Polymerization of Waveguide Structures within Three-Dimensional Photonic Crystals, *Adv. Mater.* **14**, 271–274 (2002).

Loncar, M., Yoshie, T., Scherer, A., Gogna, P., and Qiu, Y., Low-Threshold Photonic Crystal Laser, *Appl. Phys. Lett.* **81**, 2680–2682 (2002).

Luo, C., Johnson, S. G., Joannopoulos, J. D., and Pendry, J. B., All-Angle Negative Refraction without Negative Effective Index, *Phys. Rev. B.* **65**, 201104-1–201104-4 (2002).

Markowicz, P., Friend, C. S., Shen, Y., Swiatkiewicz, J., Prasad, P. N., Toader, O., John, S., and Boyd, R. W., Enhancement of Two-Photon Emission in Photonic Crystals, *Opt. Lett.* **27**, 351–353 (2002).

Markowicz, P. P., Tiryaki, H., Pudavar, H., Prasad, P. N., Lepeshkin, N. N., and Boyd, R. W., Dramatic Enhancement of Third-Harmonic Generation in 3D Photonic Crystals, *Phys. Rev. Lett.* (in press).

Masuda, H., and Fukuda, K., Ordered Metal Nanohole Arrays Made by a Two-Step Replication of Honeycomb Structures of Anodic Alumina, *Science* **268**, 1466–1468 (1995).

Masuda, H., Ohya, M., Nishio, K., Asoh, H., Nakao, M, Nohitomi, M., Yokoo, A., and Tamamura, T., Photonic Band Gap in Anodic Porous Alumina with Extremely High Aspect Ratio Formed in Phosphoric Acid Solution, *Jpn. J. Appl. Phys.* **39**, L1039–L1041 (2000).

Mekis, A., Chen, J. C., Kirland, I., Fan, S., Villeneuve, P. R., and Joannopoulos, J. D., High Transmission Through Sharp Bends in Photonic Crystal Waveguides, *Phys. Rev. Lett.* **77**, 3787–3790 (1996).

Megens, M., Wijnhoven, J. E. G. J., Lagendijk, A., and Vos, W. L., Light Sources Inside Photonic Crystals, *J. Opt. Soc. Am. B* **16**, 1403–1408 (1999).

Mizeikis, V., Juodkazis, S., Marcinkevičius, A., Matsuo, S., and Misawa, H., Tailoring and Characterization of Photonic Crystals, *J. Photochem. Photobiol. C: Photochem. Rev.* **2**, 35–69 (2001).

Nelson, R., and Haus, J. W., One-Dimensional Photonic Crystals in Reflection Geometry for Optical Applications, *Appl. Phys. Lett* **83**, 1089–1091 (2003).

Okamoto, K., Loncar, M., Yoshie, T., Scherer, A., Qiu, Y., and Gogna, P., Near-Field Scanning Optical Microscopy of Photonic Crystal Nanocavities, *Appl. Phys. Lett.* **82**, 1676–1678 (2003).

Painter, O., Vuckovic, J., and Scherer, A., Defect Modes of a Two-Dimensional Photonic Crystal in an Optically Thin Dielectric Slab, *J. Opt. Soc. Am. B.*, **16**, 275–285 (1999).

Park, S. H., and Xia, Y., Assembly of Mesoscale Particles over Large Areas and Its Application in Fabricating Tunable Optical Filters, *Langmuir* **15**, 266–273 (1999).

Parker, G., and Charlton, M, Photonic Crystals, *Phys. World* **August,** 29–30 (2000).

Pendry, J. B., and MacKinnon, A., Calculation of Photon Dispersion Relations, *Phys. Rev. Lett.* **69**, 2772–2775 (1992).

Pendry, J. B., Calculating Photonic Band Structure, *J. Phys.* (Condensed Matter) **8**, 1085–1108 (1996).

Pradhan, R. D., Tarhan, I. I., and Watson, G. H., Impurity Modes in the Optical Stop Bands of Doped Colloidal Crystals, *Phys. Rev. B* **54**, 13721–13726 (1996).

Prasad, P. N., and Williams, D. J., *Introduction to Nonlinear Optical Effects in Molecules and Polymers,* John Wiley & Sons, New York, 1991.

Purcell, E. M., Spontaneous Emission Probabilities at Radio Frequencies, *Phys. Rev.* **69**, 681–681 (1946).

Ranka, J. K., Windeler, R. S., and Stentz, A. J., Visible Continuum Generation in Air Silica Microstructure Optical Fibers with Anomalous Dispersion at 800 nm, *Opt. Lett.* **25**, 25–27 (2000).

Reese, C. E., Baltusavich, M. E., Keim, J. P., and Asher, S. F., Development of an Intelligent Polymerized Crystalline Colloidal Array Colorimetric Reagent, *Anal. Chem.* **73**, 5038–5042 (2001).

Reynolds, A. L., TMM Photonic Crystals & Virtual Crystals program, http://www.elec.gla.ac.uk/groups/opto/photoniccrystal/Software/SoftwareMain.htm.

Romanov, S. G., Fokin, A. V., Alperovich, V. I., Johnson, N. P., and DeLaRue, R. M., The Effect of the Photonic Stop-Band upon the Photoluminescence of CdS in Opal, *Phys. Status Solidi A Appl. Res.* **164**, 169–173 (1997).

Schnitzer, I., Yablonovitch, E., Caneau, C., Gmitter, T. J., and Scherer, A., 30% External Quantum Efficiency from Surface Textured, Thin-Film Light-Emitting Diodes, *Appl. Phys. Lett.* **63**, 2174–2176 (1993).

Shen, Y. R., *The Principles of Nonlinear Optics,* John Wiley & Sons, New York, 1984.

Shoji, S., and Kawata, S., Photofabrication of Three-Dimensional Photonic Crystals by Multibeam Laser Interference into a Photopolymerizable Resin, *Appl. Phys. Lett.* **76**, 2668–2670 (2000).

Slusher, R. E., and Eggleton, B. J., eds., *Nonlinear Photonic Crystals,* Springer-Verlag, Berlin, 2003.

Temelkuran, B., Ozbay, E., Kavanaugh, J. P., Tuttle, G., and Ho, K. M., Resonant Cavity Enhanced Detectors Embedded in Photonic Crystals, *Appl. Phys. Lett.* **72**, 2376–2378 (1998).

Vlasov, Y. A., Bo, X. Z., Sturm, J. C., and Norris, D. J., On-Chip Natural Assembly of Silicon Photonic Bandgap Crystals, *Nature (London)* **414**, 289–293, (2001).

Wang, X., Zhang, X. G., Yu, Q., and Harmon, B. N., Multiple-Scattering Theory for Electromagnetic Waves, *Phys. Rev. B* **47**, 4161–4167 (1993).

Wostyn, K., Zhao, Y., de Schaetzen, G., Hellemans, L., Matsuda, N., Clays, K., and Persoons, A., Insertion of a Two-Dimensional Cavity into a Self-Assembled Colloidal Crystal, *Langmuir* **19**, 4465–4468 (2003a).

Wostyn, K., Zhao, Y., Yee, B., Clays, K., and Persoons, A., Optical Properties and Orientation of Arrays of Polystyrene Spheres Deposited Using Convective Self-Assembly, *J. Chem. Phys.* **118**, 10752–10757 (2003b).

Yablonovitch, E., Inhibited Spontaneous Emission in Solid-State Physics and Electronics, *Phys. Rev. Lett.* **58**, 2059–2062 (1987).

Yamasaki, T. and Tsutsui, T., Spontaneous Emission from Fluorescent Molecules Embedded in Photonic Crystals Consisting of Polystyrene Microspheres, *Appl. Phys. Lett.* **72**, 1957–1959 (1998).

Zhu, J., Li, M., Rogers, R., Meyer, W. V., Ottewill, R. H., Russel, W. B., and Chaikin, P. M., Crystallization of Hard Sphere Colloids in Microgravity, *Nature* **387**, 883–885 (1997).

CHAPTER 10
Nanocomposites

Nanocomposites are random media containing domains or inclusions that are of nanometer size scale. These nanoscopic domains or inclusions are also referred to as mesophases, whereas the bulk phase that they constitute is called the macroscopic phase. Optical nanocomposites can be of two different types, depending on the size of domains or inclusions. In one type, the size of the domains or inclusions are significantly smaller than the wavelength of light. These nanocomposites can be prepared as very high optical quality fibers, films, or bulk in which each domain/inclusion can perform a specific photonic or optoelectronic (combined electronic and photonic) function. This permits introducing multifunctionality, and each of these functions can independently be optimized. The other type of nanocomposite contains domains/inclusions comparable to or larger than the wavelength of light. In this case, the nanocomposite is a scattering medium through which light transmission can be manipulated to produce various photonic functions. Both types of nanocomposites are described here.

Section 10.1 provides a discussion of the merits of the nanocomposites as photonic media. Section 10.2 describes nanocomposites as media suitable for optical waveguiding. It describes how a glass:polymer composite can exploit the advantages of both polymer and glass. The incorporation of nanocrystal inclusions in such composites provides the benefit of manipulation of the refractive index of the overall macroscopic phase. Section 10.3 describes the usage of scattering-type nanocomposites as lasing medium. This approach has given rise to terms such as random lasers and laser paints, which are introduced in this section.

Section 10.4 introduces the concept of electric field enhancement in a specific domain, by judicious selection of the constituent nanodomains. This field enhancement can be utilized to produce a significant increase in the strength of optical interactions—in particular, nonlinear optical effects, which strongly depend on the local electric field. Section 10.5 describes multiphasic nanocomposites that provide the exciting prospect of designing optical materials with many nanoscopic phases (mesophases). The optical interactions within and among mesophases can be controlled to derive a specific photonic response or multifunctionality. Examples of photonics applications are provided. Section 10.6 is an extension of the concept of multiphasic nanocomposite, applied to photorefractivity. Photorefractivity is a multifunctional property that involves two optoelectronic functionalities: photoconductivity and the linear electro-optic effect. The section illustrates how the use of nanocomposites can produce strong photorefractivity at judiciously selected wavelengths, including those used for optical telecommunications. The latter is conveniently achieved by using organic:inorganic hybrid nanocomposites that are formed by using inorganic quantum dots as inclusions in a hole transporting polymer matrix. This type of nanocomposite is also promising for broadband electroluminescence and broad-band solar cells, which are included in Section 10.6.

Section 10.7 covers another type of multiphasic nanocomposite. These polymer-dispersed liquid crystal (often abbreviated as PDLC) nanocomposites contain liquid crystal droplets. Depending on the droplet size, the PDLC can be scattering type or optically transparant. The applications of PDLCs in electrically switchable filters, as well as in photorefractivity, are discussed.

Section 10.8 describes the nanocomposite approach that mixes very dissimilar (mutually incompatible) materials at nanometer lengths to produce metamaterials. The metamaterials discussed in this chapter are materials that are trapped in a thermodynamically metastable or unstable phase. They are kinetically unable to phase separate beyond the nanometer scale in any practically significant time scale. Nanoscale processing of a composite containing up to 50% by weight of an inorganic oxide glass and a polymer is presented as an example, along with some of the unique features exhibited by such a composite. Section 10.9 provides highlights of the chapter.

Suggested general references are:

Nalwa, ed. (2003): *Handbook of Organic–Inorganic Hybrid Materials and Nanocomposites*

Zhang et al. (1996): *Photorefractive Polymers and Composites*

Beecroft and Ober (1997): *Nanocomposite Materials for Optical Applications*

10.1 NANOCOMPOSITES AS PHOTONIC MEDIA

Nanocomposites discussed in this chapter refer to a random medium consisting of different domains, nanometers in size. Thus, a nanocomposite is different from an ordered nanostructure, such as a photonic crystal discussed in Chapter 9. The nanocomposite approach offers the potential benefits of utilizing many nanodomains in which optical interactions and excited electronic state dynamics can be independently and judiciously manipulated to derive photonic multifunctionality. The two types of nanocomposites discussed are:

- Nanocomposites in which each domain/inclusion is separated from the other on length scale significantly smaller than the wavelength of light. For this reason, light propagation through such a composite experiences an effective refractive index, and the scattering of light is consequently minimized. These nanocomposites can be very high optical quality bulk media. Most of the chapter deals wtih this type of nanocomposites.

- Nanocomposites consisting of domains with sizes comparable to or even larger than the wavelength of light. In this case, light propagation senses the heterogeneity of the nanocomposites, whereby scattering dominates. In this case, an effective refractive index can not be used to describe light propagation. These nanocomposites will be called scattering random media to distinguish them from the type with significantly smaller nanodomains, which will be referred to as optical nanocomposites. These media can be utilized to manipulate the propagation or localization of photons as discussed later.

Nanocomposites offer the opportunity to be prepared as hybrid optical materials such as inorganic:organic blends—for example, inorganic glass:organic polymer hybrids. The examples of nanocomposites provided in this chapter are inorganic nanoparticles dispersed in polymers, inorganic glass:organic polymer mixed phase, and composites containing many nanodomains. Some of the chief merits of the various nanocomposites as photonic media, discussed in this chapter, are:

- *Manipulation of Effective Refractive Index.* Nanoparticles of materials with high refractive index can be dispersed in a glass or a polymer to increase the effective refractive index of the medium (Yoshida and Prasad, 1996). This approach is helpful in producing optical waveguides where higher refractive index leads to better beam confinement. This manipulation is described in Section 10.2.
- *Anderson-Type Photon Localization.* Anderson localization refers to localization of electrons by multiple scattering in a disordered medium (Anderson, 1958; Lee and Ramakrishnan, 1985). A similar phenomenon occurs for light in a scattering random medium where the domain sizes are comparable to or larger than the wavelength of light, and the refractive index difference (refractive index contrast) between domains is large to produce strong scattering (John, 1991; Wiersma et al., 1997). Photon localization is manifested as an exponential decrease of the transmission coefficient, instead of linear decrease, with the thickness (optical path length) of the medium. A useful application of scattering random media-type nanocomposites is for random lasers, as discussed in Section 10.3. Another example is a polymer dispersed liquid crystal used as light valves, a topic covered in Section 10.7.
- *Enhancement of Local Field.* By judicious selection of the relative dielectric constants of the two media, one can enhance the local field to produce significant gain in the nonlinear optical effects (Sipe and Boyd, 2002). Enhancement of the effective third-order nonlinear optical coefficient can be utilized to produce a nonlinear optical phase shift for optical switching, gating, and bistability functions. This feature of a nanocomposite is described in Section 10.4.
- *Control of Optical Communications Between Domains.* Judicious control of interaction between constituents of different nanodomains produces the opportunity to localize the excitation dynamics in each domain and minimize

the excitation energy transfer from one constituent to another (Ruland et al., 1996). The excitation energy transfer process involves nanoscopic interactions, as discussed in Chapter 2. This feature of minimizing energy transfer between domains, as demonstrated in Section 10.5, can be utilized to produce nanocomposites for multidomain lasing from the same bulk optical nanocomposite.

- *Introduction of Multifunctionality.* Nanostructure control allows one to design multiphase optical nanocomposites in which each nanodomain can perform a specific photonic or optoelectronic function and also derive a new manifestation resulting from the combined functions of various nanodomains. The example provided in Section 10.6 illustrates a hybrid nanocomposite approach for photorefractivity. Section 10.7 illustrates a photorefractive nanocomposite involving quantum dots (quantum dots are discussed in Chapter 4), liquid crystal nanodroplets, and a hole transporting polymer to produce the function of photorefractivity (Winiarz and Prasad, 2002). The photorefractive effect requires the combined action of photocharge generation and electro-optic effect, with the latter property producing a change of refractive index by the application of an electric field (Yeh, 1993).

- *Metamaterials.* An active area of advanced materials development is that of metamaterials. Metamaterials are usually defined as engineered structures possessing certain macroscopic properties unavailable in constituent materials and not readily observed in nature. In the context of this chapter, we consider metamaterials as materials in either a thermodynamically metastable state (a shallow local minimum) or an unstable state (no minimum). The major challenge is to stabilize these metamaterials by kinetically slowing down their conversion to the stable thermodynamic phases. These metamaterials allow one to harvest properties, derived from the phase that is thermodynamically not stable. Nanostructure processing offers this prospect as is illustrated in Section 10.8, where the formation of a metastable oxide glass:organic polymer optical nanocomposite is described. This class of mtamaterials should not be confused with left-handed metamaterials that exhibit a negative refractive index (Shvets, 2003).

10.2 NANOCOMPOSITE WAVEGUIDES

Increasing interest in optical integrated circuit has stimulated studies on optical waveguide materials. Many researchers have spent a great deal of effort to realize useful materials for optical waveguides. The sol–gel processing is one approach because of (a) high optical quality of materials produced and (b) freedom to impregnate them with a variety of additives to modify their optical characteristics. The sol–gel reaction is a technique to make metal oxide and non-metal oxide glasses through chemical reactions, without high-temperature processing (Brinker and Scherer, 1990). The precursor solution is an alkoxide, such as silicon alkoxide or titanium alkoxide, that is reacted with water containing an acid or a base catalyst. A hydrolysis reaction takes place, followed by a polycondensation reaction, resulting in a three-dimensional oxide network.

Pure inorganic sol–gel materials, such as SiO_2 and TiO_2, have been investigated for optical applications, including optical waveguides (Klein, 1988; Schmidt and Wolter, 1990; Gugliemi et al., 1992; Motakef et al., 1992; Yang et al., 1994; LeLuyer et al., 2002). Weisenbach and Zelinski (1994) produced sol–gel processed SiO_2/TiO_2 (50/50 wt %) slab waveguides, typically having losses of less than 0.5 dB/cm at the wavelength of 633 nm. The thickness of the waveguide was 0.18 μm. These materials have excellent optical quality and provide the freedom to change their indices. However, pure inorganic sol–gel materials are difficult to be produced as films thicker than 0.2 μm in a single coating, because a large shrinkage of the material during the thermal densification process causes cracking in thicker films. Such films are too thin to guide modes in the waveguides, so several repeated coatings are necessary. Also, silica has a low refractive index, not ideal for channel waveguides where a strong confinement of light is needed. Polymer materials have been used for waveguiding (Chen, 2002). The advantage offered by polymeric materials is the ease of fabrication. Polymers can readily be produced in the form of a thin film by relatively simple techniques such as spin coating. They have mechanical strength and flexibility so that films do not crack, and sufficiently thick films of waveguiding dimensions (~1 μm or even thicker) can be produced. Polymer films can readily be patterned and processed to produce channel waveguides. They can serve as passive waveguides or active waveguides by appropriate choice and selective doping of photonically active molecules (e.g., dyes or molecules with high electro-optic coefficient).

However, polymers exhibit higher optical losses compared to silica. They also do not exhibit high surface quality. Polymers possess a large coefficient of thermal expansion. Hybrid optical nanocomposites, containing both polymeric and glass phases, utilize the relative merits of both inorganic glass (e.g., silica) and polymers and offer a number of advantages (Burzynski and Prasad, 1994). Some of them are:

- Improved mechanical strength derived from polymer
- Improved optical quality derived from inorganic glass
- Manipulation of refractive index by using higher refractive index glass or its nanoparticles

Sol–gel processed organic:inorganic composite materials are capable of producing thick enough films by single coating, without cracking. Some researchers have used organically modified alkoxysilanes and tetraalkoxysilanes to produce organically modified silicates, called ORMOSILS or ORMOCERs (Schmidt and Wolter, 1990; Krug et al., 1992; Motakef et al., 1994) to achieve an optical propagation loss of 0.15 dB/cm at 633 nm with a slab waveguide. This consisted of a composite material of poly(dimethylsiloxane), SiO_2 and TiO_2, with a thickness of 1.55 μm. Another approach is mixing a prepolymerized material (polymer) into the sol–gel precursors (Wung et al., 1991; Zieba et al., 1992; Yoshida and Prasad, 1996). In this case, it is important to avoid large extent of phase separation throughout the sol–gel reaction, by a careful selection of solvents and optimized material processing conditions.

The example provided here is the sol–gel-processed SiO_2/TiO_2/poly(vinylpyrrolidone) (Yoshida and Prasad, 1996). Poly(vinylpyrrolidone) (PVP) is one of the ideal pre-polymerized materials because of its solubility in polar solvents and its thermal cross-linking characteristics. The solubility in polar solvents is a preferable characteristic for mixing with the sol–gel precursors, since the sol–gel reaction uses polar solvent(s) and water and produces alcohol in the reaction. Therefore, PVP is expected to provide a homogeneous solution in sol–gel processing, avoiding any phase separation. The cross-linking of PVP is another advantage, because it is expected to have better thermal stability than many un-cross-linked polymers. The cross-linked PVP also allows to employ some etching techniques, using water-based etchants, to pattern channel waveguide structures from slab waveguides, so a conventional photolithographic technique can be applied. Sheirs et al. (1993) studied the cross-linking mechanism of PVP. PVP is cross-linked by a radical reaction originated by residual hydrogen peroxide which is used to initiate the polymerization reaction. The cross-linking takes place during a 120°C 24-hr thermal treatment.

TiO_2 is known as a high refractive index material, so it was used to increase the refractive index of the system. We succeeded to impregnate TiO_2 into the PVP/SiO_2 system to realize the manipulation of refractive index, which provides an extended freedom for designing optical waveguides (Yoshida and Prasad, 1996). Optical propagation losses were 0.62 and 0.52 dB/cm for 40% and 50% TiO_2 waveguides, respectively. Table 10.1 summarizes the results of optical propagation losses in titanium 2-ethylhexoxide-derived composite waveguides, together with the results of measured refractive indexes for different compositions. The refractive indices are also plotted in Figure 10.1, as a function of concentration of TiO_2. The data for a 0% TiO_2 waveguide was taken from our earlier work on PVP/SiO_2 composite materials (Yoshida and Prasad, 1995). For all the titania concentrations, propagation losses of 0.62 dB/cm or lower were achieved by avoiding both scattering and absorption losses. Additionally, a manipulation of refractive index in the range 1.39–1.65 was accomplished by changing the ratio of SiO_2 and TiO_2. The variation of the refractive index was proportional to the concentration of TiO_2.

Table 10.1. Summary of Optical Propagation Loss and Refractive Index Values for Different Composition SiO_2/TiO_2/PVP Composite Waveguides at the Wavelength of 633 nm

TiO_2 wt %	0	10	20	30	40	50
SiO_2 wt %	50	40	30	20	10	0
Optical propagation loss (db/cm)	0.20	0.20	0.43	0.45	0.62	0.52
Refractive index	1.49	1.52	1.55	1.58	1.62	1.65

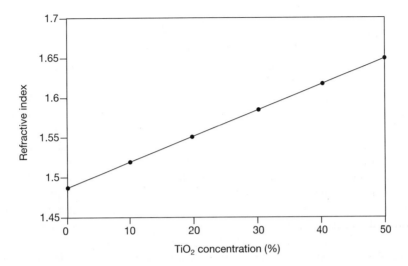

Figure 10.1. Refractive indices of the SiO$_2$/TiO$_2$/PVP composite as a function of TiO$_2$ concentration.

10.3 RANDOM LASERS: LASER PAINTS

Generally, light scattering by particulates in a lasing medium is considered to be a strong loss mechanism for laser operation, often being detrimental. However, under favorable conditions, strong scattering in a random medium can provide an optical feedback, instead of the feedback from the mirrors used in a conventional laser cavity. This feedback can produce significant gain that more than offset optical loss due to scattering. In such a case, the medium can be used to effect a lasing action and produce what is called a random laser (Cao et al., 1999).

Letokhov (1968) theoretically predicted that light amplification was possible in a scattering medium. Lawandy et al. (1994) reported lasing action from a colloidal solution containing rhodamine 640 perchlorate dye and TiO$_2$ particles in methanol. To prevent flocculation, the TiO$_2$ particles were coated with a layer of Al$_2$O$_3$ and had a mean diameter of 250 nm. The emission from the dye exhibited spectral narrowing and temporal characteristics of a multimode laser oscillator, even though no external cavity was present. They also reported that the threshold excitation energy for laser action was surprisingly low.

This result has led to the concept of a laser paint, where a suspension containing submicron particles of high refractive index, which produce highly scattering centers, can be used for lasing. The field of random lasers is expanding, with lasing having been observed from both organic dye infiltrated systems and inorganic gain media. Ling et al. (2001) used films of polymers, such as polymethylmethacrylate (PMMA), containing laser dye and TiO$_2$ particles of size ~ 400 nm to produce random lasing from the dye. In such a scattering random medium, the gain medium (the dye) and the scattering centers (TiO$_2$) are separated. They reported that, as the TiO$_2$ particle density in the polymer increased, the lasing threshold decreases. Two different types of thresholds observed were: one for spectral narrowing and another for observing individual lasing modes arising from amplification of light along certain closed loops (optical path length).

Williams et al. (2001) have reported continuous wave laser action in strongly scattering rare-earth-metal-doped dielectric nanospheres. This is the first report of electrically pumped random media to provide stimulated emission. Wiersma and Cavaleri (2002) considered an interesting system that can be temperature-tuned due to the simultaneous presence of a liquid crystal and a dye inside sintered glass powders. Such tunable random laser systems may be useful in photonic applications—for example, in temperature sensing.

Anglos et al. (2004) have demonstrated random laser action in organic/inorganic hybrid random media consisting of ZnO semiconductor nanoparticles dispersed in a polymer matrix such as PMMA, polydimethylsiloxane (PDMS), epoxy, or polystyrene (PS). For example, upon pumping a series of ZnO/PDMS hybrid composite with 248-nm radiation, laser-like narrow emission at ~ 384 nm was observed, with bandwidth narrowing from 15–16 nm (observed at low pump intensity) to ~ 4 nm.

10.4 LOCAL FIELD ENHANCEMENT

As described in Section 10.1, a nanocomposite medium offers the prospect to redistribute the field among various domains (also referred to as mesophases) and enhance them selectively in/near a specific domain (Haus et al., 1989a, 1989b; Gehr and Boyd, 1996; Sipe and Boyd, 2002). Boyd, Sipe, and co-workers have conducted extensive theoretical and experimental studies of the local field effects on the linear and nonlinear optical properties of the nanocomposites (Gehr and Boyd, 1996; Nelson and Boyd, 1999; Yoon et al., 2000). They theoretically treated the problem using an effective medium picture in which the bulk (macroscopic) optical properties of a nanocomposite are described by an effective dielectric constant ε_{eff} and are related to the dielectric constants of the individual domains (mesophases). A simple model involves a mesophase (nanodomains) of an inclusion with the dielectric constant ε_i dispersed in a host of dielectric constant ε_h. Hence an assumption is made that the domain sizes are significantly larger than the interatomic spacing, so that the values of ε_i and ε_h can be used for nanodomains. Yet the domain sizes are much less than the wavelength of light, so that the concept of an effective dielectric constant to describe the average macroscopic property is applicable.

Two topologies (geometries) of a random nanocomposite can be considered. They are: (i) the Maxwell Garnett geometry in which spherical nanoparticles or small nanodomains of inclusion are randomly dispersed in a host medium and (ii) the Bruggeman geometry in which the inclusion has a larger fill fraction, and the two constituent phases (host and inclusions) are interdispersed. These two topologies are schematically represented in Figure 10.2. The Maxwell Garnett geometry derives its name from the pioneering work by Maxwell Garnett on metallic col-

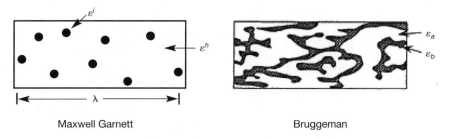

Figure 10.2. Schematic representations of the Maxwell Garnett and Bruggeman geometries.

loidal solutions. Using microscopic polarizabilities to describe the local fields, ε_{eff} can be related to ε_i and ε_h for the Maxwell Garnett topology as

$$\frac{\varepsilon_{\text{eff}} - \varepsilon_h}{\varepsilon_{\text{eff}} + 2\varepsilon_h} = f_i \left(\frac{\varepsilon_i - \varepsilon_h}{\varepsilon_i + 2\varepsilon_h} \right) \quad (10.1)$$

where f_i is the fill fraction (fractional volume) of the inclusion. In the case of Maxwell Garnett topology, f_i is significantly smaller than f_h for the host. The case of Bruggeman topology is the limit of where there is no minor (inclusion) phase. In this case an extension of (10.1) can be made by assuming $\varepsilon_h \approx \varepsilon_{\text{eff}}$, because the host medium now is derived from an interdispersed form of the two constituents, 1 and 2. Hence the effective susceptibility in the Bruggeman topology is given by the equation

$$0 = f_1 \frac{\varepsilon_1 - \varepsilon_{\text{eff}}}{\varepsilon_1 + 2\varepsilon_{\text{eff}}} + f_2 \frac{\varepsilon_2 - \varepsilon_{\text{eff}}}{\varepsilon_2 + 2\varepsilon_{\text{eff}}} \quad (10.2)$$

where ε_1 and ε_2 are the dielectric constants and f_1 and f_2 are the fill fractions of constituents 1 and 2. The local electrostatic field, derived from the effective dielectric constant, shows that it becomes concentrated in the host region near a spherical inclusion, if $\varepsilon_i > \varepsilon_h$, as is shown by the field lines in Figure 10.3. This has important consequences for the nonlinear optical properties of nanocomposites, such as the intensity-dependent refractive index, described by third-order nonlinear optical interactions that depend on the fourth power of the local field. Thus for the case of $\varepsilon_i \gg \varepsilon_h$, the addition of a small amount of a linear material to a nonlinear host can lead to an increase in the effective optical nonlinearity of the nanocomposite, because of the field enhancement (concentration) in the host region around the inclusion. However, no clear demonstration of enhancement of optical nonlinearities in a nanocomposite is currently available. But such enhancement of optical nonlinearities in an ordered multilayer nanocomposite has been reported (Fischer et al., 1995; Nelson and Boyd, 1999). Even though these are not disordered nanocomposites, the underlying principle of enhancement of nonlinear optical effects in a layered nanocomposite can be the same—that is, enhancement (concentration) of field in

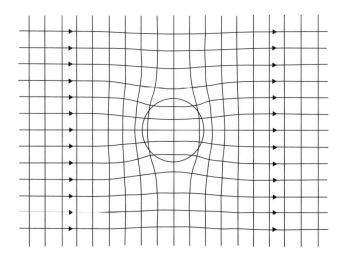

Figure 10.3. Concentration of local electrostatic field in the host region near a spherical inclusion.

the layer with lower dielectric constant. These layered nanocomposites are different from the one-dimensional photonic crystal discussed in Chapter 9, because the layer thickness here is significantly smaller than the wavelength of light. Thus no photonic band structure results, and the medium can be adequately described by an effective optical medium model. In contrast, the layer spacing in a one-dimensional photonic crystal, as described in Chapter 9, is comparable to the wavelength of light.

10.5 MULTIPHASIC NANOCOMPOSITES

An exciting prospect is provided by the concept of a multiphasic optical nanocomposite. An example of a nanocomposite is shown schematically by a cartoon diagram in Figure 10.4 (Gvishi et al., 1995; Ruland et al., 1996). It consists of a sol–gel processed glass medium, which has nanosize pores running through the glass, because sol–gel processing produces a porous structure. By controlling the temperature of the glass, one can control the size of these pores. For example, the conditions of our processing produce glass containing 5-nm-size pores. These pores can be infiltrated to create layers of structures formed at the pore walls to produce nanodomains. One can also use chemistry—for example, polymerization chemistry—to fill these pores. One can thus produce organic polymers in the pores of a glass, together with an interfacial phase separating the glass and the polymer. Since the phase separation is on a nanometer size scale, much smaller than the wavelength of light, the material is optically transparent These multiphaseic composites can be prepared in various bulk forms such as monolith, films, and fibers (Gvishi et al., 1997).

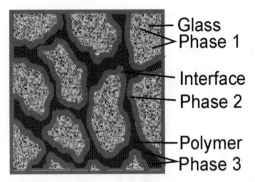

Figure 10.4. Schematic representation of a multiphasic nanocomposite sol–gel glass.

A multiphasic nanocomposite offers the very exciting prospect of controlling the excitation dynamics to derive a specific photonic response from each domain, thus achieving multifunctionality. One example is an optical nanocomposite in which an ionic dye is deposited on the surface of the pores. The other dye, rhodamine 6G, is mixed in the monomer, methylmethacrylate, which is then filled in the pores and polymerized as polymethylmethacrylate, commonly known as PMMA. One thus has an ionic dye, ASPI, which emits in the red color and is at the walls of the pores; the other dye, R6G, which emits in the yellow region, is dispersed in the polymer PMMA phase. In the composite medium, emission from both dyes is observed. Furthermore, one can use both these dyes as gain media for lasing to cover a broad wavelength range, including ranges characteristic for both dyes, as shown by the gain curves in Figure 10.5. This is in contrast to the case when one uses these two dyes in a common solvent and looks at the emission; in such a case, only emission due to ASPI is observed. The emission from R6G is quenched, because it transfers energy to the narrower energy gap red-emitting dye, ASPI. Thus only lasing from the red dye ASPI, as shown by the curve on the left, is observed in the solution phase (Figure 10.5a). The width of the lasing tunability curve for the ASPI dye is ~21 nm, while that for the R6G dye is ~12 nm. On the other hand, for a composite glass containing both these dyes, it is ~37 nm (Figure 10.5b). Hence, the laser emission from the composite glass containing both dyes is tunable across a wide range (560–610 nm). In solution, the quenching is a result of Förster energy transfer (discussed in Chapter 2). In a multiphasic glass, as described here, the extremely large surface-to-volume ratio allows the two dyes to be separated to minimize the energy transfer. Therefore, using optical nanocomposites, one can have many domains lasing separately and obtain multiwavelength lasing from the same bulk medium, utilizing the radiative features of different nanodomains.

Another example of the benefit offered by a nanocomposite is provided by that in the poly-*p*-phenylene vinylene (PPV) polymer: glass system. Poly(*p*-phenylene vinylene) (PPV) is a promising π-conjugated polymer for applications in photonics, because of its large nonresonant optical nonlinearity (Pang et al., 1991; Singh et al.,

288 NANOCOMPOSITES

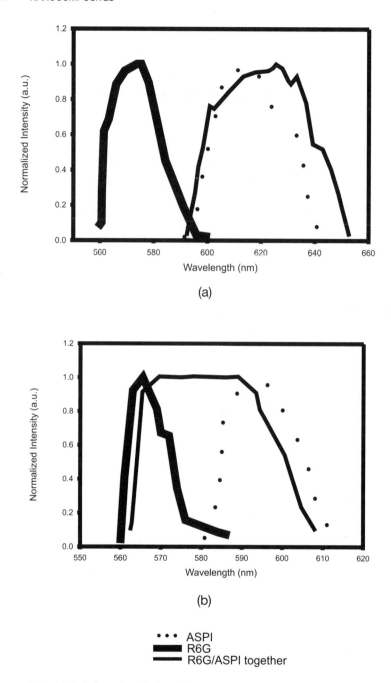

Figure 10.5. (a) Solution phase lasing. (b) Lasing in the glass matrix. In part b, ASPI and R6G are in phase 2 and phase 3, respectively.

1988). Photoexcitation across the π–π* energy gap creates electron–hole pairs that localize themselves in the polymer chain, giving rise to photoconductivity and interesting electrical properties (Lee at al., 1993). The large photoluminescence quantum efficiency of PPV and the discovery of electroluminescence in PPV and its derivatives generated a great deal of interest in this class of materials (Burroughes et al., 1990). Researchers have demonstrated optical amplification in PPV and its analogs in solutions (Moses, 1992; Brouwer et al., 1995) and in pure thin films (Hide et al., 1996), when PPV is excited with an appropriate light source.

However, the optical losses in pure PPV as solid bulk phase are high owing to its large photoinduced absorption and scattering. This problem is severe for applications in which thicker samples are required. On the other hand, an optical nanocomposite of PPV with an inorganic material (glasses) has superior optical quality and, therefore, could be useful for a number of applications in photonics. The first report on such a PPV–silica composite created by the sol–gel process was published by Wung et al. (1991), who reported the preparation of films of thicknesses in the micrometer range. Since then there have also been other reports on nonlinear optical properties (Davies et al., 1996) and fluorescence studies (Faraggi et al., 1996) of these composites.

The example discussed here is the fabrication and lasing properties of a PPV–silica conjugated polymer–glass composite bulk, formed using *in situ* polymerization of a PPV monomer within the pores of a commercially available porous glass, Vycor glass (30% pore volume), by the base-catalyzed polymerization reaction (Kumar et al., 1998). The base-catalyzed polymerization reaction eliminates the need for an extended heat treatment under vacuum for full conversion of PPV. In this way, high-optical-quality PPV–silica composites of thicknesses as large as a few millimeters could be prepared. The extremely small pore size (30 Å) of the Vycor glass enables nanometer-size distribution of the PPV network within the glass. Since the phase separation between the glass and the polymer is on a few-nanometer scale in the nanocomposite, optical scattering losses caused by refractive-index mismatch between the glass ($n = 1.48$) and the polymer ($n \sim 2.0$) are negligible.

The nanocomposite prepared by polymerization of PPV on the surface of the pores also provides for the polymer a topology with reduced interchain coupling. In the bulk phase of this polymer, interchain coupling reduces the emission from the polymer significantly, because the optical loss introduced by the photoinduced absorption of the interchain coupled states dominates. Therefore, no net optical gain can be achieved to produce lasing. When the PPV polymer is produced *in situ* by polymerization in the pores of this glass, the polymer forms only on the surface of the pores, thereby minimizing the interchain interaction between the two polymer chains. Consequently, the photoinduced absorption due to interchain band is minimized to enhance optical gain. This method has been utilized to produce a high-optical-quality nanocomposite medium, which is shown as a disk in Figure 10.6. It has the PPV polymer dispersed on a nanometer scale, in porous glass. Figure 10.7 shows the lasing behavior. On the left-hand side of Figure 10.7 the dotted curve is the lasing curve and one can see the spectral narrowing characteristic of lasing, compared to the solid line, which is a fluorescence curve. The fluorescence curve is

Figure 10.6. Photograph of a thick disk made of the nanocomposite in which PPV is formed in the pores of a porous glass.

spread out and no spectral narrowing is observed until the cavity lasing takes place. The curve on the right-hand side in Figure 10.7 shows the input–output relation between the lasing output and the pump power.

Another example of use of optical nanocomposite for lasing is provided by the work of Beecroft and Ober (1995). They achieved optical amplification in a composite containing nanocrystals of the lasing material Cr-Mg_2SiO_4 (chromium-doped forsterite) in a copolymer made of tribromostyrene/naphthylmethacrylate. Forsterite crystal is an attractive lasing medium for coverage of the near-IR region (1167–1345 nm) (Petricevic et al., 1988). However, growth of a large monolithic crystal is a combersom process. It is desirable to deal with a flexible polymer matrix that can be fabricated as films and fibers for various photonic applications. Therefore, optical nanocomposites of polymers, containing nanocrystallites of lasing medium, are attractive alternatives. The refractive index of the copolymer used was adjusted to be close to that of forsterite by varying the composition of the copolymers.

10.6 NANOCOMPOSITES FOR OPTOELECTRONICS

The optical nanocomposite approach offers opportunities to produce high-performance and relatively low-cost optoelectronic media, suitable for many applications. A photorefractive medium is an excellent example of a multifunctional optoelectronic medium, as the property of photorefractivity is derived from a combination of two functionalities: photoconductivity and electro-optic effect (Yeh, 1993; Moerner and Silence, 1994; Zhang et al., 1996; Würthner et al., 2002). Table 10.2 shows "important steps" involved in photorefractivity. The first step is photoinduced charge carrier generation at centers which are called photosensitizers. The second step is the involvement of a charge transporting medium that allows the carriers to move or migrate. As a result, there is a separation of charges (even in the absence of an external electric field, such separation can take place by diffusion of charges out of irradiated regions). The third step is the trap-

10.6 NANOCOMPOSITES FOR OPTOELECTRONICS   291

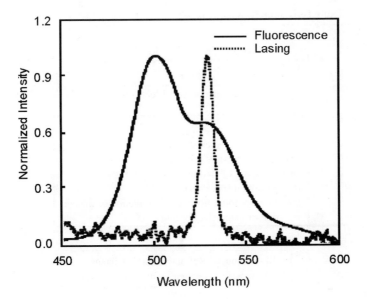

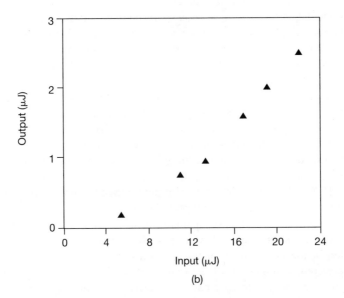

Figure 10.7. (a) Fluorescence and lasing spectra obtained from the PPV–silica nanocomposite. (b) Laser output energy per pulse plotted as a function of the pump energy for the PPV–silica nanocomposite.

Table 10.2. Important Steps Involved in Producing Photorefractivity

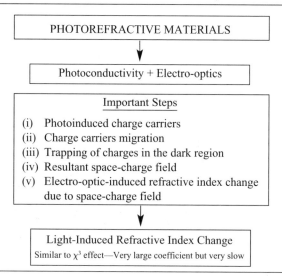

ping of charges in the dark (unirradiated) regions. Consequently, a space-charge field is produced by this charge separation. If the medium also exhibits electro-optic effects, its refractive index is dependent on the field strength applied. The net result is a change in the refractive index due to the space-charge field. But, it is a light-induced refractive index change, because it is light that generates a space-charge field within the material, which results in the change of the refractive index of the medium. It is very similar to a third-order nonlinear optical effect (due to the presence of the third-order susceptibility, $\chi^{(3)}$, discussed in Chapter 2) that also produces an optically induced change in refractive index. Since photorefractivity results from the combined action of photoconductivity and electro-optics, it is a fairly slow process compared to the $\chi^{(3)}$ processes, because it relies on spatial migration of charges.

Figure 10.8 shows the schematics of a photorefractive hologram generated by crossing of two coherent beams in a photorefractive polymer composite medium. When a nonuniform light-intensity pattern, resulting from the interference of two mutually coherent laser beams, illuminates such a material, charge carriers are photogenerated at the bright areas of the sinusoidal intensity profile. Subsequently, there is preferential migration of the more mobile charge carrier species into the darker regions of the interference pattern by drift and diffusion mechanisms (holes in Figure 10.8). Since most polymeric materials with sufficient optical transparency are good insulators, drift under an applied bias is the dominant mechanism in charge transportation. Next, the mobile charges are trapped by impurities and defect sites, and a nonuniform space-charge distribution is established. This results in an internal space-charge electric field. If the material is also capable of exhibiting an

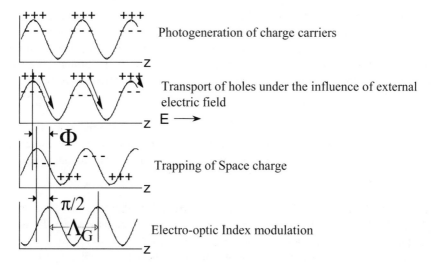

Figure 10.8. Schematic representations of the buildup of photorefractivity, Λ_G, is the grating spacing.

electro-optic response, the periodic space-charge field induces a modulation of the material's refractive index through a linear Pockels effect and/or an orientational birefringence effect (due to reorientation of molecules in electrical field).

A unique feature of the photorefractive effect is the existence of a phase shift between the illuminating intensity pattern and the resulting refractive index grating (Moerner and Silence, 1994). This is due to the fact that the electric field is a space-derivative of the space-charge distribution that predicts a $\pi/2$ phase shift between them. Such a nonlocal grating cannot occur by any of the local mechanisms (namely, photochromism, thermochromism, thermorefraction) of grating formation, because charge transport over a macroscopic distance is involved. The photorefractive effect not only provides an opportunity by which localized and nonlocalized electronic and optical processes may be studied, it is also expected to play an important role in many practical applications such as high-density optical data storage, optical amplification, and dynamic image processing.

An important manifestation of the photorefractive phase shift is that in this kind of medium, when two beams are overlapped and form a photorefractive grating, "coupling" of the two beams takes place. One beam transfers energy to the other and amplifies it at the cost of its own intensity. This two-beam coupling and power transfer proceeds in a defined direction (i.e., only from one beam to the other), depending on how the electro-optic tensor is aligned in the medium (Moerner and Silence, 1994). Thus, the two effects produced in a photorefractive medium are diffraction from a holographic grating and two-beam asymmetric coupling (power transfer). The grating can be used to steer beams or diffract to form holograms for dynamic holography. By uniformly shining the light, one can erase this holographic grating. The second application utilizes a photorefractive medium for beam amplification. For example, one can take a beam, which may contain optical noise, and couple it with a clean beam and transfer power from it to clean beam and amplify the clean beam. In optical communications, one sends the beam through a long optical fiber where the polarization can get scrambled. Asymmetric two-beam coupling in a photorefractive medium can be used to purify the polarization of such a beam.

A hybrid photorefractive nanocomposite containing inorganic semiconductor quantum dots, dispersed together with electro-optically active centers (organic molecules) in a hole transporting polymer matrix, offers a number of advantages, some of which are:

Selective Photosensitization over a Broad Spectral Range, from Blue to IR. By using quantum dots as photosensitizers, one can judiciously select the wavelength of photosensitization (Winiarz et al., 1999a, 1999b, 2002). This can be accomplished by the choice of the size of quantum dots and their composition which control their energy gap and consequently optical absorption. The optical transitions in quantum dots are discussed in detail in Chapter 4. Hence, we can make a photorefractive medium that has a tunable optical response. One can make a photorefractive medium applicable to communication wavelengths at 1.3 μm and 1.5 μm, which is very difficult to do with a bulk photorefractive medium or using organic photorefractive medium with organic photosensitizers, because organics photosensitizing in IR are generally not stable. This approach was used with lead sulfide or mercury sulfide nanocrystals as photosensitizers at 1.3 μm, a primary wavelength for optical communications (Winiarz et al., 2002). They were dispersed in a polymer matrix containing an electro-optically active organic structure. Photorefractivity at 1.3 μm was realized. A similar approach can be used to achieve photorefractivity at 1.5 μm. This is a nice example of how one can use Q-dots to photosensitize.

Enhanced Photocharge Generation. The quantum dots efficiently produce photogeneration of charge carriers. If the barrier for hole transfer from the quantum dot to the hole transporting polymer is small, this will produce enhanced photoconductivity and also increase space-charge fields. The barrier for hole transfer can be minimized by creating a direct interface between quantum dots and the polymers. The photogeneration process in a polymeric medium is often described by the Onsager model, in which the electrons and holes initially form a thermalized pair of an average distance r_0, from where they can dissociate under the influence of a field to produce free carriers (Mort and Pfister, 1976; Pope and Swenberg, 1999). Thus a strong field dependence of photogeneration efficiency is observed. Figure 10.9 compares the photogeneration efficiency and photosensitivity of the CdS-doped PVK host, compared to those for C_{60}-doped PVK (Winiarz et al., 1999a). The CdS quantum dots (labeled in the figure as Q-CdS) are of two different sizes. Q-CdS of size 1.6 nm have the bandgaps to absorb strongly at the excitation wavelength of 514.5 nm; the Q-CdS of size <1.4 nm have the bandgap shifted to higher energy so

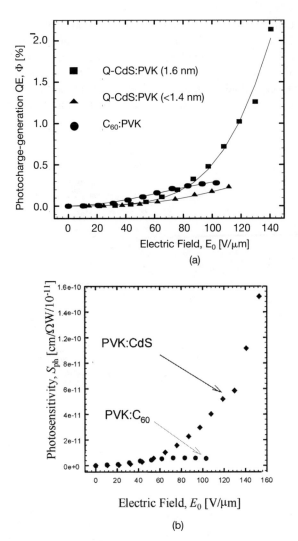

Figure 10.9. Enhanced photocharge generation in inorganic quantum dot:organic polymer HYBRID nanocomposite. (a) The photocharge generation quantum efficiency as a function of applied electric field. The solid line corresponds to the best fit using the Onsager formalism. (b) Photosensitivity plotted as a function of the applied field. The wavelength used is 514.5 nm.

that absorption at 514.5 nm is significantly less. The nanocomposite Q-CdS:PVK containing 1.6-nm quantum dots exhibit considerably higher photogeneration efficiency at 514.5 nm, compared to that involving the smaller-size quantum dot, as well as to that utilizing C_{60}. The electric field dependence of the photogeneration efficiency is well described by the Onsager model, because the solid lines represent this theoretical fit.

Enhanced Mobility. A surprising result is that the mobility of the hole carriers in the PVK matrix is enhanced by introduction of quantum dots (Roy Choudhury et al., 2003, 2004). This feature is evident from the time-of-flight mobility measurement. The mobility data for various concentrations of CdS quantum dots are plotted as a function of electric field in Figure 10.10. A significant increase in the mobility is observed with an increase in the quantum dot doping concentration. The strong

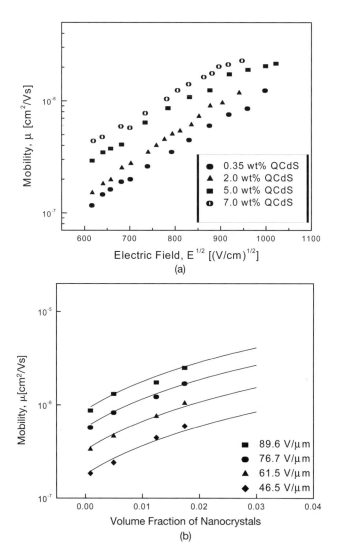

Figure 10.10. Enhanced charge carrier mobility in an inorganic quantum dot:organic polymer HYBRID nanocomposite. (a) Comparison of field dependences of hole mobilities in samples having varying concentrations of CdS nanoparticles. (b) Hole mobility as a function of varying concentration of nanoparticles at four different applied electric fields.

electric field dependence is indicative of a field induced drift mobility (versus diffusion of carriers). The nearly linear dependence with respect to $E^{1/2}$ and the Arrhenius-type thermal behavior fits the dispersive model of continuous-time random walk (Roy Choudhury et al., 2004).

The broad spectral coverage, together with enhanced photocharge generation and carrier mobility, suggests that the quantum dot:polymer nanocomposites can be a very useful medium to produce efficient, broadband solar cells. If one mixes nanocrystals that can absorb over a broad range of wavelengths, one can harvest photons over a wide spectral range to make a broadband solar cell.

The quantum dot:polymer nanocomposite approach has also been used for another optoelectronic effect, electroluminescence (Bakueva et al., 2003) light-emitting diode, LED, function (electroluminescence in quantum dots and in polymers have been discussed in Chapter 4 and Chapter 8, respectively). Bakueva et al. used nanocomposites that consisted of PbS nanocrystals dispersed in a derivative MEH-PPV of the PPV polymer or another conjugated polymer, CN-PPV. They reported photo- and electroluminescence that were tunable over a broad spectral range, from 1000 to 1600 nm. With the use of quantum dots of different sizes, internal quantum efficiency for electroluminescence reached values of up to 1.2%.

10.7 POLYMER-DISPERSED LIQUID CRYSTALS (PDLC)

Polymer-dispersed liquid crystals, often abbreviated as PDLC, are composites produced from solution phase to incorporate liquid crystal domains in a polymer matrix (Drzaic, 1995; Bunning et al., 2000). The liquid crystal droplets are produced by evaporating the solvent and subsequent thermal annealing. They are droplets dispersed in the polymer matrix. Two types of PDLC have been studied:

1. A polymer matrix containing domains of liquid crystals that are comparable to or significantly larger than the wavelength of light (e.g., micron-sized liquid crystal domains). These are scattering random media, defined earlier in this Section 10.1, because of a refractive index mismatch between the polymer and the liquid crystal domain. These composites are the traditional PDLCs used for light valve operation to be described later. In these composites, the relative domain size of the liquid crystal is optimized to enhance multiple scattering events, in order to maximize the contrast between the two domains.
2. Polymers containing significantly smaller size ($\leq$ 100 nm in diameter) nanodroplets of liquid crystals where light scattering is minimized. Here, an effective refractive index describes light propagation and the phase of propagating lightwave. These PDLCs can be labeled here as nano-PDLCs to distinguish them from the first type discussed above.

Liquid crystals are anisotropic. They have two refractive indices: extraordinary and ordinary refractive indices. By applying the field, one can reorient the director of a liquid crystal droplet and consequently change its refractive index. Figure 10.11 shows one of the older applications of using this PDLC as a light valve. It schematically shows this light valve function. The ordinary refractive index of the liquid crystal droplet is fairly close to that of the polymer, but the extraordinary refractive index is fairly high. When the liquid crystal droplets (of type 1) are randomly oriented, the light passing through this medium experiences, on the average, a large difference of the refractive index between the liquid crystal droplets and the polymer. This refractive index mismatch creates scattering, resulting in loss of transmission (off state). When an electric field is applied, the liquid crystal droplets reorient and the light now experiences the ordinary refractive index which is well matched with that of the polymer. The light can now be transmitted through the medium without scattering. Thus, the application of an electric field switches the medium, from a light blocking to the light transmissive state.

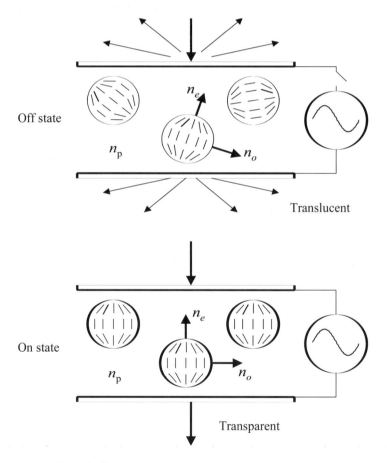

Figure 10.11. Schematic illustration of the light valve function in the off state (without applied electric field) and in the on state (with applied electric field).

In the case of nano-PDLC, an application of an electric field simply changes the effective refractive index of the nanocomposites. An application of nano-PDLC is in the context of photorefractivity, where an effective electro-optic effect is derived from the liquid crystal (Golemme et al., 1997). If the polymer matrix is chosen to be hole conducting and photosensitized with Q-dots, as described above, photorefractivity can be achieved. In an approach taken by us, a polymer PMMA, not hole conducting by itself, was doped with ECZ (ethylcarbazole) to make it hole conducting (Winiarz and Prasad, 2002).

Figure 10.12 shows schematic representation of a PDLC system containing nanodroplets of liquid crystal and ~ 10-nm-size CdS quantum dots dispersed in a PMMA ploymer. CdS produced *in situ,* used for photogeneration of charge carriers, absorbs the 532-nm laser light. The electrons remain trapped in the nanoparticle. The holes move and get trapped in the dark region. There is a resulting space-charge field that reorients the liquid crystal nanodroplet to change the effective refractive index. In a two-beam crossing geometry described earlier, a grating is now formed by this nanocrystal/nanodroplet liquid crystal to produce photorefractive effect. The change of effective refractive index, produced by the liquid crystal, can be very large, because the extraordinary index of 1.71 and ordinary index of 1.5 are widely different. One can form a grating with a large refractive index change, Δn. The result is that one can achieve very high diffraction efficiency, even in a thin film. Figure 10.13 shows the diffraction efficiency as a function of the applied electric field. A maximum net diffraction efficiency of ~70% is achieved (Winiarz and Prasad, 2002). The figure also shows the overall transmission of the sample as a function of the applied field. If one takes the internal optical losses into account, the diffraction efficiency is ~100%. The downside at this stage is the response time to reorient the nanodroplets in a plastic medium. The formation of the grating is mainly limited by the reorientation of these nanodroplets of liquid crystal, which is in seconds. By controlling the nanodroplet size and the interfacial interaction, one

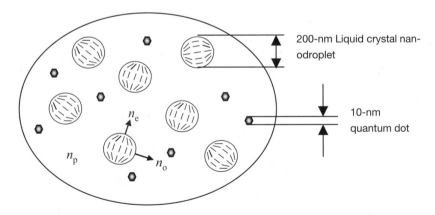

Figure 10.12. Schematic representation of polymer-dispersed liquid crystals containing quantum dots.

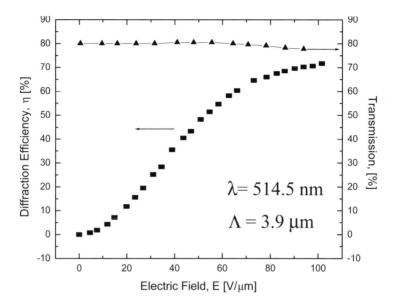

Figure 10.13. Diffraction efficiency (■) and optical transmission (▲) of the polymer-dispersed liquid crystal:ECZ:quantum dot photorefractive composite as a function of electric field.

may be able to produce faster response. This is a good example of a multifunctional multicomponent system, with inorganic Q-dots as photosensitizers, liquid crystal nanodroplets for electro-optics, and the polymer matrix for charge conduction. A plastic medium also allows for fabrication of large-area free-standing photorefractive sheets.

Wu and co-workers (Ren and Wu, 2003; Ren et al., 2003) have utilized another variation of polymer-dispersed liquid crystals to produce electrically tunable lens and prism gratings. They created tunable negative and positive lenses by exposing the UV-curable monomer in a liquid crystal host, to UV light through a patterned mask. This mask produced a continuously variable optical output, decreasing from the periphery to the center, for a negative (concave) lens formation. The peripheral (outer) area, with a stronger UV light, produced a higher concentration of the polymer, while the center, being exposed to a lower level of UV light, produced a lower concentration of the polymer. When a uniform electric field is applied, it produces a high degree of reorientation of the liquid crystal directors in the center, which has a lesser network (less polymer concentration). This reorientation creates a concave refractive index profile, forming a negative lens. The curvature of this gradient lens can be tuned by changing the applied voltage. However, at high field, all liquid crystal directors are aligned in the field direction. Consequently, the gradient no longer exists, and the lens effect vanishes in the high-voltage regime. To create a prism grating, a photomask with periodically varying optical density was used. The intensity distribution for each period formed a saw-tooth pattern, starting from a low value to a large value at the end of the period. As a result, a polymer network with a periodic gradient morphology is produced. As the voltage is applied to reorient the liquid crystal directors, a periodic gradient refractive index pattern, scalable with the field, results, producing a prism grating.

10.8 NANOCOMPOSITE METAMATERIALS

As described in Section 10.1, metamaterials are defined here as materials in a thermodynamically metastable state. In other words, there is a minimum in the free energy, but it is not the lowest energy minimum which defines the thermodynamically stable state. However, for our discussion, we shall extend the definition of a metamaterial to also include a phase that is thermodynamically unstable, but kinetically too slow to transform to a stable phase on any practical time scale. Often, when two incompatible materials such as an inorganic glass and organic polymer are mixed together, an unstable or a metastable phase results which eventually phase separates completely. Nanoscale processing of a composite of such incompatible materials provides an opportunity to limit the phase separation in nanometers. Further phase separation beyond this size scale becomes kinetically too slow to be of any practical consequence. Since the resulting phase separation is on the scale of nanometers, significantly smaller than the wavelength of light, the material is optically transparent.

An example of nanocomposite metamaterials is a glass–polymer composite containing up to 50% of an inorganic oxide glass (e.g., silica and vanadium oxide) and a conjugated polymer, PPV (Wung et al., 1991, 1993; He et al., 1991). The PPV polymer, already discussed in Section 10.4, exhibits many interesting optical properties as described there. It is a rigid polymer that, without any flexible side chain, is insoluble once formed. In our work, the nanoprocessing of this composite involved mixing of a sol–gel precursor of silica (or vanadium oxide) which is commonly used for chemical processing of glass through solution chemistry discussed in Section 10.4. The PPV polymer was produced from its monomer by an acid-catalyzed reaction in the present case. The sol–gel precursor was mixed with the PPV monomer precursor in the solution phase and hydrolyzed to form the silica network. Simultaneously the PPV monomer was polymerized. By controlling the processing condition, the composite was made to form a rigid structure before the phase separation proceeded beyond several nanometers. The TEM study of the composite confirmed the nanoscale phase separation (Embs et al., 1993). As the structure becomes very rigid, the phase separation becomes kinetically too slow to be of any significance. It should be noted that this nanocomposite containing the interdispersed polymer and glass phases is different from the PPV composite described in Section 10.4 where PPV was formed by polymerization in the pores of an already formed porous silica glass (post-doping). The PPV:glass nanocomposite described here, in contrast, is produced by mixing them in the precursor stage in the solution phase (premixing).

The nanocomposite PPV:glass even at the high level (~50%) of mixing is optically transparent and exhibits very low optical loss, compared to a film of pure PPV, a feature clearly derived from the presence of an inorganic glass (Wung et al., 1991).

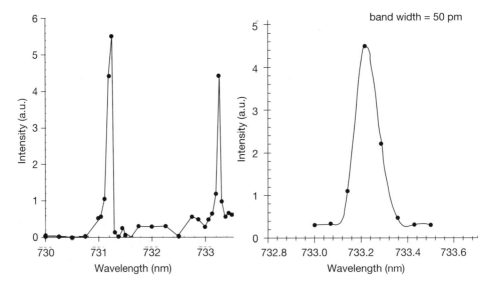

Figure 10.14. Wavelength multiplexing and narrow bandwidth filtering in the sol–gel processed PPV:silica composite.

The composite exhibits many features derived from inorganic glass (e.g., silica) such as high surface quality and easily polishable ends for end-fire coupling of light into the waveguide. It also exhibits features derived from the polymer PPV, such as mechanical strength and high third-order optical nonlinearity (Pang et al., 1991).

For dense wavelength division multiplexing (DWDM) function, an important part of optical communication technology, holographic gratings have been recorded in the PPV/sol–gel glass nanocomposite. The left-hand panel in Figure 10.14 shows two gratings recorded in the same volume, used to multiplex/demultiplex two wavelength separated by 2 nm. The right-hand panel in Figure 10.14 shows a single holographic grating used to filter-out light of an extremely narrow, 50-pm bandwidth.

10.9 HIGHLIGHTS OF THE CHAPTER

- Nanocomposites are media containing randomly distributed nanometer size domains or inclusions, sometimes also referred to as mesophases.
- Nanocomposites can serve as multifunctional photonic media, because each domain/inclusion can be optimized to perform a specific photonic or optoelectronic function.
- The two types of optical nanocomposites are (i) scattering random media with domains comparable to or larger than the wavelength of light and (ii) highly transparent optical nanocomposites with domains significantly smaller than the wavelength of light.

10.9 HIGHLIGHTS OF THE CHAPTER

- Multifunctional optical nanocomposites can have high optical quality, when the nanodomains are significantly smaller than the wavelength of light, so that they do not scatter light.
- Optical nanocomposite media, containing high refractive index nanoparticles dispersed in a glass or a polymer, can be utilized to produce optical waveguides, where a higher refractive index leads to better beam confinement.
- Additional structural flexibility, together with manipulation of refractive index, can be achieved in fabrication of optical waveguides, using glass:polymer:nanoparticle composites of varying compositions.
- Local scattering by larger nanoparticles of high refractive index in a scattering random medium can be used to produce local lasing in a nanocomposite, an approach giving rise to terms such as random lasers or laser paints.
- By judicious selection of the relative dielectric constants of the two media, one can achieve enhancement of local electric field in a particular domain of a nanocomposite to produce significant amplification of nonlinear optical effects.
- The two topologies of optical nanocomposites, from the perspectives of field enhancement, are (i) the Maxwell Garnett geometry in which spherical nanoparticles or small nanodomains are randomly dispersed in a host medium and (ii) Bruggeman geometry in which the inclusion has a larger fill fraction, with the two consistuent phases interdispersed.
- Multiphasic optical nanocomposites consists of multiple nanodomains where the optical functions of each domain can be isolated or combined to produce new manifestations and thereby achieve multifunctionality.
- Judicious control of interactions between constituents of different nanodomains of a multiphasic optical nanocomposite offers the prospect of minimizing the excitation energy transfer among constituents. This feature is illustrated by an approach to obtain multidomain, multiwavelength lasing.
- Another application of a multiphasic optical nanocomposite is provided by enhanced lasing of a conjugated polymer, poly-*p*-phenylene vinylene (PPV). This occurs at an interfacial region in the pores of a silica glass of controlled porosity.
- A hybrid optical nanocomposite provides a new-generation optoelectronic media for solar energy conversion, and photorefractivity. It contains (a) inorganic quantum dots as photosensitizers that absorb light to produce charge-carriers (photocharge generation) and (b) a hole transporting polymer for charge conduction in a flexible plastic medium,
- Photorefractivity is a multifunctional property combining two funcationalities: photoconductivity and the electro-optic effect. Light absorption creates electron–hole pairs, and the relative migration of these pairs creates an internal electric field that changes the refractive index of the medium through the electro-optic effect.
- The use of quantum dots in a hybrid optoelectronic medium provides the op-

portunity to tailor photoresponse to a specific wavelength (by choice of the type of semiconductor and quantum dot size, as discussed in Chapter 4), this can also produce a broad spectral coverage (by using many quantum dots, together covering a broad range) that is highly desirable for broadband solar cells.

- The use of quantum dots in a hybrid polymer nanocomposite also produces enhanced photocharge generation as well as enhanced mobility of carriers.
- The use of quantum dots of a narrow bandgap semiconductor allows one to achieve photorefractivity at communication wavelengths. An example presented is of PbS or HgS quantum dots:poly-N-vinyl carbazole nanocomposites containing electro-optic chromophores. They exhibit photorefractivity at 1.3 μm.
- Polymer-dispersed liquid crystals (PDLCs) are nanocomposites that incorporate liquid crystal nanodroplets into a polymer matrix.
- PDLCs containing liquid crystal droplets, comparable to or larger than the wavelength of light, can be used as dynamically switchable optical elements (such as light valve), where the optical transmission is influenced by the electric-field-induced reorientation of liquid crystal droplets leading to a change of its effective refractive index.
- Polymer-dispersed liquid crystals, containing quantum dots, have been used as a very effective photorefractive media. These media utilize photocharge generation in quantum dots to produce an internal space-charge field, thereby causing a change in the refractive index of the liquid crystal nanodroplets through the electro-optic effect in liquid crystal.
- Other applications of polymer-dispersed liquid crystals are for electrically tunable lenses, prisms, and gratings.
- Nanocomposite metamaterials, produced by controlled nanoscale phase separation of two thermodynamically incompatible phases, such as glass and plastic, offer the opportunities to combine the merits of diverse optical materials.
- An example of a metamaterial presented here is a 50:50 composite of glass and a conjugate polymer, poly-p-phenylene vinylene (PPV).
- The PPV:glass composite exhibits high optical quality derived from the glass, along with mechanical strength and high third-order optical nonlinearity derived from the PPV polymer.

REFERENCES

Anderson, P. W., Absence of Diffusion in Certain Random Lattices, *Phys. Rev.* **105**, 1492–1501 (1958).

Anglos, D., Stassinopoulos, A., Dos, R. N., Zacharakis, G., Psylkai, M., Jakubiak, R., Vaia, R. A., Giannelis, E. P., and Anastasiadris, S. H., Random Laser Action in Organic/Inorganic Nanocomposites, *J. Opt. Soc. Am. A* **12**, 208–212 (2004).

Bakueva, L., Musikhin, S., Hines, M. A., Chang, T.-W. F., Tzolov, M., Scholes, G., D., and

Sargent, E. H., Size-Tunable Infrared (1000–1600 nm) Electroluminescence from PbS Quantum Dot Nanocrystals in a Semiconductor Polymer, *Appl. Phys. Lett.* **82,** 2895–2897 (2003).

Beecroft, L. L., and Ober, C. R., Novel Ceramic Particle Synthesis for Optical Applications: Dispersion Polymerized Preceramic Polymers as Size Templates for Fine Ceramic Powders, *Adv. Mater.* **7,** 1009–1009 (1995).

Beecroft, L. L., and Ober, C. R., Nanocomposite Materials for Optical Applications, *Chem. Mater.* **9,** 1302–1317 (1997).

Brinker, C. J., and Scherer, G. W., *Sol–Gel Science: The Physics and Chemistry of Sol–Gel Processing,* Academic Press, New York, 1990.

Brouwer, H. J., Krasnikov, V. V., Hilberer, A., Wildeman, J., and Hadziioannou, G., Novel High Efficiency Copolymer Laser Dye in the Blue Wavelength Region, *Appl. Phys. Lett.* **66,** 3404–3406 (1995).

Bunning, T. J., Natarajan, L. V., Tondiglia, V. P., and Sutherland, R. L., Holographic Polymer-Dispersed Liquid Crystals, *Annu. Rev. Mater. Sci.* **30,** 83–115 (2000).

Burroughes, J. H., Bradley, D. D. C., Brown, A. R., Marks, R. N., Mackay, K., Friend, R. H., Burns, P. L., and Holmes, A. B., Light-Emitting Diodes Based on Conjugated Polymers, *Nature* **347,** 539–541 (1990).

Burzynski, R., and Prasad, P. N., Photonics and Nonlinear Optics with Sol–Gel Processed Inorganic Glass: Organic Polymer, in *Sol–Gel Optics—Processing and Application,* L. C. Klein, ed., Kluwer, Norwell, MA, 1994, pp. 417–449.

Cao, H., Zhao, Y. G., Ho, S. T., Seeling, E. W., Wang, Q. H., and Chang, R. P. H., Random Laser Action in Semiconductor Powder, *Phys. Rev. Lett.* **82,** 2278–2281 (1999).

Chen, R., Integration Glass Plastic, *SPIE's OE Magazine* **November,** 24–26 (2002).

Dabbousi, B. O., Bawendi, M. G., Onitsuka, O., and Rubner, M. F., Electroluminescence From CdSe Quantum Dot/Polymer Composites, *Appl. Phys. Lett.* **66,** 1316–1318 (1995).

Davies, B. L., Samoc, M., and Woodruff, M., Comparison of Linear and Nonlinear Optical Properties of Poly(*p*-phenylene vinylene)/Sol–Gel Composites Derived from Tetramethoxysilane and Methyltrimethoxysilane, *Chem. Mater.* **8,** 2586–2594 (1996).

Drzaic, P. S., *Liquid Crystal Dispersions,* World Scientific, Singapore, 1995.

Embs, F. W., Thomas, E. L., Wung, C. J., and Prasad, P. N., Structure and Morphology of Sol–Gel Prepared Polymer–Ceramic Composite Thin Films, *Polymer* **34,** 4607–4612 (1993).

Faraggi, E. Z., Sorek, Y., Levi, O., Avny, Y., Davidov, D., Neumann, R., and Reisfeld, R., New Conjugated Polymery Sol–Gel Glass Composites: Luminescence and Optical Waveguides, *Adv. Mater.* **8,** 833–839 (1996).

Fischer, G. L., Boyd, R. W., Gehr, R. J., Jenekhe, S. A., Osaheni, J. A., Sipe, J. E., and Weller-Brophy, L. A., Enhanced Nonlinear Optical Response of Composite Materials, *Phys. Rev. Lett.* **74,** 1871–1874 (1995).

Gehr, R. J., and Boyd, R. W., Optical Properties of Nanostructured Optical Materials, *Chem. Mater.* **8,** 1807–1819 (1996).

Golemme, A., Volodin, B. I., Kippelen, B., and Peyghambarian, N., Photorefractive Polymer-Dispersed Liquid Crystals, *Opt. Lett.* **22,** 1226–1228 (1997).

Gugliemi, M., Colombo, P., Mancielli, D. E., Righini, G. C., Pelli, S., and Rigato, V., Characterization of Laser-Densified Sol–Gel Films for the Fabrication of Planar and Strip Optical Wave-Guides, *J. Non-Cryst. Solids* **147,** 641–645 (1992).

Gvishi, R., Bhawalkar, J. D., Kumar, N. D., Ruland, G., Narang, U., and Prasad, P. N., Multiphasic Nanostructured Composites for Photonics: Fullerene-Doped Monolith Glass, *Chem. Mater.* **7,** 2199–2202 (1995).

Gvishi, R., Narang, U., Ruland, G., Kumar, D. N., and Prasad, P. N. Novel, Organically Doped, Sol–Gel Derived Materials for Photonics: Multistructured Composite Monoliths and Optical Fibres, *Appl. Organomet. Chem.* **11,** 107–127 (1997).

Haus, J. W., Inguva, R., and Bowden, C. M., Effective-Medium Theory of Nonlinear Ellipsoidal Composites," *Phys. Rev. A* **40,** 5729–5734 (1989a).

Haus, J. W., Kalyaniwalla, N., Inguva, R., Bloemer, M., and Bowden, C. M., Nonlinear Optical Properties of Conductive Spheroidal Composites," *J. Opt. Soc. Am. B* **6,** 797–807 (1989b).

He, G. S., Wung, C. J., Xu, G. C., and Prasad, P. N., Two-Dimensional Optical Grating Produced on a Poly-*p*-phenylene vinylene/V2O5-Gel Film by Ultrashort Pulsed Laser Radiation, *Appl. Opt.* **30,** 3810–3817 (1991).

Hide, F., Diaz-Garcia, M. A., Schwartz, B. J., Andersson, M. R., Pei, Q., and Heeger, A. J., Semiconducting Polymers: A New Class of Solid-State Laser Materials, *Science* **273,** 1833–1835 (1996).

John, S., Localization of Light, *Phys. Today* **44,** 32–40 (1991).

Klein, L. C., *Sol–Gel Technology for Thin Films, Fibers, Preforms, Electronics, and Specialty Shapes,* Noyes Publications, Park Ridge, NJ, 1988.

Krug, H., Merl, N., and Schmidt, H., Fine Patterning of Thin Sol–Gel Films, *J. Non-Cryst. Solids* **147,** 447–450 (1992).

Kumar, D. N., Bhawalkar, J. D., and Prasad, P. N., Solid-State Cavity Lasing from Poly(*p*-phenylene vinylene)–Silica Nanocomposite Bulk, *Appl. Opt.* **37,** 510–513 (1998).

Lawandy, N. M., Balachandran, R. M., Gomes, A. S. L., and Sauvain, E., Laser Action in Strongly Scattering Media, *Nature* **368,** 436–438 (1994).

Lee, C. H., Yu, G., and Heeger, A. J., Persistent Photoconductivity in Poly(*p*-phenylene Vinylene): Spectral Response and Slow Relaxation, *Phys. Rev. B* **47,** 15, 543–15,553 (1993).

Lee, P. A., and Ramakrishnan, T. V., Disordered Electronic Systems, *Rev. Mod. Phys.* **57,** 287–337 (1985).

LeLuyer, C., Lou, L., Bovier, C., Plenet, J. C., Dumas, J. G., and Mugnier, J., A Thick Sol–Gel Inorganic Layer for Optical Planar Waveguide Fabrication, *Opt. Mater.* **18,** 211–217 (2002).

Letokhov, V. S., Light Generation by a Scattering Medium with a Negative Resonant Absorption, *Sov. Phys.-JETP* **16,** 835–840 (1968).

Ling, Y., Cao, H., Burin, A. L., Ratner, M. A., Liu, X, and Chang, R. P. H., Investigation of Random Lasers with Resonant Feedback, *Phys. Rev. A* **64,** 063808-1–063808-7 (2001).

Moerner, W. E., and Silence, S. M. Polymeric Photorefractive Materials, *Chem. Rev.* **94,** 127–155 (1994).

Mort, J., and Pfister, G., *Electronic Properties of Polymers,* John Wiley & Sons, New York, 1976.

Moses, D. High Quantum Efficiency Luminescence from a Conducting Polymer in Solution: A Novel Polymer Laser Dye, *Appl. Phys. Lett.* **60,** 3215–3216 (1992).

Motakef, S., Boulton, J. M., Teowee, G., Uhlmann, D. R., Zelinski, B. J., and Zanoni, R.,

Polyceram Planar Waveguides and Optical Properties of Polyceram Films, *J. Proc. SPIE* **1758,** 432–445 (1992).

Motakef, S., Boulton, J. M., and Uhlmann, D. R., Organic–Inorganic Optical Materials, *Opt. Lett.* **19,** 1125–1127 (1994).

Nalwa, H. S., ed., *Handbook of Organic–Inorganic Hybrid Materials and Nanocomposites,* Vols. 1 and 2, American Scientific Publishers, Stevenson Ranch, CA, 2003.

Nelson, R. L., and Boyd, R. W., Enhanced Electro-Optic Response of Layered Composite Materials, *Appl. Phys. Lett.* **74,** 2417–2419 (1999).

Pang, Y., Samoc, M., and Prasad, P. N., Third Order Nonlinearity and Two-Photon-Induced Molecular Dynamics: Femtosecond Time-Resolved Transient Absorption, Kerr Gate, and Degenerate Four-Wave Mixing Studies in Poly(p-phenylene vinylene)/Sol–Gel Silica Film, *J. Chem. Phys.* **94,** 5282–5290 (1991).

Petricevic, V., Geyen, S. K., Alfano, R. R., Yamagishi, K., Anzai, H., and Yamaguchi, Y., Laser Action in Chromium-Doped Forsterite, *Appl. Phys. Lett.* **52,** 1040–1042 (1988).

Pope, M., and Swenberg, C. E., *Electronic Processes in Organic Crystals and Polymers,* Oxford University Press, Oxford, England, 1999.

Ren, H., and Wu, S.-T., Tunable Electronic Lens Using a Gradient Polymer Network Liquid Crystal, *Appl. Phys. Lett.* **82,** 22–24 (2003).

Ren, H., Fan, Y.-H., and Wu, S.-T., Prism Grating Using Polymer Stablilzed Nematic Liquid Crystal, *Appl. Phys. Lett.* **82,** 1–3 (2003).

Roy Choudhury, K., Samoc, M., and Prasad, P. N., Charge Carrier Transport in poly(N-vinylcarbazole):CdS Quantum Dot Hybrid Nanocomposite, *J. Phys. Chem. B,* in press.

Roy Choudhury, K., Winiarz, J. G., Samoc, M., and Prasad, P. N., Charge Carrier Mobility in an Organic–Inorganic Hybrid Nanocomposite, *Appl. Phys. Lett.,* **92,** 406–408 (2003).

Ruland, G., Gvishi, R., and Prasad, P. N., Multiphasic Nanostructured Composites: Multi-dye Tunable Solid-State Laser, *J. Am. Chem. Soc.* **118,** 2985–2991 (1996).

Schmidt, H., and Wolter, H., Organically Modified Ceramics and Their Applications, *J. Non-Cryst. Solids* **121,** 428–435 (1990).

Sheirs, J., Bigger, S. W., Then, E. T. H., and Billingham, N. C., The Application of Simultaneous Chemiluminescence and Thermal Analysis for Studying the Glass Transition and Oxidative Stability of Poly(N-vinyl–2-pyrrolidone), *J. Polym. Sci., Polym. Phys. Ed.* **31,** 287–297 (1993).

Shvets, G., Photonic Approach to Making a Material with a Negative Index of Refraction, *Phys. Rev. B* **67,** 035109-1-035109-8 (2003).

Singh, B. P., Prasad, P. N., and Karasz, F. E., Third-Order Non-Linear Optical Properties of Oriented Films of Poly(p-phenylene vinylene) Investigated by Femtosecond Degenerate Four Wave Mixing, *Polymer* **29,** 1940–1942 (1988).

Sipe, J. E., and Boyd, R. W., Nanocomposite Materials for Nonlinear Optics in *Nonlinear Optics of Random Media, Topics in Applied Physics,* V. M. Shalaev, ed., Springer, Berlin, 2002.

Weisenbach, L., and Zelinski, B. J., Attenuation of Sol–Gel Waveguides Measured as a Function of Wavelength and Sample Age, *J. Proc. SPIE* **2288,** 630–639 (1994).

Wiersma, D. S., and Cavaleri, S., Temperature-Controlled Random Laser Action in Liquid Crystal Infiltrated Systems, *Phys. Rev. E* **66,** 056612-1–056612-5 (2002).

Wiersma, D., and Lagendijk, A., Laser Action in Very White Paint, *Phys. World* **10,** 33–37 (1997).

Wiersma, D. S., Bartolini, P., Lagendijk, A., and Righini, R., Localization of Light in a Disordered Medium, *Nature* **390,** 671–673 (1997).

Williams, G. R., Bayram, S. B., Rand, S. C., Hinklin, T., and Laine, R. M., Laser Action in Strongly Scattering Rare-Earth-Metal-Doped Dielectric Nanospheres, *Phys. Rev. A* **65,** 013807-1–013807-6 (2001).

Winiarz, J. G., and Prasad, P. N. Photorefractive Inorganic-Organic Polymer-Dispersed Liquid Crystal Nanocomposite Photosensitized with Cadmium Sulfide Quantum Dots, *Opt. Lett.* **27,** 1330–1332 (2002).

Winiarz, J. G., Zhang, L., Lal, M., Friend, C. S., and Prasad, P. N. Photogeneration, Charge Transport, and Photoconductivity of a Novel PVK/CdS–Nanocrystal Polymer Composite, *Chem. Phys.* **245,** 417–428 (1999a).

Winiarz, J. G., Zhang, L., Lal, M., Friend, C. S., and Prasad, P. N., Observation of the Photorefractive Effect in a Hybrid Organic–Inorganic Nanocomposite, *J. Am. Chem. Soc.* **121,** 5287–5298 (1999b).

Winiarz, J. G., Zhang, L., Park, J., and Prasad, P. N., Inorganic:Organic Hybrid Nanocomposites for Photorefractivity at Communication Wavelengths, *J. Phys. Chem. B* **106**(5), 967–970 (2002).

Wung, C. J., Pang, Y., Prasad, P. N., and Karasz, F. E., Poly(*p*-phenylene vinylene)–Silica Composite: A Novel Sol–Gel Processed Non-Linear Optical Material for Optical Waveguides, *Polymer* **32,** 605–608 (1991).

Wung, C. J., Wijekoon, W. M. K. P., and Prasad, P. N., Characterization of Sol–Gel Processed Poly(*p*-phenylenevinylene) Silica and V_2O_5 Composites Using Waveguide Raman, Raman and *FTIR* Spectroscopy, *Polymer* **34,** 1174–1178 (1993).

Würthner, F., Wortmann, R., and Meerholz, K., Chromophore Design for Photorefractive Organic Materials, *Chem. Phys. Chem.* **3,** 17–31 (2002).

Yang, L., Saaredra, S. S., Armstrong, N. R., and Hayes, J., Fabrication and Characterization of Low-Loss, Sol–Gel Planar Wave-Guides, *J. Anal. Chem.* **66,** 1254–1263 (1994).

Yeh, P., *Introduction to Photorefractive Nonlinear Optics,* John Wiley & Sons, New York, 1993.

Yoon, Y. K., Bennink, R. S., Boyd, R. W., and Sipe, J. E., Intrinsic Optical Bistability in a Thin Layer of Nonlinear Optical Material by Means of Local Field Effects, *Opt. Commun.* **179,** 577–580 (2000).

Yoshida, M., and Prasad, P. N., Sol–Gel Derived PVP:SiO2 Composite Materials and Novel Fabrication Technique for Channel Waveguides, *Mater. Res. Symp. Proc.* **392,** 103–108 (1995).

Yoshida, M., and Prasad, P. N., Sol–Gel-Processed SiO_2/TiO_2/Poly(vinylpyrrolidone) Composite Materials for Optical Waveguides, *Chem. Mater.* **8,** 235–241 (1996).

Zhang, Y., Burzynski, R., Ghosal, S., and Casstevens, M. K., Photorefractive Polymers and Composites, *Adv. Mater.* **8,** 115–125 (1996).

Zieba, J., Zhang, Y., Prasad, P. N., Casstevens, M. K., and Burzynski, R., Sol–Gel Processed Inorganic Oxide: Organic Polymer Composites for Second-Order Nonlinear Optics Applications, *Proc. SPIE* **1758,** 403–409 (1992).

CHAPTER 11
Nanolithography

This chapter describes some of the nanofabrication processes used in nanophotonics. These processes utilize light-induced photochemical or photophysical changes to produce nanostructures. By using nonlinear optical processes, near-field excitation, and plasmonic field enhancement, nanostructures with feature sizes significantly smaller than the wavelength of light used to fabricate them can be obtained. In addition, some other processes that do not utilize photonics to produce nanostructures, but produce useful structures for nanophotonic applications, are also described.

This chapter also includes nanofabrication processes such as photobleaching and photopolymerization in direct writing, which are not lithographic processes in the traditional sense. However, they are included since they utilize similar photonic patterning as the one involving traditional lithography; thus both are being presented together.

Section 11.1 discusses two-photon lithography, where two-photon excitation, being quadratically dependent on the intensity, produces the resulting photochemical or photophysical changes that are spatially localized near the high intensity region of the focal point. This feature, together with greater depth penetration enabled by use of long-wavelength exciting light (often in the near-IR) in most bulk material (e.g. polymers), allows one to fabricate many types of three-dimensional micro- and nanostructures that are useful for three-dimensional optical circuitry, optical microelectromechanical systems (optical MEMS), and three-dimensional data storage schemes. Several examples of two-photon lithography are presented. Another variation of two-photon microfabrication, described here, utilizes two-photon holography to produce complex three-dimensional structures by interference of coherent beams.

Section 11.2 describes near-field lithography in which near-field excitation, both by using an aperture (e.g., tapered fiber) or an apertureless geometry (using local field enhancement), is used to induce photophysical or photochemical changes and, thereby, produce nanostructures. The near-field interactions and various geometries used for near-field excitation are described in Chapter 3. A combination of two-photon excitation and near-field geometry allows one to create nanostructures with even smaller feature sizes. Examples provided include that of near-field lithography on a self-assembled monolayer template (see Chapter 8 for a discussion of self-assembled monolayers).

Nanophotonics, by Paras N. Prasad
ISBN 0-471-64988-0 © 2004 John Wiley & Sons, Inc.

Section 11.3 introduces the concept of using a phase-mask in soft lithography, where a phase change through a mask is used to produce intensity modulation in the near-field (contact mode) pattern. Soft lithography, described in this chapter, refers to lithography with soft organic materials such as elastomeric polymers, which can be used to produce an elastomeric stamp from a master (made of hard materials such as a metal). The intensity variation, introduced by this phase mask, produces nanostructures in an underlying photoresist.

Section 11.4 describes the use of plasmon guided evanescent waves through metallic nanostructures to effect photochemical and photophysical changes in a layer below the metallic structure. The principles of plasmonics and plasmonic guides are discussed in Chapter 5. This approach, referred to as plasmonic printing, is illustrated with an example.

Section 11.5 describes nanosphere lithography that utilizes self-assembled close-packed nanospheres in a monolayer as the mask. Infiltration through the voids in this close-packed structure leads to nanostructure deposition on the substrate. Removal of the nanospheres leaves nanostructures. A specific example provided here is the fabrication of organic light-emitting "nanodiode" arrays using nanosphere lithography.

Section 11.6 presents "dip-pen nanolithography," which utilizes the tip of an atomic force microscope (AFM) as a pen to lay down molecules, acting as an ink, on a substrate. The use of an AFM tip can provide resolutions in a few nanometers. Several examples of this method are discussed.

Section 11.7 describes nanoimprint lithography, which involves printing of a mold with nanostructure on a soft (or molten) materials. The mold is removed, leaving the corresponding negative contrast (embossed) structure in the material.

Section 11.8 introduces the concept of alignment of nanoparticles that are driven by photon absorption. The light absorption is used to produce chemically modified domains that force the migration of included nanoparticles. This light-driven migration of nanoparticles can be used to produce grating structures as well as other complex nanostructures.

The highlights of the chapter are presented in Section 11.9.

Considering the diverse range of lithographic techniques described here, it is not possible to provide a general reference for the subject. Some references on the various topics are provided here for general reading.

Bhawalkar et al. (1996): *Nonlinear Multiphoton Processes in Organic and Polymeric Materials*

Rogers et al. (2000): *Printing, Molding, and Near-Field Photolithographic Methods for Patterning Organic Lasers, Smart Pixels and Simple Circuits*

Piner et al. (1999): Dip-Pen Nanolithography

Haynes and VanDuyne (2001): *Nanosphere Lithography: A Versatile Nanofabrication Tool for Studies of Size-Dependent Nanoparticle Optics*

Kik et al. (2002): *Plasmon Printing—A New Approach to Near-Field Lithography*

11.1 TWO-PHOTON LITHOGRAPHY

In recent years, three-dimensional microfabrication by direct laser writing or laser rapid prototyping has been attracting a great deal of interest for its various applications to microelectromechanical systems (MEMS), 3D optical waveguide circuitry, 3D optical data storage, and so on. Up to now, most of microelectromechanical devices such as sensors and actuators were fabricated by conventional lithography and etching process on silicon substrate, which includes many complicated and time-consuming steps. On the other hand, laser rapid prototyping enables computer-aided single step generation of three-dimensional structures by laser-induced photopolymerization or photocrosslinking. Use of linear absorption of the laser light by the monomer for photopolymerization generally limits the resolution of the feature to the wavelength of the light. Also, deep penetration into the bulk material is not typically possible due to strong linear absorption, so that the fabrication is necessarily a layer-by-layer process.

The use of two-photon excitation, instead of conventional UV curing, can enhance the resolution and allow access to the volume of a precast bulk material (Bhawalkar etal, 1996). The three-dimensional microfabrication capability achieved by two-photon excitation was demonstrated by the early pioneering work of Rentzepis's group on two-photon optical memory (Parthenopoulos and Rentzepis, 1989; Liang et al., 2003) and by Webb's group on two-photon lithography (Wu et al., 1992). In a two-photon process, the transition probability is quadratically proportional to the light intensity. If the laser light is focused using a lens with high numerical aperture, the two-photon absorption is largely restricted to the vicinity of the focal point, of which spatial size can be less than 1 μm. Narrower definition of structures can be achieved utilizing the feature that two-photon induced photochemistry may require a certain threshold of intensity to reach. Thus sub-wavelength nanoscopic structures can be fabricated by using two-photon-induced lithography. Because the photons used to excite the monomers or initiators are half the nominal energy required to excite the absorbing material, it is possible to excite at wavelengths wherein the material is transparent to one-photon excitation. But at the focus, where the intensity is high, two-photon excitation occurs. In this manner, two-photon excitation can provide a mechanism to circumvent the problems of linear absorption process in such a way that the laser light can penetrate deeply into the material without loss, to provide access to the desired volume element. In addition, because the absorption profile scales quadratically with intensity, if the sample is excited with a beam that has a Gaussian spatial profile, the absorption profile will be narrower than the beam profile. This, in turn, provides the opportunity to fabricate 3D nanoscopic (subwavelength) structures.

Several groups have demonstrated fabrication of subdiffraction-limited features (Kawata and Sun, 2003; Kuebler et al., 2003; Prasad et al., 2000). For example, at our Institute, two-photon-induced photo-polymerization has been used to produce subwavelength 3D structures within the volume of a polymer block.

The monomers most commonly used for photopolymerization can be classified broadly as epoxy resins acrylates and urethane acrylates. In the case of epoxy-resin-type monomers, light-generated photo acids initiate the polymerization while, in the case of acrylate-type monomers, photogenerated free radicals initiate the polymerization.

In our laboratory, several kinds of two-photon absorbing organic chromophores with very large two-photon absorption cross sections have been developed. The photon energy absorbed by a molecule can be emitted as energy up-converted fluorescence with photon energy significantly larger than that of the excitation light. This fluorescence can be utilized as a light source to induce various photochemical reactions such as photopolymerization, photocrosslinking and photodissociation. For various photochemical reactions, which can be activated at different wavelengths, several two-photon chromophores, whose linear absorption and fluorescence spectra span the visible spectrum from 400 nm to 600 nm, have been developed. The multiple wavelengths of absorption and emission can allow one to activate different species at different wavelengths to produce multiple structures of varied functionally. Our two-photon chromophores, which usually have a donor–acceptor character, can be tuned to achieve a broad spectral range of emission by varying the donor and acceptor parts. Another approach is to have a photoinitiator with high two-photon absorption cross section (Cumpston et al., 1999 Zhou et al., 2002). This photoinitiator can be of free radical generator type or photoacid generator type, depending on the type of monomer to be used.

We have demonstrated the two-photon curing of a commercial urethane acrylate-based monomer using a two-photon initiator synthesized by the group of Dr. Seng Tan at the Air Force Research Laboratory in Dayton. An example of such a photoinitiator, called AF281, is shown in Figure 11.1. These photoinitiators were designed to have strong two-photon absorption at 800 nm. A series of these com-

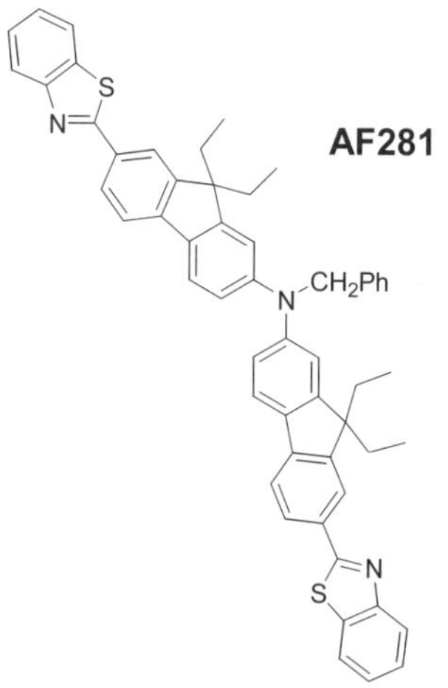

Figure 11.1. Chemical structure of AF-281 two-photon photoinitiator.

pounds have been synthesized with tailor-made properties for curing different monomers using two-photon excitation.

Using these two-photon photoinitiators, fabrication of submicron-size optical circuitry such as channel waveguides and gratings produced within solid bulk samples were demonstrated. The bulk solid was prepared from a mixture of UV curable and thermally curable epoxies, where thermal and photochemical processes initiated by two-photon absorption could control its refractive index. The experimental setup for microfabrication is shown in Figure 11.2. A stepper-motor-controlled XY stage was used to translate the sample onto which a beam was focused with a microscopic objective lens of high numerical aperture (NA). The resolution can be improved by increasing the NA of the objective lens. The laser source for these experiments was a mode-locked Ti:sapphire laser with an output wavelength of 800 nm, a pulse duration of 90 fs, a repetition rate of 84 MHz, and an average power of 200 mW.

We also used a series of two-photon photoinitiators (free radical generators) from Air Force Research Laboratory (e.g., AF281, Figure 11.1) to fabricate microstructures using a commercial acrylate-based monomer Deso-105 from Desotech.

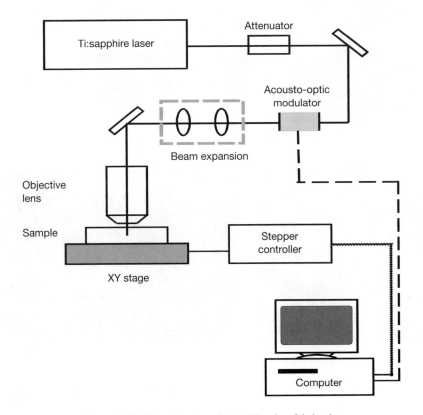

Figure 11.2. Setup for two-photon 3D microfabrication.

Various types of microstructures and submicron structures were fabricated using this process. Examples are 3D optical waveguide structures, $1 \times N$ (one splitting in to N) optical splitters, 3D multistack optical data storage, volume gratings, and 3D security identification/barcodes. Some of these fabricated structures are shown in Figure 11.3.

Two-photon lithography is witnessing a rapid growth of activities worldwide. Marder, Perry, and co-workers developed efficient two-photon-activated photoacid generators and used them to initiate the polymerization of epoxides by two-photon activation (Zhou et al., 2002; Kuebler et al., 2003). They used the photoacid in conjunction with a positive-tone chemically amplified resist to fabricate three-dimensional microchannel structures. In yet another earlier report, they produced a 3D photonic crystal using two-photon activated photoinitiation to induce polymerization (Cumpston et al., 1999). This work is already described in Chapter 9. Perry, Marder, and co-workers have also used two-photon excitation to induce growth of nanoparticles for 2D and 3D metal patterning (Stellaci et al., 2002). In this work,

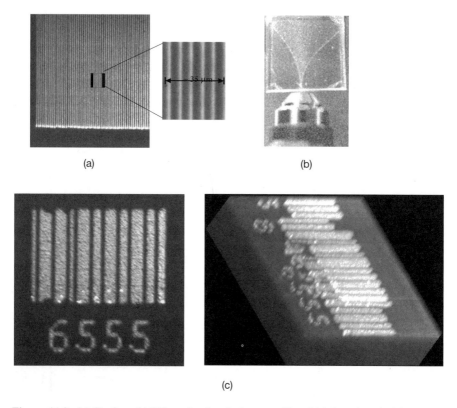

Figure 11.3. (a) Grating, (b) Y-branch of optical waveguide which is written inside a polymer bulk, and (c) planar and three-dimensional views of 3D barcode where two sets of barcodes are vertically stacked.

photoreduction of silver ions by a dye sensitizer was used to grow ligand-coated silver nanoparticles, which, upon growth, coalesced to form continuous metal features in a matrix or on a surface.

Servin et al. (2003) used two-photon polymerization of photosensitive high-optical-quality inorganic–organic hybrid materials, ORMOCER–1, with femtosecond pulses at ~ 780 nm from a Ti:sapphire laser, to produce many three-dimensional structures. The medium is a liquid resin that was polymerized using two-photon absorption by the initiator molecules. The nonirradiated liquid resin was dissolved in alcohol. They also produced photonic crystals that consisted of individual rods of 200-nm diameter with a 250-nm spacing.

The work of Kawata and Sun (2003) is an example of subdiffraction three-dimensional spatial resolution achieved by using photoproduced radical quenching effect to produce a thresholding function in two-photon-induced polymerization. In other words, only above a certain threshold of intensity (achieved in the central portion of the focused spot) the polymerization occurred. Another illustrative fabrication of subdiffraction structures presented here is from our laboratory. As shown in Figure 11.4, a structural definition of ~200 nm was produced by two-photon polymerization, using 100-fs pulses from a mode-locked Ti:sapphire laser.

For the nano-patterning with high spatial resolution, understanding the performance of individual voxels is also critical. Sun, Lee, Kawata, and co-workers (Sun et al., 2003) found that at near-threshold exposure condition, voxel shape scaling follows different laws by means of varying laser power and changing exposure time (Figure 11.5). The aspect ratio of vertical and lateral lengths of a voxel monotonically increases upon increased exposure and reaches a saturation stage for the two processes, which are conventionally regarded as equivalent for voxel volume tuning

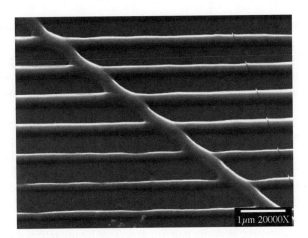

Figure 11.4. SEM image of structures produced by two-photon polymerization of Deso-Bond 956-105. A resolution of ~200 nm was achieved.

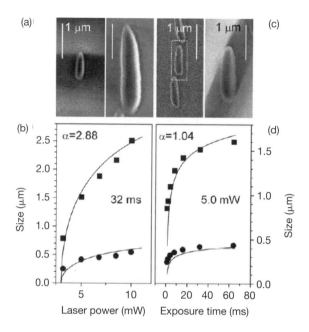

Figure 11.5. Exposure time (T)- and power (P)-dependent voxels. (a) SEM images of voxels obtained under (3.2 mW, 32 ms) and (10 mW, 32 ms); (b) P dependence; (c) voxels produced with varied exposure time (5 mW, 1 ms) and (5 mW, 64 ms); and (d) corresponding T dependence [vertical (■) and lateral (●) sizes]. A 780-nm, 80-fs, mode-locked Ti:sapphire laser, operating at 80 MHz, was employed for two-photon photopolymerization. SCR resin was used which consists of urethane acrylate monomer and oligomers. Courtesy of K.-S. Lee, HanNam University, Korea.

(Sun et al., 2003). Based on these findings, they were able to produce voxels with a near 100-nm lateral spatial resolution under optimized conditions. In addition, by employing two-photon polymerization at low laser power and short period of exposure time, Lee et al. (2003) have successfully fabricated a wide variety of two-dimensional and three-dimensional structures with high resolution, as shown in Figure 11.6.

Kirkpatrick et al. (1999) used another variation of two-photon induced polymerization, two-photon holographic recording, to produce micro- and nanostructures. They called it holographic two-photon-induced polymerization (H-TPIP). This approach utilizes overlap of two coherent ultrashort pulses, resulting in a spatially modulated intensity pattern that can be used to produce polymerized grating structures. They used a commercially available resin, mixed with a dye possessing a large two-photon absorption cross section. The advantage of the holographic approach is that a complex intensity pattern, generated by interference of multiple coherent beams, can be used to produce complex 3D nanostructure, without the need to translate the sample or raster scan the laser beam.

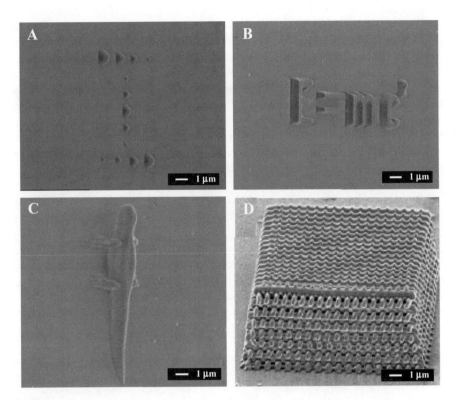

Figure 11.6. SEM images of various two-dimensional (2D) and three-dimensional (3D) micro-patterns fabricated by the two-photon-induced polymerization. (A, B) 2D patterns of voxels and letters. (C, D) 3D patterns of a lizard sculpture and a photonic crystal. For the micro-fabrication, a laser system of mode-locked Ti:sapphire laser of 780-nm wavelength, 80-fs pulsewidth, and 82-MHz repetition rate was employed. The laser focal spot was scanned in three dimensions by a piezoelectric stage to polymerize the desired part of the photoactive resin, which is a mixture of SCR-500 and two-photon absorbing chromophores. Courtesy of K.-S. Lee, HanNam University, Korea.

11.2 NEAR-FIELD LITHOGRAPHY

Nanoscale optical interactions involving near-field excitation can be used to fabricate nanoscale structures using either photochemistry or photophysics, such as domain inversion. This approach has contributed to the emerging field of near-field lithography, whereby lithographic techniques, in conjunction with near-field excitation, can be used to fabricate nanoscale structures. A number of photochemical and photophysical processes have been employed for near-field nanofabrication. Numerous examples of near-field nanofabrication have appeared in the literature, and it is not possible to list them all. Some of the examples of near-field lithography are provided here.

The first example provided is our use of photobleaching of an organic dye to produce nanopatterns (Pudavar et al., 1999; Pudavar and Prasad, unpublished results). Figure 11.7 shows an example of photobleaching of an organic dye in a plastic medium, used to write patterns such as little pixels for memory application or a grating for the coupling of light. This example utilized permanent photobleaching. In other words, the dye is chemically altered permanently. The dye used for this purpose can be activated by one-photon absorption as well as by two-photon absorption. If photobleaching is accomplished with one-photon absorption using light at 400 nm which emanates from a tapered fiber, the photobleached spot size is ~ 120 nm. The pixel size of 120 nm provides a high number of pixel arrays in one place.

Use of two-photon excitation (see Section 11.1) in a near-field geometry allows one to achieve even narrower nanoscale resolution. If two-photon absorption is used with a pulsed laser at 800 nm to photobleach the dye, the spot size produced is much narrower (~ 70 nm). Photobleaching occurs only near the focal point and on the sides, when the energy falls below a certain threshold level, no photobleaching occurs. Therefore, the pixels can be much more spatially localized and narrower. The feature is shown on the top right where 70-nm size spots are produced by two-photon excitation. Similarly, the bottom two curves illustrate writing of gratings.

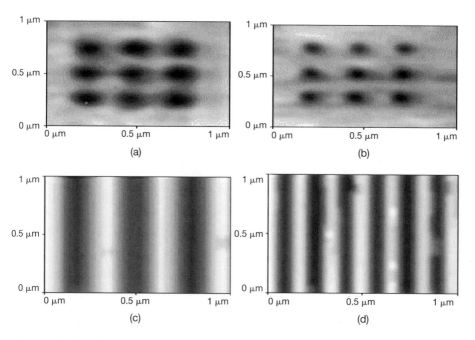

Figure 11.7. Optical memory (a, b) and grating patterns (c, d) written by NSOM. (a, c) written with one-photon excitation and readout with two-photon excitation. (b, d) Both written and readout with two-photon excitation. From Shen et al. (2001), reproduced with permission.

With one-photon absorption, one can write gratings that are 150 nm wide and the period is 290 nm. The right bottom of Figure 11.7 shows the recorded gratings using two-photon absorption, which are 70 nm wide and the period is 160 nm. Therefore, one can achieve better spatial resolution with two-photon excitation.

Ramanujam et al. (2001) fabricated narrow surface relief features in a side-chain azobenzene polyester, using near field lithography. Irradiation at 488 nm from an argon-ion laser at intensity levels of 12 W/cm^2 produced topographic features (surface relief) as narrow as 240 nm and as high as 6 nm. Very little anisotropy due to photoisomerization of azobenzene was induced in the film, and the dominant effect was topographic modification.

Sun and Leggett (2002) utilized a self-assembled monolayer (SAM) (described in Chapter 10) as a template for near-field lithography to produce nanostructures. Their approach is illustrated in Figure 11.8. They used a UV fiber with an aperture of 50 nm to couple 244-nm light that was used to irradiate and scan over the surface of a gold film covered (by solution coating) with a self-assembled monolayer consisting of alkanethiolate molecules. The thiol (-SH) group exhibits strong binding affinity to a gold surface and is often used to produce an SAM on a gold surface.

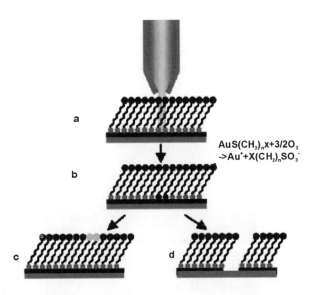

Figure 11.8. A schematic illustration of the scanning near-field photolithography (SNP) process and procedures used to prepare chemical patterns on substrate surfaces and three-dimensional nanostructures in gold films. (a) A UV probe carrying 244-nm light is scanned close to the surface (i.e., operating in the near field regime). (b) SAMs in the exposed region are photochemically oxidized. (c) Oxidized adsorbate molecules are displaced by immersion into an alcoholic thiol solution. (d) Alternatively, selective etching of Au may be initiated in the regions where the SAMs have been oxidized. From Sun and Leggett (2002), reproduced with permission.

320 NANOLITHOGRAPHY

The alkanethiolate molecules in the exposed region of the SAM are photooxidized to alkane sulfonate, which is only weakly bonded to the gold surface. Thus the chemically modified product can be removed (displaced) by another thiol by dipping into a solution containing this different thiol. Alternatively, a suitable etching solution could then remove gold from the exposed area. Sun and Leggett were able to produce three-dimensional gold structures (useful for plasmonics, see Chapter 5), less than 100 nm in size.

Yin et al. (2002) used another variation of near-field two-photon optical lithography that utilized an apertureless near-field optical microscope. The principle of an apertureless near-field scanning optical microscope has already been discussed in Chapter 3. Yin et al. used the field enhancement at the extremity of a metallic probe to induce nanoscale two-photon excitation, leading to polymerization in a commercial photoresist, SU-8. They produced lithographic features with a ~ 70-nm resolution.

Figure 11.9 shows another application where femtosecond laser pulses are used to ablate materials in order to produce nanostructures. Using a near-field geometry, one can ablate nanometer domains and remove defects. The figure shows a photomask that, in the top picture, exhibits a small structure protruding from the surface. This protruding structure is ablated away using femtosecond pulses in near-field propagation. The ablated structure is shown by the bottom picture in Figure 11.9.

Figure 11.10 shows a liquid crystal where one can change the domains optically and can define the optical contrast. Figure 11.11 provides an example of surface modification of an organic crystal, anthracene, using near-field excitation. Basically, by interaction with laser, one can create nanometer-size pits with modified sur-

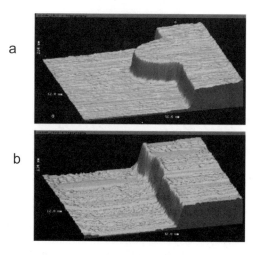

Figure 11.9. AFM image of a programmed defect prior to (a) and after repair. From Lieberman et al. (1999), reproduced with permission.

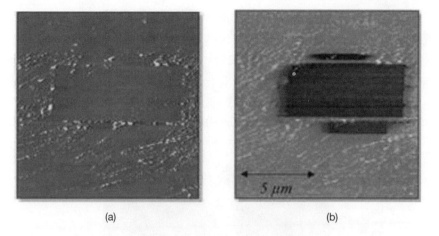

Figure 11.10. Topography (a) and optical image (b) of modified liquid crystal (8CB) by using the NSOM tip. The modification can be visualized in reflection mode using cross-polarization. From Moyer et al. (1995), reproduced with permission.

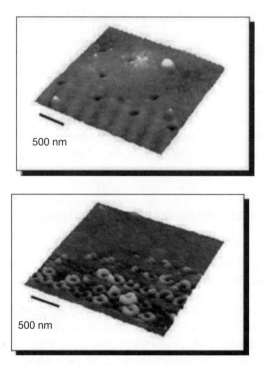

Figure 11.11. Near-field surface modification. Top: the holes in the organic crystal anthracene, indicating a true local vaporization of material on a scale of smaller than 100 nm. Bottom: craters formation. From Zeisel et al. (1997), reproduced with permission.

322 NANOLITHOGRAPHY

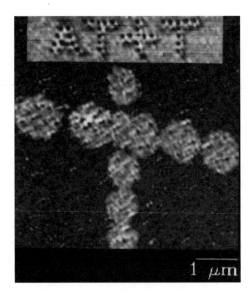

Figure 11.12. Magneto-optic domains as small as 60 nm are optically written and imaged in a 14-nm-thick Co/Pt multilayer magnetooptic film with NSOM, presenting the possibility of 45 Gbits/in.^{2} storage densities. From Betzig et al. (1992), reproduced with permission.

face quality. The craters created by laser ablation can be used as templates to build structures, by techniques such as self-assembly, on the modified surface.

Figure 11.12 shows an example where one can use near-field microscopy on magneto-optic materials to change the magnetization in nanometer-size domains. Then magneto-optical properties can be used to change the light polarization in different ways. This approach can be used for magnetic memory, because one can store information by changes in magnetic domains.

11.3 NEAR-FIELD PHASE-MASK SOFT LITHOGRAPHY

Soft lithography refers to the use of elastomeric (soft) photomasks that can conveniently be produced from a master and easily removed (lifted off) from the developed photoresist. Whitesides and co-workers (Rogers et al., 1998, 2000) have demonstrated that many organic electronic and integrated optical device structures can be produced using elastomeric stamps, molds, and conformable photomasks. One specific application described here is that of use of an elastomeric phase mask, with near-field contact-mode photolithography, to generate ~90-nm features. The principle utilized is that the surface relief on the elastomeric stamp modulates the phase of light passing through it and thus acts as a phase mask. If the depth of the relief is chosen to introduce a π-phase shift, then nulls appear in the near-field pattern of the transmitted intensity at each edge in the relief. The widths of these nulls can be of the order of ~100 nm when ~365-nm light is used. These nulls in the intensity can be used to produce ~100-nm lines in a positive photoresist, after exposure and development. Whitesides' group has also demonstrated that the method can utilize even a broadband incoherent light source, such as a mercury lamp.

A schematic of the experimental procedure utilized by Whitesides and coworkers is shown in Figure 11.13. A prepolymer of polydimethylsiloxane (PDMS) film of appropriate thickness is cast on a master, generated by photolithography. The prepolymer film is cured to produce an elastomeric mask, which is then removed from the master. This mask is now brought into conformal contact with the photoresist and is exposed to UV light where the photoresist senses the near-field intensity distribution determined by the π-phase shift. Subsequent development of the photoresist after exposure through the elastomeric phase mask produced structures of width ~100 nm. Figure 11.14 shows an AFM picture of the parallel lines formed in the photoresists. These lines have a width of ~100 nm and are ~300 nm in height.

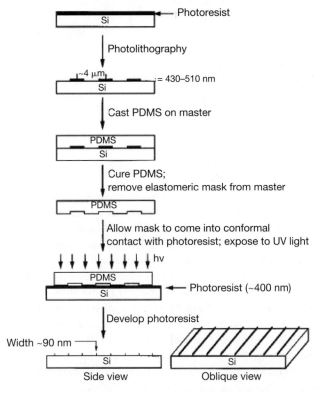

Figure 11.13. Schematic illustration of the fabrication of nanometer-size structures by near-field contact-mode photolithography. From Rogers et al. (1998), reproduced with permission.

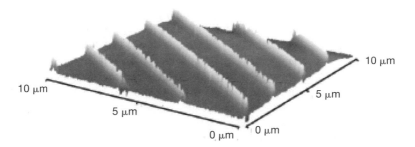

Figure 11.14. Parallel lines formed in photoresist using near-field contact-mode photolithography have widths on the order of 100 nm and are 300 nm in height as imaged by AFM. From Rogers et al. (1998), reproduced with permission.

This method can conveniently be used to produce slits in metals that can be used to define narrow separation between source and drain electrodes for organic electronics (Rogers et al., 1999). Other types of nanostructures can be produced by multiple exposures with different photomasks. For example, nanometer-size dots can be generated by exposing photoresist twice with a periodic binary phase mask, rotating the mask by 90° between the two exposures (Rogers et al., 1998). The resolution of the contact mode lithography using elastomeric phase masks is determined not only by the wavelength of light used, but also by the refractive index of the photoresist. Hence simply by using photoresist of high refractive index, the resolution of photolithography can be increased.

11.4 PLASMON PRINTING

Plasmon printing is a method introduced by Atwater and co-workers which allows replication of patterns with a resolution considerably below the diffraction limit, with the use of a standard photoresist and a broad beam light illumination (Kik et al., 2002). The method utilizes surface plasmon resonances of a metallic nanostructure, discussed in Chapter 5, to considerably enhance the electric field near it, when the light wavelength used corresponds to the plasmon resonance frequency.

A schematic of the plasmon printing method is shown in Figure 11.15. Here, metallic nanostructures in the form of an array of metal nanoparticles are used. Glancing angle illumination by light at the wavelength of plasmon resonance of the metal nanoparticles produces a strongly enhanced field near the particle. As discussed in Chapter 5, the enhanced field is dipolar in nature for small metal particles. This enhanced field, localized nanoscopically near the nanoparticle, is used to produce nanoscopic exposure of a thin layer of photoresist directly below the metal nanostructure on the mask layer. Upon development, the resist layer produces a nanostructure pattern. In order for local field intensity to be enhanced directly below the particle, the illumination beam should be polarized approximately normal

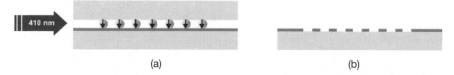

Figure 11.15. (a) Schematic representation of plasmon printing, showing glancing angle illumination using polarized visible light, producing enhanced resist exposure directly below the metal nanostructures on the mask layer. (b) The resulting pattern in the resist layer after development. From Kik et al. (2002), reproduced with permission.

to the resist layer; that is, it is p-polarized. For that purpose, a glancing incidence exposure is utilized.

For effective plasmon printing, the following considerations are important:

- Nanostructures of metals with long carrier relaxation time, in order to achieve strong field enhancement. For this reason, gold ($\tau_{relax} \approx 4$ fs) and silver ($\tau_{relax} \approx 10$ fs) are reasonable choices.
- Metallic nanostructures (nanoparticles) with sufficiently small size to achieve strong spatially confined dipolar enhancement of field below the particle. When the particle size is large, multipolar oscillations, as discussed in Chapter 5, reduce the spatial confinement and produce a reduction of the field enhancement. For this reason, gold or silver nanoparticles of 30- to 40-nm diameter are desirable.
- Metallic nanostructures with plasmon resonance in the wavelength range of 300–450 nm where most photoresists exhibit maximum sensitivity. For this reason, silver nanoparticles (plasmon resonance at ~410 nm) may be preferrable.

In their demonstration of plasmon printing, Atwater and co-workers (Kik et al., 2002) utilized spray-deposition to produce ~41-nm-diameter silver nanoparticles. A standard g-line resist (AZ 1813 from Shipley) was used which has maximum sensitivity in the 300- to 450-nm wavelength range. It was spin-coated as a ~75-nm-thick film. The wavelength of 410 nm from a 1000-W Xe arc lamp was selected by a monochromator and passed through a polarizer to obtain polarization normal to the sample surface. To increase power density, the beam was compressed in the vertical direction by a cylindrical lens and was incident on the sample at a glancing angle. Subsequent to development, AFM images of the film showed circular depressions of 50 nm in diameter, with a depth of 12 nm.

11.5 NANOSPHERE LITHOGRAPHY

Nanosphere lithography involves polymer or silica nanospheres that are self-assembled to produce a close-packed monolayer on a substrate and then infiltrated by the materials to be deposited to produce a two-dimensional nanopattern on the substrate

(Deckman and Dunsmuir, 1982; Haynes and VanDuyne, 2001). The first step in nanosphere lithography involves forming a close-packed monolayer of submicron size nanospheres, similar to the one used for the fabrication of colloidal photonic crystals described in Chapter 9. In the present case a monolayer deposition may simply utilize spin coating from a dispersion of the nanospheres in an appropriate solvent. The speed of spin coating is optimized to produce a close-packed monolayer, a periodic particle array of nanospheres. The film is dried subsequently by thermal or laser pulse deposition to produce a colloidal crystal mask. Figure 11.16A provides schematic illustration of the colloidal crystal mask. The material to be patterned (such as a metal) is deposited on the substrate through the interstitial holes (voids) between the nanospheres. A solvent wash is then used to remove the polymer nanosphere, leaving behind a nanopatterned material. In the case of silica nanospheres, they can be etched using an HF solution. In the case of hexagonally packed nanospheres, the final patterned periodic array of the deposited material (e.g., Ag) is triangular in shape whose spacing can be varied by varying the size of the nanospheres. Figure 11.16B shows AFM image of the array of the triangular shape Ag nanoparticles. The metallic nanoarrays thus produced can be used for plasmonics, a topic discussed in Chapter 5.

A nice example of the use of nanosphere lithography is provided by the work of Marks and co-workers (Veinot et al., 2002) to produce organic light-emitting "nanodiode" arrays. Their fabricated multilayer structure for nanodiode is illustrated in Figure 11.17. They utilized deposition of carboxy-terminated polystyrene nanobeads (sizes 400 nm and 160 nm) on a clean, ultra-smooth ITO/glass substrate, coated with a self-assembled triarylamine monolayer. This monolayer coating acts to facilitate hole injection into, and cohesion of, the subsequent hole transporting layers of 1,4-bis(1-naphthyl-phenylaminobiphenyl) (NPB). After formation of a close-packed polystyrene nanosphere monolayer, and subsequent drying, they were used as a shadow mask to vacuum deposit a 50-nm-thick hole-transporting layer of NPB. This was followed by deposition of a ~40-nm-thick electron-transporting layer of aluminum

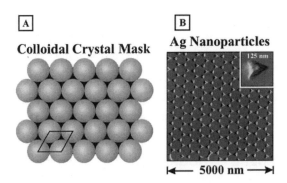

Figure 11.16. (A) schematic illustration of colloidal crystal mask containing with $D = 542$-nm nanospheres. (B) AFM image of triangular-shaped silver nanoparticles. From Haynes and VanDuyne (2001), reproduced with permission.

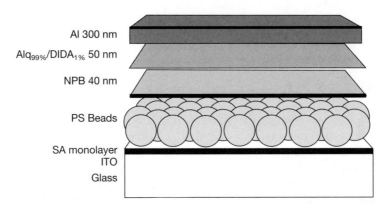

Figure 11.17. Schematic of nano-OLED fabrication. From Veinot et al. (2002), reproduced with permission.

tris(8-hydroxyquinolate) (Alq$_3$) containing the emissive dopant, N,N'-di-isoamylquinacridone (DIQA). Finally, a 300-nm-thick Al layer was deposited. Following tape lift-off of the beads, a regular pattern of trigonal prismatic posts of light-emitting multilayer heterostructure with 90-nm sides was obtained in the case of 400-nm-size beads. In the case of 160-nm size, the sides of the trigonal prismatic structures were estimated to be ~43 nm. Figure 11.18 (inset) shows an AFM image of the array of trigonal prismatic diodes obtained with the 400-nm beads. Also displayed in this figure are the electroluminescence intensity-applied voltage curves for

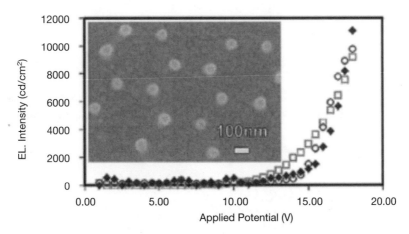

Figure 11.18. Emission intensity-V profiles of 90-nm nanodiode (◆), 43-nm nanodiode arrays (○), and macrodiode (□). Nanodiode data are corrected for fill factor. **Inset:** SEM image of an array of trigonal prismatic diodes (90 nm/side). From Veinot et al. (2002), reproduced with permission.

both the 90-nm and the 43-nm nanodiode arrays. The results also indicate minimal size-dependent compromise of either rectification or emissive characteristics.

11.6 DIP-PEN NANOLITHOGRAPHY

Dip-pen nanolithography (DPN) is another type of soft lithography developed by Mirkin and co-workers (Piner et al., 1999; Hong et al., 1999) which allows nanopatterning of molecules to produce structures which are one molecule thick. It is a direct-write method which utilizes the tip of an atomic force microscope (AFM) as a pen, with molecules to be laid down on a substrate acting as an ink. The water condensation between the tip and the substrate forms a water-filled capillary that moves with the tip and helps the transport and subsequent chemisorption of organic molecules with low water solubility to be deposited on the substrate. Mirkin and co-workers demonstrated the deposition of alkanethiol (octadecanethiol) in a pattern on a gold (111) substrate. The process is schematically represented in Figure 11.19.

By the choice of the molecule transported, one can control the nature of the surface. For example, the use of 16-mercaptohexadecanoic acid produces a hydrophilic surface, while octadecanethiol yields a hydrophobic surface. Figure 11.20 illustrates nanoscale letter writing using DPN, where the generated nanostructure is in the form of "AFOSR" with approximately 70-nm line widths (Mirkin, 2000). One can use DPN to produce multicomponent nanostructure. For example, one can write an initial nanostructure with octadecanethiol and then write, within this initial structure, another nanostructure with a different type of ink (e.g., 16-mercaptohexadecanoic acid).

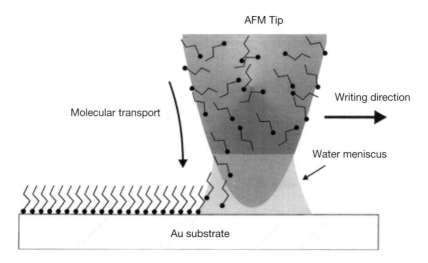

Figure 11.19. Schematic representation of dip-pen nanolithography. From Piner et al. (1999), reproduced with permission.

Figure 11.20. The word AFOSR written using dip-pen nanolithography. From Mirkin (2000), reproduced with permission.

DPN holds promise for producing nanostructures for photonic applications. For example, DPN can be used to produce photonic crystals. Mirkin and co-workers (Zhang et al., 2003) utilized DPN to fabricate arrays of sub-50-nm gold dots and line structures which can be utilized for plasmonic applications. For this purpose, they used DPN to pattern the etch resist, 16-mercaptohexadecanoic acid, on Au/Ti/SiO$_x$/Si substrates and then used wet-chemical etching to remove the exposed gold. For etching out the exposed gold, they used an aqueous mixture of 0.1 M Na$_2$S$_2$O$_3$; 1.0 M KOH, 0.01 M K$_3$Fe(CN)$_6$, and 0.001 M K$_4$Fe(CN)$_6$. The distance between the gold dots produced were 100 nm. Figure 11.21 shows an AFM picture of the gold dot nanoarrays.

Figure 11.21. AFM topographic images of an etched MHA/Au/Ti/SiO$_x$/Si dot nanoarray. From Zhang et al. (2003), reproduced with permission.

11.7 NANOIMPRINT LITHOGRAPHY

Nanoimprint lithography is a high-throughput, low-cost, and nonconventional lithographic technique. It involves an imprint step in which a mold, with nanostrucures on its surface, is pressed into a thin resist (or soft material) that is on a substrate. This step is followed by removal of the mold. Thus the nanostructure in the mold is embossed in the resist film. Further thickness contrast can be created by using an anisotropic etching process, such as reactive ion etching, to remove the residual resist in the compressed areas. The resolution of this nonoptical method is not diffraction-limited. Nanostructuring with 25-nm feature size and 90° corners have been demonstrated, and the method has successfully been used to fabricate nanoscale photodetectors, silicon quantum dots, quantum wire, and ring transistors.

Chou et al. (2002) used laser-assisted direct imprint nanolithography, for fabrication of silicon nanostructures. The method is schematically represented in Figure 11.22. A mold made of quartz is pressed against the silicon substrate. A thin surface layer of silicon, which is in contact with the quartz mold, is melted by a laser pulse at 308 nm from an XeCl excimer laser, which passes through the quartz mold without being absorbed. The molten silicon layer can be as deep as 300 nm and remains in the molten state for hundreds of nanoseconds. The molten layer is then embossed by the quartz mold. The mold is separated from the imprinted silicon after it solidi-

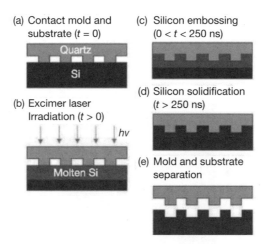

Figure 11.22. Schematic of laser-assisted direct imprint (LADI) of nanostructures in silicon. (a) A quartz mold is brought into contact with the silicon substrate. A force presses the mold against the substrate. (b) A single XeCl (308-nm wavelength) excimer laser pulse (20-nm pulse width) melts a thin surface layer of Si. (c) The molten silicon is embossed while the silicon is in the liquid phase. (d) The silicon rapidly solidifies. (e) The mold and silicon substrate are separated, leaving a negative profile of the mold patterned in the silicon. From Chou et al. (2002), reproduced with permission.

fies. A variety of silicon structures, with a resolution better than 10 nm, were generated by this method.

11.8 PHOTONICALLY ALIGNED NANOARRAYS

A novel promising photonic approach for nanofabrication is ordering of nanoparticles that is produced by their light-driven migration (Bunning et al., 2000; Vaia et al., 2001). A schematic of this approach is shown in Figure 11.23.

The nanoparticles represented by black dots are randomly distributed in a matrix that can be a monomer of appropriate structure or a polymer that allows these nanoparticles to be randomly distributed. When two beams are crossed to generate an intensity grating (holographic geometry), as shown in the middle, intensity modulation producing alternate bright and dark regions occurs. This intensity modulation pattern can be used to initiate a spatially modulated polymerization of the monomers. Thus, in the bright regions polymerization occurs, whereas in the dark regions the monomers remain unpolymerized. One can crosslink a polymer in the bright region, whereas in the dark regions, no crosslinking of the polymer occurs. If the nanoparticles are not compatible with the polymerized region, they will tend to migrate toward the dark unpolymerized regions. This spatial movement of the nanoparticles can cause them to align in the dark regions, producing a periodic array of nanoparticles.

One can utilize this approach to align an array of Q-dots or to produce photonic crystals where the aligned nanodots with high refractive index produce a high refractive index contrast. Alternatively, one can process the polymer-dispersed liquid crystal (Bunning et al., 2000) in a similar manner. One can have larger-size nan-

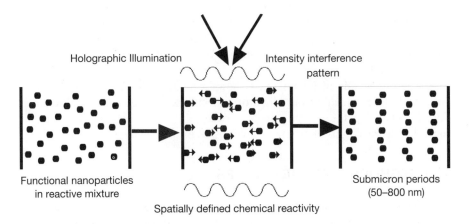

Figure 11.23. Using the holographic (laser) photopolymerization to induce movement and sequester nanoparticles into defined three-dimensional patterns. (Courtesy of T. Bunning and R. Vaia, Air Force Research Laboratory, Dayton.)

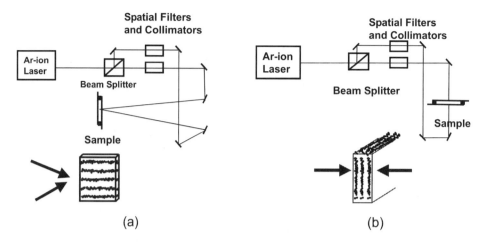

Figure 11.24. Two types of writing geometry. transmission (a) and reflection (b) gratings. Courtesy of T. Bunning, Air Force Research Laboratory, Dayton.

odroplets of liquid crystals (for scattering-type PDLC discussed in Chapter 10). These droplets are shown as black dots in Figure 11.23. Suppose one polymerizes them in a matrix, or crosslinks the matrix to squeeze out these nanodroplets; then they phase-separate in the dark region to form a grating. One will have a grating that diffracts light when there is no field. When a field is applied, the refractive index contrast between the liquid crystal nanodroplets and the polymer disappears, resulting in loss of the grating and concomitant loss of diffraction. This process thus can be used to produce a grating that is dynamically switchable with an electric field, and it can be used to steer a beam or diffract light in certain directions.

Figure 11.24 shows these PDLC gratings. A counterpropagating holographic geometry, shown on the right-hand side, produces alignment of nanodroplets of liquid crystal in the vertical planes to form a reflection grating. The diffraction efficiency achieved from this dynamic grating is 70%, and the switching time is in tens of microseconds.

The left-hand side of Figure 11.23 shows a transmission grating, formed in a horizontal plane, which diffracts light of a specific wavelength in a specific direction. By changing the crossing angle between the two beams, gratings can be formed to diffract lights of different wavelengths in specific directions.

11.9 HIGHLIGHTS OF THE CHAPTER

- Application of nonlinear optical processes, near-field excitation, and plasmonic field enhancement cause photochemical or photophysical changes that produce nanostructures significantly smaller than the wavelength of light used.

11.9 HIGHLIGHTS OF THE CHAPTER

- Two-photon lithography involves two-photon absorption-induced photoprocesses to produce nanostructures.
- The quadratic dependence of two-photon excitation on the optical intensity, together with energy threshold needed to effect photochanges, provides a means to produce photogenerated nanostructures with definitions significantly less than the wavelength of excitation.
- Examples of photogenerated nanostructures produced by two-photon lithography are complex three-dimensional optical circuitry, multilayer optical data storage, and 3D photonic crystals.
- Two-photon holography is another kind of two-photon lithography that utilizes overlap of two or more coherent short pulses. It generates a spatially modulated pattern that can photoproduce the corresponding nanostructures.
- Near-field lithography utilizes near-field excitation to photogenerate nanostructures.
- The use of two-photon excitation in a near-field geometry achieves even narrower nanoscale resolution, below 100 nm, compared to the corresponding one-photon excitation in the same geometry.
- Near-field lithography can produce nanoscale physical modification (such as physical transformation) of domains, surface modification, and patterning.
- Soft lithography refers to embossing of a nanostructure using a soft, elastomeric mask from a master and subsequently removing (lifting off) the master.
- Phase-mask soft lithography utilizes an elastomeric phase mask with near-field contact-mode photolithography to generate ~90-nm features. It utilizes the principle that the surface relief on the elastromeric stamp modulates the phase of light passing through it and thus acts as a phase mask.
- Plasmon printing involves surface plasmon resonances of a metallic nanostructure to enhance the electromagnetic field, which is subsequently used to produce nanoscopic exposure of a thin layer of photoresist directly below the metal nanostructure.
- Nanosphere lithography utilizes a close-packed monolayer of polymer or silica nanoparticles on a substrate that is infiltrated (into interstitial voids) by the materials to be deposited in a two-dimension pattern on the substrate. A solvent is then used to dissolve the polymer nanoparticle, or HF is used to etch out the silica nanoparticles.
- Dip-pen nanolithography is a direct-write method that utilizes the tip of an atomic force microscope (Chapter 7 describes this microscope) as a pen, with molecules as the ink to be laid down on a substrate.
- Nanoimprint lithography utilizes printing of a nanostructure (pattern) in a mold onto a molten film on a substrate, by pressing the mold against it.
- A novel photonic method for fabrication of nanostructures is ordering of embedded nanoparticles, by their light-driven migration in a medium.

REFERENCES

Betzig, E., Trautman, J. K., Wolfe, R., Gyorgy, E. M., Finn, P. L., Kryder, M. H., and Chang, C.-H., Near-Field Magneto-optics and High Density Data Storage, *Appl. Phys. Lett.* **61**, 142–144 (1992).

Bhawalker, J. D., He, G. S., and Prasad, P. N., Nonlinear Multiphoton Processes in Organic and Polymeric Materials, *Rep. Prog. Phys.* **59**, 1041–1070 (1996).

Bunning, T. J., Kirkpatrick, S. M., Natarajan, L. V., Tondiglia, V. P., and Tomlin, D. W., Electrically Switchable Grating Formed Using Ultrafast Holographic Two-Photon-Induced Photopolymerization, *Chem. Mater.* **12**, 2842–2844 (2000).

Chou, S. Y., Krauss, P. R, and Renstrom, P. J., Nanoimprint Lithography, *J. Vac. Sci. Technol. B* **14**, 4129–4133 (1996).

Chou, S. Y., Chris, K. , and Jian, G., Ultrafast and Direct Imprint of Nanostructures in Silicon, *Nature* **417**, 835–837 (2002).

Cumpston, B. H., Ananthavel, S. P., Barlow, S., Dyer, D. L., Ehrlich, J. E., Erskine, L. L., Heikal, A. A., Kuebler, S. M., Lee, I.-Y. S., McCord-Maughon, D., Qin, J., Röckel, H., Rumi, M., Wu, X.-L., Marder, S. R., and Perry, J. W., Two-Photon Polymerization Initiators for Three-Dimensional Optical Data Storage and Microfabrication, *Nature* **398**, 51–54 (1999).

Deckman, H. W., and Dunsmuir, J. H., Natural Lithography, *Appl. Phys. Lett.* **41**, 377–379 (1982).

Frisken S. J., Light-Induced Optical Waveguide Uptapers, *Opt. Lett.* **18**, 1035–1037 (1993).

Haynes, C. L., and VanDuyne, R. P., Nanosphere Lithography: A Versatile Nanofabrication Tool for Studies of Size-Dependent Nanoparticle Optics, *J. Phys. Chem. B* **105**, 5599–5611 (2001).

Hong, S., Zhu, J., and Mirkin, C. A., Multiple Ink Nanolithography: Toward a Multiple Pen Nano-Plotter, *Science* **286**, 523–525 (1999).

Kawata, S., and Sun, H.-B., Two-Photon Photopolymerization as a Tool for Making Micro-Devices, *Appl. Surf. Sci.* **208–209**, 153–158 (2003).

Kik, P. G., Maier, S. A., and Atwater, H. A., Plasmon Printing—A New Approach to Near-Field Lithography, *Mater. Res. Soc. Symp. Proc.* **705**, 66–71 (2002).

Kirkpatrick, S. M., Baur, J. W., Clark, C. M., Denny, L. R., Tomlin, D. W., Reinhardt, B. R., Kannan, R., and Stone, M. O., Holographic Recording Using Two-Photon-Induced Photopolymerization, *Appl. Phys. A* **69**, 461–464 (1999).

Kuebler, S. M., Braun, K. L., Zhou, W., Cammack, J. K., Yu, T., Ober, C. K., Marder, S. R., and Perry, J. W., Design and Application of High-Sensitivity Two-Photon Initiators for Three-Dimensional Microfabrication, *J. Photochem. Photobiol. A Chem* **158**, 163–170 (2003).

Lee, K. S., Yang, H. K., Kim, M. S, Kim, R. H, Kim, J. Y., Sun, H. B, and Kawata, S., private communication, 2003.

Liang, Y. C., Dvornikov, A. S., and Rentzepis P. M., Nonvolatile Read-out Molecular Memory, *Proc. Natl. Acad. Sci.* **100**, 8109–8112 (2003).

Lieberman, K., Shani, Y., Melnik, I., Yoffe, S., and Sharon, Y., Near-Field Optical Photomask Repair with a Femtosecond Laser, *J. Microsc.* **194**, 537–541 (1999).

Mirkin, C. A., Programming the Assembly of Two- and Three-Dimensional Architectures with DNA and Nanoscale Inorganic Building Blocks, *Inorg. Chem.* **39**, 2258–2274 (2000).

Moyer, P. J., Kämmer, S., Walzer, K., and Hietschold, M., Investigations of Liquid Crystals and Liquid Ambients Using Near-Field Scanning Optical Microscopy, *Ultramicroscopy* **61,** 291–294 (1995).

Parthenopoulos, D. A., and Rentzepis, P. M., Three-Dimensional Optical Storage Memory, *Science* **245,** 843–845 (1989).

Piner, R. D., Zhu, J., Xu, F., Hong, S., and Mirkin, C. A., Dip-Pen Nanolithography, *Science* **283,** 661–663 (1999).

Prasad, P. N., Reinhardt, B., Pudavar, H., Min, Y. H.; Lal, M., Winiarz, J., Biswas, A., and Levy, L., Polymer-Based New Photonic Technology Using Two Photon Chromophores and Hybrid Inorganic–Organic Nanocomposites, 219th ACS National Meeting, San Francisco, CA, March 26–30, 2000, Book of Abstracts (2000), POLY–338.

Pudavar, H. E., Joshi, M. P., Prasad, P. N., and Reinhardt, B. A., High-Density Three-Dimensional Optical Data Storage in a Stacked Compact Disk Format with Two-Photon Writing and Single Photon Readout, *Appl. Phys. Lett.* **74,** 1338–1340 (1999).

Ramanujam, P. S., Holme, N. C. R., Pedersen, M., and Hvilsted, S., Fabrication of Narrow Surface Relief Features in a Side-Chain Azobenzene Polyester with a Scanning Near-Field Microscope, *J. Photochem. Photobiol. A: Chem.* **145,** 49–52 (2001).

Rogers, J. A., Paul, K. E., Jackman, R. J., and Whitesides, G. M., Generating ~90 Nanometer Features Using Near-Field Contact-Mode Photolithography with an Elastomeric Phase Mask, *J. Vac. Sci. Technol. B* **16,** 59–68 (1998).

Rogers, J. A., Dodabalapur, A., Bao, Z., and Katz, H. E., Low-Voltage 0.1 μm Organic Transistors and Complementary Inverter Circuits Fabricated with a Low-cost Form of Near-field Photolithography, *Appl. Phys. Lett.* **75,** 1010–1012 (1999).

Rogers, J. A., Bao, Z., Meier, M., Dodabalapur, A., Schueller, O. J. A., and Whitesides, G. M., Printing, Molding, and Near-Field Photolithographic Methods for Patterning Organic Lasers, Smart Pixels and Simple Circuits, *Synth. Metals* **115,** 5–11 (2000).

Serbin, J., Egbert, A., Ostendorf, A., Chichkov, B. N., Houbertz, R., Domann, G., Schultz, J., Cronauer, C., Fröhlich, L., and Popall, M., Femtosecond Laser-Induced Two-Photon Polymerization of Inorganic–Organic Hybrid Materials for Applications in Photonics, *Opt. Lett.* **28,** 301–303 (2003).

Shen, Y., Swiatkiewicz, J., Jakubczyk, D., Xu, F., Prasad, P. N., Vaia, R. A., and Reinhardt, B. A., High-Density Optical Data Storage with One-Photon and Two-Photon Near-Field Fluorescence Microscopy, *Appl. Opt.* **40,** 938–940 (2001).

Stellacci, F., Bauer, C. A., Meyer-Friedrichsen, T., Wenseleers, W., Alain, V., Kuebler, S. M., Pond, S. J. K., Zhang, Y., Marder, S. R., and Perry, J. W., Laser and Electron-Beam Induced Growth of Nanoparticles for 2D and 3D Metal Patterning, *Adv. Mater.* **14,** 194–198 (2002).

Sun, S., and Leggett, G. J., Generation of Nanostructures by Scanning Near-Field Photolithography of Self-Assembled Monolayers and Wet Chemical Etching, *Nano Letters* **2,** 1223–1227 (2002).

Sun, H.-B., Takada, K., Kim, M.-S., Lee, K.-S., and Kawata, S. Scaling Laws of Voxels in Two-Photon Photopolymerization Nanofabrication, *Appl. Phys. Lett.* **83,** 1104–1106 (2003).

Vaia, R. A., Dennis, C. L., Natarajan, V., Tondiglia, V. P., Tomlin, D. W., and Bunning, T. J., One-Step, Micrometer-Scale Organization of Nano- and Mesoparticles Using Holographic Photopolymerization: A Generic Technique, *Adv. Mater.* **13,** 1570–1574 (2001).

Veinot, J. G. C., Yan, H., Smith, S. M., Cui, J., Huang, Q., and Marks, T. J., Fabrication and Properties of Organic Light-Emitting Nanodiode Arrays, *Nano Lett.* **2,** 333–335 (2002).

Wu, E.-S., Strickler, J., Harrell, R., and Webb, W. W., Two-Photon Lithography for Microelectronic Application, *SPIE Proc.* **1674,** 776–782 (1992).

Yin, X., Fang, N., Zhang, X., Martini, I. B., and Schwartz, B. J., Near-Field Two-Photon Nanolithography Using an Apertureless Optical Probe, *Appl. Phys. Lett.* **81,** 3663–3665 (2002).

Zeisel, D., Dutoit, B., Deckert, V., Roth, T., and Zenobi, R., Optical Spectroscopy and Laser Desorption on a Nanometer Scale, *Anal. Chem.* **69,** 749–754 (1997).

Zhang, H., Chung, S. W., and Mirkin, C. A., Fabrication of Sub-50-nm Solid-State Nanostructures on the Basis of Dip-Pen Nanolithography, *Nano Lett.* **3,** 43–45 (2003).

Zhou, W., Kuebler, S. M., Braun, K. L., Yu, T., Cammack, J. K., Ober, C. K., Perry, J. W., and Marder, S. R., An Efficient Two-Photon-Generated Photoacid Applied to Positive-Tone 3D Microfabrication, *Science* **296,** 1106–1109 (2002).

CHAPTER 12

Biomaterials and Nanophotonics

Biological systems provide researchers with a fertile ground with regard to materials, enabling new nanotechnologies that cover a wide range of applications. In Nature, bioprocesses yield nanostructures that are nearly flawless in composition, stereospecific in structure, and flexible. Furthermore, because of their biodegradable nature, these biomaterials are environmentally friendly. Compounds of biological origin can spontaneously organize (self-assemble) into complex nanostructures and function as systems possessing long range and hierarchical nanoscale order. In addition, chemical modification and genetic engineering can be used to modify biomaterials to enhance or engineer a specific functionality. Indeed, nanophotonics applications can utilize a number of diverse groups of biomaterials for a variety of active and passive photonic functions as discussed here. This chapter describes some important types of biomaterials which allow nanostructure control for nanophotonics. The four types of biomaterials described here are (i) bioderived materials, either in the natural form or with chemical modifications, (ii) bioinspired materials, synthesized using guiding principles of biological systems, (iii) biotemplates providing anchoring sites for self-assembling of photonic active structures, and (iv) bacteria bioreactors for producing photonic polymers by metabolic engineering. The applications of biomaterials include efficient harvesting of solar energy, low-threshold lasing, high-density data storage, optical switching, and filtering. This chapter discusses some of these applications of biomaterials.

Section 12.1 describes some selected examples of bioderived materials. A widely investigated bioderived material for photonics is bacteriorhodopsin (Birge et al., 1999). The main focus has been to utilize its excited-state properties and associated photochemistry for high-density holographic data storage. Native DNA is another example of bioderived material used as photonic media for optical waveguide and host for laser dyes. Here, nanoscale self-assemblying of photonically active groups is provided by the double-stranded helical structure of DNA. Yet another example of bioderived materials is provided by biocolloids that consist of highly structured and complex, discrete nanoscale-size biological particles that can be organized into close-packed arrays via surface-directed assembly to form photonic crystals, discussed in Chapter 9.

Bioinspired materials are synthetic nanostructured materials produced by mimicking natural processes of synthesis of biological materials. A growing field is biomimicry with a strong focus on producing multifunctional hierarchical materials

Nanophotonics, by Paras N. Prasad
ISBN 0-471-64988-0 © 2004 John Wiley & Sons, Inc.

and morphologies that mimic nature. An example of this category is a light-harvesting dentritic structure that is presented in Section 12.2.

Biotemplates, discussed in Section 12.3, refer to natural nanostructures with appropriate morphologies and surface interactions to serve as templates for creating multiscale and multicomponent photonics materials. The biotemplates can be a naturally occurring biomaterial, a chemically modified bioderived material, or its synthetic analogue. Examples discussed are oligonucleotides, polypeptides, and viruses with organized nanostructures of varied morphologies.

Metabolically engineered materials, covered in Section 12.4, refer to those synthesized by the naturally occurring bacterial biosynthetic machinery that can be manipulated to produce a family of helical polymers having a wide range of optical properties. An example provided is a wide range of biodegradable polyester polymers, the polyhydroxyalkaonates (PHA), synthesized by bacteria. The highlights for the chapter are provided in Section 12.5. Additonal suggested general reading materials are:

Prasad (2003): *Introduction to Biophotonics*
Saleh and Teich (1991): *Fundamentals of Photonics*
Birge et al. (1999): *Biomolecular Electronics: Protein-Based Associative Processors and Volumetric Memories*

12.1 BIODERIVED MATERIALS

This section presents examples of naturally occurring biomaterials or their chemically derivatized forms that have been investigated for photonics. Among those are:

- Bacteriorhodopsin
- DNA
- Biocolloids

Another bioderived photonic material, receiving a great deal of attention, is green fluorescent protein (GFP) in its wild and mutant forms (Prasad, 2003). The green fluorescent protein and its variants exhibit strong emission that can be generated by both one-photon and two-photon excitations. By choosing an appropriate variant, one can select the excitation and the emission wavelength. The optical properties of these proteins are very much dependent on their nanostructures (Yang et al., 1996; Prasad, 2003). GFP has extensively been used as biological fluorescent markers for *in vivo* imaging and fluorescence resonance energy transfer (FRET) imaging to study protein–protein and DNA–protein interactions. Other photonic applications of GFP, reported in the literature, are for photovoltaic devices and for lasing. The biophotonics book by this author (Prasad, 2003) describes these photonic applications of GFP.

Here only bacteriorhodopsin, DNA, and biocolloids are discussed.

Bacteriorhodopsin. Bacteriorhodopsin (bR) (Birge et al., 1999) is a naturally occurring protein for which a broad range of photonic applications has been proposed. Its robustness, ease of processing into optical quality films, suitable photophysics and photochemistry of the excited state, and flexibility for chemical and genetic modifications make this protein of significant interest for photonic applications (Birge et al., 1999). These applications include random access thin film memories (Birge et al., 1989), photon counters and photovoltaic converters (Marwan et al., 1988; Sasabe et al., 1989; Hong, 1994), spatial light modulators (Song et al., 1993), reversible holographic media (Vsevolodov et al., 1989; Hampp et al., 1990), artificial retinas (Miyasaka et al., 1992; Chen and Birge, 1993), two-photon volumetric memories (Birge, 1992), and pattern recognition systems (Hampp et al., 1994). The application discussed here is holographic data storage, which utilizes a change in absorption and, consequently, a change in the refractive index, when a given intermediate excited state of bacteriorhodopsin is populated. Bacteriorhodopsin consists of seven *trans*-membrane α-helices that form the secondary structure of this protein. The light absorption by the light-adapted form of this protein is due to a chromophore called all-*trans*-retinal. Light absorbed by this chromophore induces an all-*trans* to 13-*cis* photoisomerization in its structure, which is followed by a series of protein intermediates exhibiting different absorption spectra and vectoral proton transport. Ultimately, the reisomerization of the chromophore leads to regeneration of the protein's original state. For details of the photoinduced cycles, the readers are referred to the article by Birge et al. (1999) and the biophotonics book by Prasad (2003).

The excitation to a particular intermediate level, such as the M state (Birge et al., 1999) in bacteriorhodopsin, produces a change in the absorption spectrum and thus produces a change in absorbance at a given frequency (ω in angular unit), denoted as $\Delta\alpha(\omega)$. The corresponding change in refractive index, $\Delta n(\omega)$, is derived from $\Delta\alpha(\omega)$ by the Kramers–Kronig relation of optics as follows (Finlayson et al., 1989):

$$\Delta n(\omega) = \frac{c}{\pi} \text{ p.v.} \int_0^\infty \frac{d\omega' \Delta\alpha(\omega')}{\omega'^2 - \omega^2} \qquad (12.1)$$

where p.v. stands for the principal value of the integral. As described in Chapter 10, a hologram is generated as a periodically modulated refractive index grating, when two monochromatic beams (an object beam and a reference beam) of wavelength λ are crossed at an angle in a holographic medium (see Figure 10.14). Their interference produces an intensity modulation, with alternate bright and dark stripes. In the bright areas, the action of light is to induce the change in absorption, $\Delta\alpha$, due to the population of the excited state, which creates a change in refractive index, Δn, given by the Kramers–Kronig relation of Eq. (12.1). This refractive index grating is local, because it replicates the intensity modulation. In contrast, the photorefractive holographic grating, described in Chapter 10, is nonlocal—that is, phase-shifted with respect to the intensity modulation. This hologram can be read by diffraction of a weak probe (readout) beam. Thus, when the hologram is illuminated by the reference beam, a bright diffraction beam is produced in the direction of the object beam. This reproduction process is very sensitive to angle and wavelength, in accordance with the Bragg diffraction condition for a thick grating (Kogelnik, 1969).

The diffraction efficiency, η, is related to the refractive index change, Δn, for a symmetric grating as

$$\eta = \sin^2\left(\frac{\pi \Delta n d}{\lambda \cos \theta/2}\right) \qquad (12.2)$$

where d is the thickness of the sample. A large Δn produces a large diffraction efficiency, until the sine function reaches its maximum value of 1. Holographic storage allows one to store many different holograms (thousands) in the same space (volume element) by changing the angle of the writing incident beams, a process called *angular multiplexing*. Alternatively, different holograms can be recorded in the same volume by changing the wavelength of the writing beams, in which case it is called *wavelength multiplexing*.

The principle of multiplexing can be used in reverse to identify an image, a process termed *optical correlation*, among a set of images that have been stored in a medium as a hologram. To accomplish this goal, we pass the object beam (which had a fixed angle during writing) through the image and illuminate the holographic medium in which multiple holograms are recorded. If the image matches one among those stored as holograms, a bright diffracted spot is generated in the direction corresponding to the direction of the reference beam used to store that image. By determining the angle at which bright beam is diffracted, one can determine which stored image has been matched. The absence of any diffracted spot indicates no match.

The magnitude of Δn can be used as a direct measure of the material's storage capacity. A large index modulation allows a large number of holograms to be recorded with good diffraction efficiency. This property, represented by a $M_\#$ parameter (M number), is used to describe the dynamic range of the material and is defined as (Mok et al., 1996)

$$\sqrt{\eta} = (M_\#)/M \qquad (12.3)$$

where M is the number of holograms recorded in the same volume, and η is the diffraction efficiency of each hologram. In some materials with large index modulation value and/or of large thickness, η can reach values close to unity (or 100%), while $M_\#$ can be as high as 10 or more.

Most holographic applications of bacteriorhodopsin utilize the optical change ($\Delta \alpha$ and subsequently Δn) in going from the ground state, bR, to a specific intermediate excited state called the M state. The Δn change between the two states, calculated using the Kramers–Kronig transformation [Eq. (12.1)], is large because of a large difference in their absorption maximum. The quantum yield of the bR to M conversion is also high. A primary limitation, however, is the short lifetime of the M state. However, chemical and genetic manipulation have provided much improved bacteriorhodopsin analogues for long-term storage. Another approach is to use nanocomposites of bacteriorhodopsin with glass or other polymers. Shamansky et al. (2002) studied the kinetics of the M state for bacteriorhodopsin, entrapped within a dried xerogel glass. They found that the decay kinetics of the M state was slowed by a factor of ~100, when the solvent was removed from the wet-gel to form the dry xerogel glass. This dramatic reduction in the M state decay rate was attributed to the reduced water content and a decrease in the hydrogen bonding network within the protein.

DNA. Naturally occurring DNA exhibits a number of properties useful for photonics. First, the constituent nucleotide building blocks of DNA (involving the heterocyclic ring bases) are optically transparent (nonabsorbing) over a wider spectral range, from 350 nm to 1700 nm. Second, their specificity in hybridization (selectivity in base pairing to form a double strand) as well as their ability to bind on their surface through electrostatic attraction (the DNA being negatively charged), or intercalate within the double strand, allows them to incorporate various photonic active structural units. Third, their dielectric properties are suitable for them to be used as cladding layer for polymer electro-optic devices (Grote et al., 2003).

DNA is abundant in nature, and some sources of DNA (such as salmon and scallop sperms) are waste products of fishing industries. Ogata and co-workers from the Chitose Institute of Science and Technology have shown that naturally occurring DNA from salmon can be used as a photonic medium (Kawabe et al., 2000; Wang et al., 2001). The first step is the purification of DNA to minimize or eliminate the protein content with the help of enzymatic degradation. Natural DNA is not soluble in common organic solvents used for casting of films or processing of polymeric fibers. However, when complexed with a lipid, such as hexadecyltrimethyl ammonium chloride (CTMA), DNA can be made dispersable in organic solvents. Ogata and co-workers have shown that films and fibers of optical waveguiding quality can be fabricated using this readily available DNA, complexed with a lipid. Their approach is schematically represented in Figure 12.1 (Grote et al., 2003).

Grote et al. (2003) have produced films for optical waveguiding where the optical losses at 1.55 μm are anticipated to be less than 1 dB/cm. Grote et al. have also demonstrated the utility of this DNA–lipid film as a high-quality cladding medium for electric field induced effects in electro-optic devices. Ogata and co-workers have doped a laser dye into a DNA–surfactant (lipid) complex film to achieve amplified spontaneous emission (Kawabe et al., 2000). Their approach to introduce a dye into the DNA–lipid surfactant complex is shown in Figure 12.2.

Many fluorescent dyes can readily be intercalated into the helices of DNA to form self-assembled ordered nanostructures. For a number of dyes, the fluorescence intensity is greatly enhanced (Jacobsen et al., 1995; Spielmann, 1998). Ogata and co-workers found that the intercalated dye molecules can be well-aligned and stabilized (Kawabe et al., 2000; Wang et al., 2001). Dyes can be intercalated in high concentration without showing any concentration quenching effect derived from aggregation.

This procedure was utilized to make rhodamine 6G-doped film for lasing (Kawabe et al., 2000). First, the DNA–surfactant complex was prepared by mixing a salmon DNA solution with hexadecyltrimethylammonium chloride (surfactant) aqueous solution to which rhodamine 6G was added. Then a rhodamine 6G-doped DNA-surfactant film was cast from an ethanol solution (with a DNA base pair to

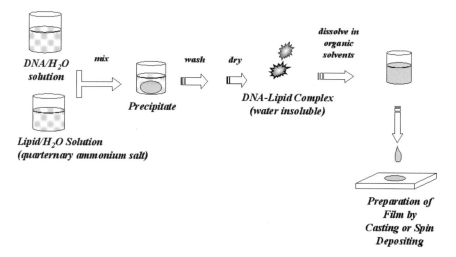

Figure 12.1. Schematics for preparation of optical materials from DNA Solution. From Grote et al. (2003), reproduced with permission.

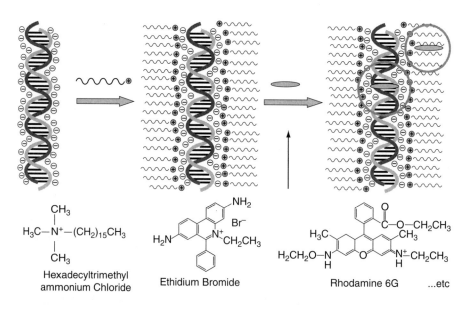

Figure 12.2. Schematics representation for the preparation method and a possible structure of a dye-doped DNA–surfactant film. Courtesy of N. Ogata and J. Grote.

dye ratio of 25:1) on a glass slide in a closed chamber (55% humidity), and a solid film was formed by slow evaporation. Above a certain threshold energy (20 μJ) and power density (300 kW/cm^2) of the pump beam (532 nm; 7 ns; 10 Hz), a narrowing of the lineshape, together with a superlinear dependence on the pump intensity, was taken as an indication of light amplification by stimulated emission from this film.

Kawabe et al. (2000) have suggested that rhodamine 6G may not be intercalating DNA structure in the strict sense because of its chemical structure. But even then, no aggregation of the dye takes place, because the film shows strong fluorescence and stimulated emission even at a high dye concentration of 1.36 weight %.

Bioobjects and Biocolloids. Many unique forms of nanosize bioobjects have highly precise geometries (plates, rods, icosahedral, etc.), some even existing in nature as monodispersed systems which can be used as building blocks for nanophotonics. Examples are viruses, sponges, sea urchin needles, and platelets from abalone shell. For example, virus particles are comprised of a capsid consisting of arranged protein subunits that form a hollow particle (with diameter in the range 20–300nm) enclosing the genome. The genomic material in the core of a virus particle can be replaced by other functional interiors to produce novel photonic functions. In addition, by use of appropriate protein chemistry, the surface of a virus particle can also be exploited to induce different photonic functions.

The monodispersed nanosize bioobjects can be used as building blocks for photonics crystals. The topic of photonics crystals has been discussed in Chapter 9. As described in Chapter 9, an important approach to fabricate a 3D photonic crystal is self-assemblying of monodispersed colloidal crystals of silica or polystyrene of appropriate diameters to close-pack in a face-centered cubic (fcc) periodic array. Biological objects, such as virus particles of sizes 100–300nm and varied shapes (icosahedral, rod, etc.) can assemble in both fcc and non-fcc packing to produce a wide range of self-assembled photonics crystals. When dispersed in an appropriate solvent media, these bioobjects form biocolloids that can self-assemble into a close-packed structure exhibiting a photonic crystal behavior. The research group at the Polymer Branch of the Air Force Research Laboratory (United States; R. Vaia) in collaboration with those at MIT (United States; E. L. Thomas) and Otago University (New Zealand; V. K. Ward) have used iridovirus, consisting of an icosahedral capsid with a diameter of ~200 nm, to produce a photonic crystal (Vaia et al., 2002; Juhl et al., 2003). They applied a strong gravitational force of ~11,000g from a centrifuge for 15 minutes, to an aqueous viral suspension containing 4% formaldehyde to cross-link the sedimented particles together. Figure 12.3 shows the local regions of the fcc packing of the iridovirus particles. It is very similar to Figure 9.14, showing the packing of the colloidal polystyrene spheres. Furthermore, McPherson and co-workers (Kuznetsov et al., 2000) have shown that viruses can be packed not only in a fcc structure, but also in other lattices such as orthorhombic and monoclinic systems. Ha et al. (2004) have used another bioobject, a sea urchin exoskeleton, to fabricate a high dielectric contrast 3D photonic crystal exhibiting a stop gap in the mid-IR range.

Another potential application of a virus is derived from its ability to include other materials such as high-refractive-index nanoparticles within its capsid, allowing

344 BIOMATERIALS AND NANOPHOTONICS

Figure 12.3. SEM micrograph of close packing of iridoviruses in a periodic structure. From Vaia (2002), unpublished results.

manipulation of its refractive index, which, in turn, can be utilized to enhance the dielectric contrast of a photonic crystal (see Chapter 9). Douglas and Young (1998) have shown that under appropriate conditions, the viral capsid can be temporarily opened to allow the transport of various substances inside and, therefore, trapping it when the capsid closes.

12.2 BIOINSPIRED MATERIALS

Biological systems utilize complex multifunctional structures to perform the functions of bio-recognition, multilevel processing, self-assembly, and templating at nanometer scale. There is a growing recognition that, based on the understanding of structure–function relation and effected cellular processes, multifunctional suprastructures can be designed. These materials are called bioinspired or nature-inspired. Some examples of bioinspired supramolecular structures for photonics have already been provided in Chapter 8. An example is light-harvesting dendrimers, developed by Frechet and co-workers (Andronov et al., 2000), which have been modeled after a naturally occurring photosynthetic system, a chlorophyll assembly. This photosynthetic system consists of a large array of chlorophyll molecules that surround a reaction center (Prasad, 2003). This chlorophyll array acts as an efficient light-harvesting antenna system to capture photons from the sun and to transfer the absorbed energy to the reaction center. The reaction center utilizes this energy to produce charge separation, eventually forming ATP and NADPH.

Frechet and co-workers prepared light-harvesting dendrimers that consist of a number of nanometer-size antennas in a hyperbranch arrangement (Andronov et al.,

2000). Figure 12.4 is a schematic representation of such a dendritic system that exhibits a nanoscale light-harvesting dendrimer. It consists of multiple peripheral sites (light-absorbing chromophores, represented by spheres) that can absorb and thus harvest sunlight. The absorbed energy is eventually transferred (funneled) quantitatively by a Förster-type energy transfer, a process discussed in Chapter 2, to an excitation energy acceptor at the center of the dendrimer where it can be "reprocessed" into a monochromatic light of a different wavelength (by emission from the core) or be converted into electrical or chemical energy. It represents a case where the dendritic antenna is based on a single molecule about 3 nm in size.

The energy harvesting antenna approach is even more useful for two-photon excitation processes. Here the antenna molecules are strong two-photon absorbers that funnel the absorbed energy to the core which may be a narrower energy gap molecule, but a more efficient emitter to yield a low threshold two-photon lasing. By using dendritic systems of this type, the number density of efficient two-photon absorbers is significantly increased for energy harvesting. Also, by isolating the function of net photoactivity at the core (e.g., lasing, optical limiting, or photochemistry), one can independently optimize two-photon absorption strength and an excited-state process. Frechet, Prasad, and co-workers demonstrated two-photon excited efficient light harvesting in novel dendrite systems (Brousmiche et al., 2003; He et al., 2003). Here the antennas are efficient two-photon absorbers (and green emitters) that absorb near-IR photons at ~800 nm and transfer the excitation energy quantitatively to the core molecule (a red emitter). In another approach, Frechet, Prasad, Tan, and co-workers used strong two-photon excitation by the pe-

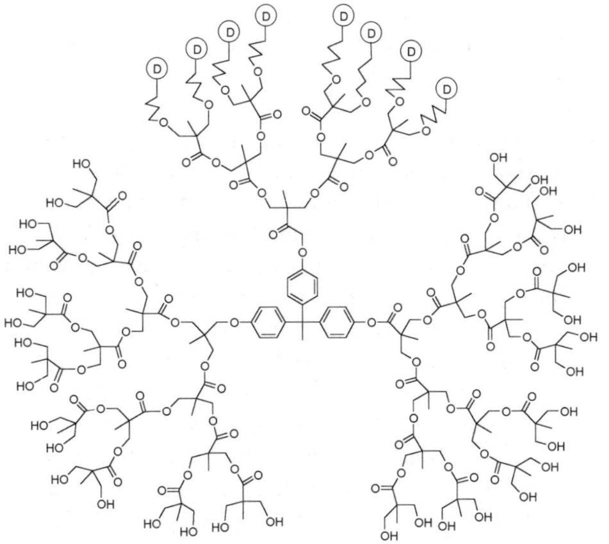

Figure 12.4. Light-harvesting antenna-based dendritic structure. Courtesy of J. M. J. Frechet.

ripheral antenna molecules, along with subsequent efficient energy transfer to a photosensitizing porphyrin-type molecule at the core to effect photodynamic therapy action (Dichtel et al., 2004). The subject of photodynamic therapy is discussed in Chapter 13.

12.3 BIOTEMPLATES

Natural microstructures can be used, in the pristine form or a surface functionalized form of a natural or bioinspired material, as templates to produce multiscale, multicomponent materials through iterative mesophase synthesis and processing. Such biotemplates can be used for the development of new assembling and processing techniques to produce periodic, aperiodic, and other engineering nanoscale architectures for photonics applications. The advantages of a biomolecular structure for templating are derived from their natural self-assemblying and self-recognition capability. The two biomolecular structures widely used for biotemplating for nanotechnology are (a) polypeptides and (b) DNA (whether natural or synthesized) or its oligomeric forms synthesized in a predetermined sequence. Peptide-driven formation and assemblying of nanoparticles have been demonstrated for a number of materials (Sarikaya et al., 2003). Polypeptides can be genetically engineered to specifically bind to selected inorganic compounds and assemble them to produce functional nanostructures. These genetically engineered polypeptides are characterized by a specific sequence of amino acids. The specificity of binding is derived from both chemical (e.g., hydrogen bonding, polarity and charge effects) and structural (e.g., size and morphology) recognition of inorganic nanostructures (e.g., nanoparticles). From a combinatorial library of millions of peptides, Whaley et al. (2000) selected specific peptide sequences that could distinguish among different crystallographic planes of GaAs. The peptide structure can thus be used to control positioning and assemblying of nanostructures.

DNA has been used extensively as a biotemplate to grow inorganic quantum confined structures (quantum dots, quantum wires, metallic nanoparticles as discussed in Chapter 4) and to organize nonbiological building blocks into extended hybrid materials (for review, see Storhoff and Mirkin; 1999). A DNA template has also been described as a "smart glue" for assembling nanoscale building blocks (Mbindyo et al., 2001). DNAs that can be used as templates can be naturally occurring or synthesized with appropriate length and base sequence (polynucleotides). A major advantage derived from the use of DNA is the ability of DNA strands to hybridize selectively its complementary strand. Coffer and co-workers were the first to utilize DNA as a template for CdS nanoparticles (Coffer et al., 1992; Coffer, 1997). In their approach, CdS nanoparticles were formed by first mixing an aqueous solution of calf thymus DNA with Cd^{2+} ions, followed by the addition of Na_2S. The formed nanoparticles of CdS exhibited the optical properties of a CdS quantum dot with an approximate size of 5.6 nm.

Mirkin and co-workers assembled 13-nm colloidal Au nanoparticles, modified with thiolated single-stranded DNA, for calorimetric DNA sensor application

(Mirkin et al., 1996; Mucic et al., 1998). They have also assembled hybrid materials composed of Au and CdSe nanoparticles (Mitchell et al., 1999). Alivisatos' group (Loweth et al., 1999) used single-stranded DNA as a template for the directed self-assembly of nanoparticles, modified with single-strand DNA that is complementary to a particular section of a DNA template. Harnak et al. (2002) and Mertig et al. (2002), respectively, fabricated metallic nanowires of gold and platinum using DNA as templates.

Khomutov et al. (2003) have utilized monolayers and multilayers of DNA/polycation complex Langmuir–Blodgett films (described in Chapter 8) as templates to generate inorganic nanostructures such as semiconductor quantum dots and ion oxide nanoparticles as quasi-linear arrays (nanorods). An example is presented from their work producing CdS quantum dots. For this purpose they prepared a film containing complexes of DNA, Cd^{2+}, and a polycation polymer, poly-4-vinilpyridine with 16% cetylpyridinium group (PVP-16). The DNA has a net negative charge and thus binds to the positively charged ions. The CdS nanoparticles were prepared *in situ* by incubation of the film in the H_2S atmosphere. By this method they obtained planar polymeric complex films containing CdS nanowires of diameter ~ 5 nm.

At our Institute (Suga, Prasad, and co-workers), the focus is on 2-D and 3-D DNA periodic arrays that can be synthesized by introducing an intermolecular bridge, bridger-DNA, between two strands. A bridger DNA has two functions. First, it can bridge two strands in head-to-tail fashion. Second, it can bridge two strands in parallel fashion. More details of this approach can be found in the monograph on biophotonics (Prasad, 2003).

Another example of a biotemplate is a virus structure. Viruses with well-defined morphology, flexible microstructures, and surfaces can be modified to serve as suitable templates for producing novel photonic materials. An example is the use of cowpea mosaic virus (CPMV) particles, which are 30-nm-diameter icosahedra, as a template to attach dye molecules or gold nanoparticles (Wang et al., 2002). Wang et al. (2002) used functionalized mutant CPMV particles and reacted them with dye maleimide reagent to attach up to 60 dye molecules per CPMV particle. Similarly, by using monomaleimido-nanogold, they were able to attach gold nanoclusters on the surface of the CPMV particles. This approach provides an opportunity to produce a high local concentration of the attached species.

Lee at al. (2002) showed that engineered viruses can recognize specific semiconductor surfaces through the method of selection by combinatorial phage display. They used this specific recognition property of a virus to organize quantum dots of ZnS, forming ordered arrays over the length scale defined by liquid crystal formation. Highly oriented and self-supporting films were self-assembled by this method.

12.4 BACTERIA AS BIOSYNTHESIZERS

The biosynthetic ability of bacteria provides a novel approach of metabolic engineering to prepare unique nanostructured polymers for photonics (Aldor and

Keasling, 2003). The production of a family of polyhydroxyalkanoate (PHA) polymers synthesized by the bacteria *Pseudomonas oleovorans* is an example of such biosynthesis (Figure 12.5). This organism has the capacity to synthesize various PHAs containing C_6 to C_{14} hydroxyalkanoic acid, dependent on the 3-hydroxyl alkanoate monomer present.

Polyhydroxyalkanoates (PHA) are a class of natural thermoplastic polyesters, produced by bacteria to store carbon in response to conditions of stress such as shortage of essential nutrient. PHAs are linear polyesters composed of 3-hydroxy fatty acid monomers; the basic structure is shown in Figure 12.6. This naturally occurring polymer (crystalline thermoplastic, resembling isostatic polypropylene) consists of left-handed 2_1 helices ("all R" confirmation), grouped into crystals of one chirality with a 5.96-Å repeat structure. Isolation and crystallization of these polymers result in a multilaminate "lath" structure that may have significant applications in the development of new photonic materials (Nobes et al., 1996). The details of PHA biosynthesis are presented elsewhere (Prasad, 2003).

Over 100 different PHAs have been isolated from various bacteria in which the 3-hydroxyalkanoate (HA) monomer units range from 3-hydroxypropionic acid to 3-hydroxyhexadecanoic acid (Steinbüchel and Valentin, 1995). There are PHAs in which unsaturated 3-hydroxyalkenoic acids occur with one or two double bonds in the R group. Other PHAs have 3 hydroxyl-hexanoic acids with a methyl group at various positions of the R group. In addition, there are HAs in which the R group contains various functional groups such as halogens (-Br, -Cl, -F), olefin, cyano, and hydroxy groups (Curley et al., 1996a, 1996b). The use of metabolic engineering with a bacterial machinery offers the following advantages in the production of nanoscale photonic materials:

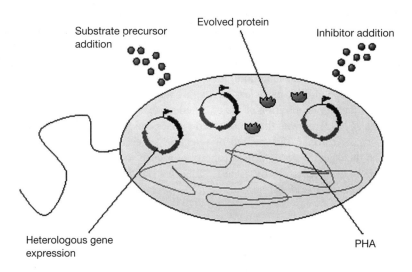

Figure 12.5. Diagrammatic representation of a BioReactor for bacteria synthesis of PHA.

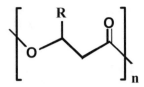

Figure 12.6. Repeating unit structure of PHA.

- Nanostructured photonic polymers with multifunctionality can be prepared, which are at the same time biodegradable.
- The PHA polymers, synthesized by bacteria, provide backbone structures and chirality for novel self-ordered photonic polymers and provide unique properties that are difficult to obtain synthetically.
- The native structure of the natural polymers is a helix, with a strong tendency to form a beta sheet.
- The biosynthesized polymers are more heat- and light-stable than other biopolymeric structures (i.e., protein and nucleic acids).
- Structural variability, accomplished by control of monomeric composition, chain length, and copolymer microstructure, provides a unique opportunity to investigate various polymer structure–function relationships.
- Chromophores having an appropriate β-hydroxyalkanoic acid side chain can be fed to the bacteria, under conditions that stimulate nutritional limitations (excess carbon, limited nitrogen), to produce PHA polymers with side groups to confer specific optical and mechanical properties.
- These polymers can be produced at low cost in relation to costs of synthetic production of specialty polymers.
- Genetic manipulation can be utilized to tailor biopolymer synthesis.

The helical and β-sheet structure of materials can be designed or modified by incorporating cross-linking functionalities and established poling techniques to develop polymers with a well-defined and temporally and thermally stable noncentrosymmetric structures. Similar materials have been designed for nonlinear optical applications, such as high-frequency E-O modulators (Prasad and Williams, 1991, also see Chapter 8). Electroactive or photosensitive moieties could be introduced to effect structural changes in the polymers which can lead to applications in transducers, smart materials, and actuators.

To test the applicability of the bacteria synthesized PHA as thin film optical media, thin films of a bacteria-synthesized PHA polymer were fabricated and their optical properties were evaluated (Bergey and Prasad, unpublished work). Both dip-coating and spin-coating methods were successful in depositing optical quality films on a glass substrate. Also, an important chromophore, APSS, developed at our Institute, was successfully doped in this polymer film. This chromophore exhibits important nonlinear optical properties, such as electro-optic activities (under electrically poled conditions) and two-photon excited up-converted fluorescence.

12.5 HIGHLIGHTS OF THE CHAPTER

- Biomaterials are potentially attractive multifunctional materials for many nanophotonics applications.
- Important classes of biomaterials for nanophotonics are (i) bioderived materials, (ii) bioinspired or nature-inspired materials, (iii) biotemplates, and (iv) metabolically engineered materials using bacterial synthesis.
- Bioderived materials are naturally occurring biomaterials or their chemically modified forms.
- Bioinspired materials are those synthesized on the basis of governing principles of biological systems.
- Biotemplates refer to the use of natural microstructures as suitable templates for self-assembling of photonic active structures.
- Metabolically engineered materials are those synthesized by using bacteria as biosynthesizers to produce photonic polymers.
- Examples of naturally occurring biomaterials for nanophotonics are (i) bacteriorhodopsin for holographic memory, (ii) DNA, for optical waveguiding as well as the host media for laser dyes, and (iv) bio-objects and biocolloids for photonic crystal media.
- Light-harvesting dendrimers, designed following the principles of the antenna effect in natural photosynthetic units, provide an example of bioinspired or nature-inspired materials.
- DNA is a biotemplate utilized to organize nonbiological blocks, such as organic dyes, semiconductor quantum dots, and metallic nanoparticles, into extended hybrid structures.
- Another example of biotemplate is a virus that has a well-defined morphology, flexible microstructures, and easily modifiable surfaces.
- Polyhydroxyalkanoates are a class of natural thermoplastic polymers, produced by bacteria and providing an example of metabolic engineering.

REFERENCES

Aldor, I. S., and Keasling, J. D., Process Design for Microbial Plastic Factories: Metabolic Engineering of Polyhydroxyalkanoates, *Curr. Opin. Biotechnol.* **14,** 475–483 (2003).

Andronov, A., Gilat, S. L., Frechet, J. M. J., Ohta, K., Neuwahl, F. V. R., and Fleming, G. R., Light Harvesting and Energy Transfer in Laser Dye-Labeled Poly (Aryl Ether) Dendrimers, *J. Am. Chem. Soc.* **122,** 1175–1185 (2000).

Birge, R. R., Protein-Based Optical Computing and Memories, *IEEE Comput.* **25,** 56–67 (1992).

Birge, R. R., Zhang, C. F., and Lawrence, A. F., Optical Random Access Memory Based on Bacteriorhodopsin, in *Molecular Electronics,* F. Hong, ed., Plenum, New York, 1989, pp. 369–379.

Birge, R. R., Gillespie, N. B., Izaguirre, E. W., Kusnetzow, A., Lawrence, A. F., Singh, D., Song, Q. W., Schmidt, E., Stuart, J. A., Seetharaman, S., and Wise, K. J., Biomolecular Electronics: Protein-Based Associative Processors and Volumetric Memories, *J. Phys. Chem. B* **103,** 10746–10766 (1999).

Brousmiche, D. W., Serin, J. M., Frechet, J. M. J., He, G. S., Lin, T.-C., Chung, S. J., and Prasad, P. N., Fluorescence Resonance Energy Transfer in a Novel Two-Photon Absorbing System, *J. Am. Chem. Soc.* **125,** 1448–1449 (2003).

Chen, Z., and Birge, R. R., Protein-Based Artificial Retinas, *Trends Biotechnol.* **11,** 292–300 (1993).

Coffer, J. L., Approaches for Generating Mesoscale Patterns of Semiconductor Nanoclusters, *J. Cluster Sci.* **8,** 159–179 (1997).

Coffer, J. L., Bigham, S. R., Pinizzotto, R. F., and Yang, H., Characterization of Quantum-Confined CdS Nanocrystallites Stabilized by Deoxyribonucleic Acid, *Nanotechnology* **3,** 69–76 (1992).

Curley, J. M., Hazer, B., Lenz, R. W., and Fuller, R. C., Production of Poly(3-hydroxyalkanoates) Containing Aromatic Substituents by *Pseudomonas oleovorans,* Macromolecules **29,** 1762–1766 (1996a).

Curley, J. M., Lenz, R. W., and Fuller, R. C., Sequential Production of Two Different Polyesters in the Inclusion Bodies of *Pseudomonas oleovorans, Int. J. Biol. Macromol.* **19(1),** 29–34 (1996b).

Dichtel, W. R., Serin, J. M., Ohulchanskyy, T. Y., Edder, C., Tan, L.-S., Prasad, P. N., and Frechet, J. M. J., Singlet Oxygen Generation via Two Photon Excited Fluorescence Resonance Energy Transfer, submitted to *J. Am. Chem. Soc.* (2004).

Douglas, T., and Young, M., Host–Guest Encapsulation of Materials by Assembled Virus Cages, *Nature* **393,** 152–155 (1998).

Finlayson, N., Banyai, W. C., Seaton, C. T., Stegeman, G. I., Neill, M., Cullen, T. J., and Ironside, C. N., Optical Nonlinearities in CdS_xSe_{1-x}-Doped Glass Wave-Guides, *J. Opt. Soc. Am. B* **6,** 675–684 (1989).

Grote, J. G., Ogata, N., Hagen, J. A., Heckman, E., Curley, M. J., Yaney, P. P., Stone, M. O., Diggs, D. E., Nelson, R. L., Zetts, J. S., Hopkins, F. K., and Dalton, L. R., Deoxyribonucleic Acid (DNA) Based Nonlinear Optics, in *Nonlinear Optical Transmission and Multiphoton Processes in Organics,* Vol. 5211, A. T. Yeates, K. D. Belfield, F. Kajzar, and C. M. Lawson, eds. SPIE Proceedings, 2003, Bellingham, WA, pp. 53–62.

Ha, Y.-H., Vaia, R. A., Lynn, W. F., Constantino, J. P., Shin, J., Smith, A. B., Matsudaira, P. T., and Thomas, E. L., Top-Down Engineering of Natural Photonic Crystalks: Cyclic Size Reduction of Sea Urchin Stereom, submitted to *Adv. Mater.* (2004).

Hampp, N., Bräuchle, C., and Oesterhelt, D., Bacteriorhodopsin Wildtype and Variant Aspartate-96-Asparagine as Reversible Holographic Media, *Biophys. J.* **58,** 83–93 (1990).

Hampp, N., Thoma, R., Zeisel, D., and Bräuchle, C., Bacteriorhodopsin Variants for Holographic Pattern-Recognition, *Adv. Chem.* **240,** 511–526 (1994).

Harnack, O., Ford, W. E., Yasuda, A., Wessels, J. M., Tris(hydroxymethyl)phosphine-Capped Gold Particles Templated by DNA as Nanowire Precursors, *Nanoletters* **2,** 919–923 (2002).

He, G. S., Lin, T.-C., Cui, Y., Prasad, P. N., Brousmiche, D. W., Serin, J. M., and Frechet, J. M. J., Two-Photon Excited Intramolecular Energy Transfer and Light Harvesting Effect in Novel Dendmitic Systems, *Opt. Lett.* **28**, 768–770 (2003).

Hong, F. T., Retinal Proteins in Photovoltaic Devices, *Adv. Chem.* **240**, 527–560 (1994).

Jacobsen, J. P., Pedersen, J. B., and Wemmer, D. E., Site Selective Bis-Intercalation of a Homodimeric Thiazole Organge Dye in DNA Oligonucleotides, *Nucl. Acid Res.* **23**, 753–760 (1995).

Juhl, S., Ha, Y.-H., Chan, E., Ward, V., Smith, A., Dockland, T., Thomas, E. L., and Vaia, R., BioHarvesting: Optical Characteristics of Wisenia Iridovirus Assemblies, *Polym. Mater. Sci. Eng.*, **91** (2004, in press).

Kawabe, Y., Wang, L., Horinouchi, S., and Ogata, N., Amplified Spontaneous Emission from Fluorescent-Dye-Doped DNA–Surfactant Complex Films, *Adv. Mater.* **12**, 1281–1283 (2000).

Khomutov, G. B., Kislov, V. V., Antipina, M. N., Gainutdinov, R. V., Gubin, S. P., Obydenov, A. Y., Pavlov, S. A., Rakhnyanskaya, A. A., Serglev-Cherenkov, A. N., Saldatov, E. S., Suyatin, D. B., Tolstikhina, A. L., Trifonov, A. S., and Yurova, T. V., Interfacial Nanofabrication Strategies in Development of New Functional Nanomaterials and Planar Supramolecular Nanostructures for Nanoelectronics and Nanotechnology, *Microelectron. Eng.* **69**, 373–383 (2003).

Kogelnik, H., Coupled Wave Theory for Thick Hologram Grating, *The Bell System Tech. J.* **48**, 2909–2948 (1969).

Kuznetsov, Y. G., Malkin, A. J., Lucas, R. W., and McPherson, A., Atomic Force Microscopy Studies of Icosahedral Virus Crystal Growth, *Colloids and Surfaces B: Biointerfaces* **19**, 333–346 (2000).

Lee, S.-W., Mao, C., Flynn, C. E., and Belcher, A. M., Ordering of Quantum Dots Using Genetically Engineered Viruses, *Science* **296**, 892–895 (2002).

Loweth, C. J., Caldwell, W. B., Peng, X., Alivisatos, A. P., and Schultz, P. G., DNA-Based Assembly of Gold Nanocrystals, *Angew. Chem. Int. Ed.* **38**, 1808–1812 (1999).

Marwan, W., Hegemann, P., and Oesterhelt, D., Single Photon Detection by an Archaebacterium, *J. Mol. Biol.* **199**, 663–664 (1988).

Mbindyo, J. K. N., Reiss, B. D., Martin, B. R., Keating, C. D., Natan, M. J., and Mallouk, T. E., DNA-Directed Assembly of Gold Nanowires on Complementary Surfaces, *Adv. Mater.* **13**, 249–254 (2001).

Mertig, M., Ciacchi, L. C., Seidel, R., and Pompe, W., DNA as a Selective Metallization Template, *Nanoletters* **2**, 841–844 (2002).

Mirkin, C. A., Letsinger, R. L., Mucic, R. C., and Storhoff, J. J., A DNA-Based Method for Rationally Assembling Nanoparticles into Macroscopic Materials, *Nature* **382**, 607–609 (1996).

Mitchell, G. P., Mirkin, C. A., and Letsinger, R. L., Programmed Assembly of DNA Functionalized Quantum Dots, *J. Am. Chem. Soc.* **121**, 8122–8123 (1999).

Miyasaka, T., Koyama, K., and Itoh, I., Quantum Conversion and Image Detection by a Bacteriorhodopsin-Based Artificial Photoreceptor, *Science* **255**, 342–344 (1992).

Mok, F. H., Burr, G. W., and Psaltis, D., System Metric for Holographic Memory Systems, *Opt. Lett.* **21**, 896–898 (1996).

Mucic, R. C., Storhoff, J. J., Mirkin, C. A., and Letsinger, R. L., DNA-Directed Synthesis of Binary Nanoparticle Network Materials, *J. Am. Chem. Soc.* **120**, 12674–12675 (1998).

Nobes, G. A. R., Marchessault, R. H., Chanzy, H., Briese, B. H., and Jendrossek, D., Splintering of Poly(3-hydroxybutyrate) Single Crystals by PHB-Depolymerase A from *Pseudomonas lemoignei, Macromolecules* **29,** 8330–8333 (1996).

Prasad, P. N., and Williams, D. J., *Introduction to Nonlinear Optical Effects in Molecules and Polymers,* John Wiley & Sons, New York, 1991.

Prasad, P. N., *Introduction to Biophotonics,* John Wiley & Sons, New York (2003).

Saleh, B. E. A., and Teich, M. C., *Fundamentals of Photonics,* Wiley-Interscience, New York, 1991.

Sarikaya, M., Tanerkerm, C. M., Jen, A. K.-Y., Schulten, K., and Baneyx, F., Molecular biomimetics: nanotechnology through biology, *Nature Mater.* **2,** 577–585 (2003).

Sasabe, H., Furuno, T., and Takimoto, K., Photovoltaics of Photoactive Protein Polypeptide LB Films, *Synth. Met.* **28,** C787–C792 (1989).

Shamasky, L. M., Luong, K. M., Han, D., and Chronister, E. L., Photoinduced Kinetics of Bacteriorhodopsin in a Dried Xerogel Glass, *Biosensors Bioelectronics* **17,** 227–231 (2002).

Song, Q. W., Zhang, C., Gross, R., and Birge, R. R., Optical Limiting by Chemically Enhanced Bacteriorhodopsin Films, *Opt. Lett.* **18,** 775–777 (1993).

Spielmann, H. P., Dynamics of a bis-Intercalator DNA Complex by H-1-Detected Natural Abundance C-13 NMR Spectroscopy, *Biochemistry* **37,** 16863–16876 (1998).

Steinbüchel, A., and Valentin, H. E., Diversity of Bacterial Polyhydroxyalkanoic Acids, *FEMS Microbiol. Lett.* **128(3),** 219–228 (1995).

Storhoff, J. J., and Mirkin, C. A., Programmed Materials Synthesis with DNA, *Chem. Rev.* **99,** 1849–1862 (1999).

Vaia, R., Farmer, B., and Thomas, E. L. (2002), private communications.

Vsevolodov, N. N., Druzhko, A. B., and Djukova, T. V., Actual Possibilities of Bacteriorhodopsin Application in Optoelectronics, in *Molecular Electronics: Biosensors and Biocomputers,* F. T. Hong, ed., Plenum Press, New York, 1989, pp. 381–384.

Wang, L., Yoshida, J., and Ogata, N., Self-Assembled Supramolecular Films Derived from Marine Deoxyribonucleic Acid (DNA)–Cationic Surfactant Complexes: Large Scale Preparation and Optical and Thermal Properties, *Chem. Mater.* **13,** 1273–1281 (2001).

Wang, Q., Lin, T., Tang, L., Johnson, J. E., and Finn, M. G., Icosahedral Virus Particles as Addressable Nanoscale Building Blocks, *Angew. Chem. Int. Ed.* **41,** 459–462 (2002).

Whaley, S. R., English, D. S., Hu, E. L., Barbara, P. F., and Belcher, A. M., Selection of Peptides with Semiconductor Binding Specificity for Directed Nanocrystal Assembly, *Nature* **405,** 665–668 (2000).

Yang, F., Moss, L. G., and Phillips, J. G. N., The Molecular Structure of Green Fluorescent Protein, *Nat. Biotechnol.* **14,** 1246–1251 (1996).

CHAPTER 13

Nanophotonics for Biotechnology and Nanomedicine

Nanophotonics broadly impacts biomedical research and technology for studying the fundamentals of interactions and dynamics at single cell/molecule level, as well as for applications to light-guided and light-activated therapy using nanomedicine. Nanomedicine is an emerging field that deals with utilization of nanoparticles in the development of new methods of minimally invasive diagnostics for early detection of diseases, as well as for facilitating targeted drug delivery, effectiveness of therapy, and real-time monitoring of drug action. This chapter illustrates a broad range of potential applications by providing some demonstrative examples. A fundamental understanding of drug–cell interactions, based on molecular changes at the single-cell level induced by a pre-onset state of a disease, can provide a basis for molecular recognition-based "personalized" therapeutic approaches to treat diseases. Nanophotonics enables one to use optical techniques for tracking of drug intake, elucidating its cellular pathway and monitoring subsequent intracellular interactions. For this purpose, bioimaging, biosensing, and single-cell biofunction studies, using optical probes, are proving to be extremely valuable.

In the area of nanomedicine-based molecular recognition of diseases, light-guided and light-activated therapies provide a major advancement. Nanoparticles containing optical probes, light-activated therapeutic agents, and specific carrier groups that can direct the nanoparticles to the diseased cells or tissues provide targeted drug delivery, with an opportunity for real-time monitoring of drug efficacy. This chapter provides some examples of both optical diagnostics and light-based therapy.

Section 13.1 illustrates the usage of near-field microscopy, a technique already discussed in Chapter 3, for bioimaging of microbes and biostructures that are of dimensions considerably less than the wavelengths of light used. Section 13.2 provides a general description of nanophotonic approaches for optical diagnostics and light-activated and guided therapy. Section 13.3 presents semiconductor quantum dots for bioimaging. These quantum dots and their size-dependent optical properties have been discussed in Chapter 4. Section 13.4 covers up-converting nanoparticles for bioimaging. These nanoparticles, containing rare-earth ions, absorb in the IR and emit in the visible spectral range and have been discussed in Chapter 6. Merits of these inorganic emitters over the organic fluorophores are also described in their respective sections.

Section 13.5 describes the application of nanophotonics for biosensing. The various biosensing approaches included in this section are plasmonic biosensors, photonic crystal biosensors, porous silicon microscovity biosensors, PEBBLE nanosensors, dye-doped nanoparitcle sensors, and optical nanofiber sensors. Section 13.6 describes the nanoclinic approach for optical diagnositics and targeted therapy. The usefulness of this approach for optical tracking of drug delivery and light-activated, targeted therapy is discussed. The example provided in this section is of light-guided magnetic therapy of cancer, where optical tracking of nanoparticles is used in optimization of cellular uptake. Section 13.7 describes optically trackable nanoclinic gene delivery system. Section 13.8 provides a discussion of light-activatable nanoclincs for photodynamic therapy. The use of up-converting nanoparticles allows for the usage of more penetrating IR radiation, compared to the visible radiation, for treatment of deeper tumor. Section 13.9 provides the highlights of the chapter.

A suggested reference for further reading is:

Prasad (2003): *Introduction to Biophotonics*

13.1 NEAR-FIELD BIOIMAGING

Near-field microscopy allows one to probe imaging of biological structures that are significantly smaller than the wavelength of light. Examples of such structures are chromosomes and viruses. Even some bacteria are also submicron sizes. These structures are difficult to detect using regular microscopy. But near-field microscopy allows one to easily detect these dimensions, because resolutions of 100 nm or less are readily obtainable. Bioimaging using near-field scanning optical microscopy (NSOM) and some applications of imaging of a bacteria, stained with a fluorescent dye, are discussed in the book on biophotonics by Prasad (2003). Described in the biophotonics book are examples of imaging of bacteria and viruses.

NSOM has also been utilized for imaging of various biomolecular subcellular structures. A good reference is that of Yanagida et al. (2002). Chromosomes are supercoiled structures containing DNA and are the major constituents of the nucleus of a cell. They are only a few microns long. NSOM has been used to obtain high-resolution topographic image of chromosomes. Even single DNA molecules, in which a DNA-specific fluorescent dye, YOYO-1, was intercalated, were imaged using NSOM. The result indicated a heterogeneous intercalation of the YOYO-1 dye, when the concentration of this dye is low. NSOM also provided finer structures of fibroblast in cytoskeleton, compared to the confocal microscopic images (Muramatsu et al., 2002; Betzing et al., 1993). Muramatsu et al. (1995) produced the first NSOM image in a liquid by studying the cytoskeletal structures of chemically fixed cells in an aqueous buffer. They obtained a spatial resolution of 70 nm in the transmission NSOM mode.

Another area of NSOM bioimaging is neuron imaging. The high resolution of NSOM in fluorescence imaging can provide increased understanding of the mechanisms involved in information processing in neurons (Yanagida et al., 2002).

13.2 NANOPARTICLES FOR OPTICAL DIAGNOSTICS AND TARGETED THERAPY

Nanoparticles of dimensions <50 nm, significantly smaller than the size of pores in the membrane surrounding a biological cell, offer a number of advantages for diagnostics and targeted therapy from within the cell. These nanostructures can be polymeric, ceramic, silica, dendritic, or liposome-based structures. Polymer-, ceramic-, and silica-based nanoparticles are more rigid. The dendritic or liposome structures are softer nanostructures that also offer structural flexibility to incorporate multifunctionality. A nanoparticle approach offers the following advantages:

- Nanoparticles are nonimmunogenic. Hence they do not elicit any immune response when introduced in the body's circulatory system. Such immune response can often lead to severe consequences by shutting off or clogging up the circulatory system.
- The nanoparticles can be tailored with compositions (such as silica or organically modified silica) that are resistant to microbial attack. They do not undergo enzymatic degradation and thus effectively protect the encapsulated probes or drugs.
- Nanoparticles provide three different structural platforms for diagnostics and therapy: (a) An interior volume in which various probes and therapeutic agents can be encapsulated. (b) A surface to which targeting groups can be attached to carry the nanoparticles to cells or biological sites expressing appropriate receptors. In addition, the surface can bind to specific biological molecules for intracellular delivery (such as in gene therapy, discussed later in Section 13.7). The surface can also be functionalized to introduce a hydrophilic (polar), hydrophobic (nonpolar), or amphiphilic character to enable dispersibility in a variety of fluid media. (c) Pores in the nanoparticles, which can be tailored to be specific sizes to allow selective intake or release of biologically active molecules or activate therapeutic agents.
- Nanoparticles, appropriately surface functionalized and of size ≤50 nm, can penetrate the pores of the cell membrane by endocytosis. This provides a convenient mechanism for intracellular diagnostics and therapy.
- Nanoparticles, such as silica-based nanobubbles (thin shells), are optically highly transparent. Therefore, light activation and optical probing can readily be accomplished.

Table 13.1 lists the applications of various nanoparticles for optical diagnostics and therapy, which are discussed in subsequent sections. The nanoparticles have

Table 13.1. Applications of Various Nanoparticles for Optical Diagnostics and Therapy

Nanoparticles	
Optical Diagnostics	**Therapy**
Bioimaging	Optical tracking using nanoclinic
• Quantum dots	• Magnetic therapy
• Up-converting nanophosphors	• Gene therapy
• Encapsulated dyes	
Biodetection	Light-activated therapy
• Plasmonic nanostructures	• Light-activated photodynamic therapy using nanoclinic
• Silicon microcavities	• Up-conversion photodynamic therapy using two-photon absorbing nanostructures
• Photonic crystal biosensors	
• PEBBLES	
• Dye-doped nanoparticles	
• Nanofiber sensors	

been used for bioimaging and biosensing. They have been demonstrated to be useful for targeted therapy (nanomedicine), which can be light-guided, light-activated, or both.

13.3 SEMICONDUCTOR QUANTUM DOTS FOR BIOIMAGING

As discussed in Chapter 4, appropriate selection of the semiconducting materials and their nanosizes allows one to cover a wide spectral range for bioimaging. Also, another useful feature is that many of these quantum dots can be excited at the same wavelength, even though their emissions are at widely different wavelengths. The typical line widths of emission bands are 20–30 nm, thus relatively narrow, which helps if one wants to use the quantum dots for multispectral imaging. Compared to organic fluorophores, some major advantages offered by quantum dots for bioimaging are (Prasad, 2003):

- Quantum dot emissions are considerably narrower compared to organic fluorophores that exhibit broad emissions. Thus, the complication in simultaneous quantitative multichannel detection, posed by cross-talks between different detection channels derived from spectral overlap, is significantly reduced.
- The lifetime of emission is longer (hundreds of nanoseconds) compared to that of organic fluorophores, thus allowing one to utilize time-gated detection to suppress autofluorescence, which has a considerably shorter lifetime.
- The quantum dots do not readily photobleach.

A major problem in the use of quantum dots for bioimaging is surface-induced quenching of emission efficiency due to the high surface area of nanocrystals. This requires surface passivation by encapsulation or by using core-shell type quantum dots, described in Chapter 4.

Another issue to address is dispersibility of the quantum dots in a biological medium. Various techniques have been used to make quantum dots dispersible in biological media. Some use encapsulation by a silica layer (Bruchez et al., 1998). Others have used covalent bonding of the quantum dots to a biological molecule such as a protein or a DNA segment (Chan and Nie, 1998; Mattoussi et al., 2000; Akerman et al., 2002). For details of these examples, the readers are referred to the book on biophotonics (Prasad, 2003). These examples have utilized primarily the quantum dots of II–VI semiconductors.

More recently, silicon nanoparticles as well as quantum dots of III–V semiconductors have been produced dispersible in biological fluids. These nanostructures provide emission covering longer wavelengths extending into IR, which permits deeper penetration into biological specimens (Prasad, 2003).

13.4 UP-CONVERTING NANOPHORES FOR BIOIMAGING

Another group of nanoparticles useful for bioimaging is that of rare-earth ion-doped oxide nanoparticles (Holm et al., 2002). The rare-earth ions are well known to produce IR-to-visible up-conversion by a number of mechanisms discussed in Chapter 6. These up-converting nanoparticles are also useful for light activation of therapy (photodynamic therapy), to be discussed in Section 13.8. The up-conversion processes in rare-earth ions exhibit a strong power dependence on excitation intensity (e.g., quadratic for a two-photon process). Thus, they provide better spatial resolution, since efficient emission is generated only near the focus of the beam where the excitation intensity is the highest. Thus bioimaging using these up-converting nanophores provide background-free (practially no autofluorescence) detection, because the excitation source is in the near-IR region (generally 975-nm laser diodes). Autofluorescence from biological constitutents often pose a major problem in bioimaging (Prasad, 2003).

An advantage offered by these nanoparticles over the two-photon excitable dye is that the up-conversion process in the rare-earth nanoparticles occurs by sequential multistep absorption through real states and is thus considerably stronger. Therefore, one can use a low-power continuous-wave diode laser at 975 nm (which is also very inexpensive and readily available) to excite the up-converted emission. By contrast, the two-photon absorption in organic dyes is direct (simultaneous) two-photon absorption through a virtual state that requires a high-peak-power pulse laser source. However, the emission from the rare-earth ion is a phosphorescence process with a lifetime typically in milliseconds, compared to a dye fluorescence with a lifetime in nanoseconds. Therefore, applications that require short-lived fluorescence cannot use the phosphorescence from these up-converting nanoparticles, also referred to as *nanophores* or *nanophosphors*.

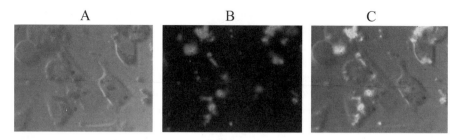

Figure 13.1. Bioimaging using up-converting nanoparticles on oral epithelial carcinoma cells (KB). KB cells were incubated with nanoparticles consisting of Er-doped Y_2O_3 nanophosphors in silica shell. (A) The light transmission image of the KB cells. (B) The fluorescent emission after excitation with 974 nm. (C) The composite of parts A and B.

A considerable amount of work on up-converting nanophores and their applications was originally done by SRI (Chen et al., 1999). More recently, our group at the Institute for Lasers, Photonics, and Biophotonics produced rare-earth-doped yttria (Y_2O_3) nanoparticles and coated them with silica to produce nanophores of size ~25 nm (Holm et al., 2002). These silica-coated nanophores are water-dispersible and extremely stable and exhibit no photobleaching. The size of these nanophores is still small enough for them to penetrate the cell and can be targeted to specific cell types by functionalizing silica nanoparticles with a carrier group attached to the surface.

These nanophores are prepared using the reverse micelle chemistry (Kapoor et al., 2000), a method described in Chapter 7. Encapsulation and functionalization, for subsequent ligand coupling of nanophosphors, is accomplished by addition of silica shell. The targeting ligand is then coupled to the —COOH groups or NH_2 groups of the spacer arms by using carbodimides. The same procedure is followed to synthesize Er/Yb co-doped Y_2O_3 and Tm/Yb co-doped Y_2O_3. The up-converted emission of these nanoparticles is red (640 nm) for the Er/Yb co-doped Y_2O_3 particles, green (550 nm) for Er-doped Y_2O_3 particles, and blue (480 nm) for Tm/Yb co-doped Y_2O_3 particles. These wavelengths of light are readily detected with standard CCD arrays and/or a CCD-coupled spectrograph.

The ability to tailor the emission wavelength, coupled with our ability to surface functionalize these nanoparticles, allows for a number of unique applications of these materials. Our initial studies were conducted in the KB cells (Holm et al., 2002). As can be seen in Figure 13.1, the infrared excitation wavelength does not induce autofluorescence in the target cells. Only the emission of the nanophores can be seen (Figure 13.1B). This signal-to-noise ratio reduction is of great benefit in the visualization of low-level luminescence signals in biological systems.

13.5 BIOSENSING

Plasmonic Biosensors. Metallic nanostructures have been used for biosensing. A very widely used biosensing method utilizes surface plasmon resonance in a

Kretchmann geometry, discussed in Chapter 2. The thin-film surface plasmon sensors are described in detail in the book on biophotonics (Prasad, 2003). These types of surface plasmon sensors are already in market. More recent activities have focused on metallic nanoparticles and metallic shells. The plasmon resonances in these nanostructures have been discussed in Chapter 5. In each case, the principle involved utilizes enhancement of fluorescence, or a change in the plasmon resonance as a result of binding of the analyte, to be detected, on the surface of a metallic nanostructure, or modification of interparticle interactions created by the analyte. Some examples of the approaches using nanoparticles and nanoshells are presented here.

Lakowicz and co-workers (Malicka et al., 2003) used metal-enhanced fluorescence for detection of DNA hybridization which forms a basis for a wide range of biotechnology and diagnostic applications. They used thiolated oligonucleotide to bind to silver nanoparticles on a glass surface. Addition of a complementary fluorescein-labeled oligonucleotide produced a dramatic time-dependent 12-fold increase in fluorescence intensity, as the hybridization process proceeded, bringing fluorescein close to silver nanoparticles for surface plasmon enhancement. This enhancement effect has been discussed in Chapter 5, Section 5.6. Thus this type of approach can be used to increase the sensitivity of DNA detection.

Storhoff and Mirkin (1999) linked a single-stranded DNA, modified with a thiol group at one terminal, to a gold nanoparticle of ~15 nm in diameter via strong gold–sulfur interactions. The 15-nm-diameter gold particles exhibit a well-defined surface plasmon resonance. Due to this resonance, the individual gold particles, even when attached to DNA, exhibit a burgundy-red color. When this DNA attached to the gold particle hybridizes with its complementary DNA in the test sample, the duplex formation leads to aggregation of the nanoparticles, shifting the surface plasmon resonance and, thus, changing the color to blue black. The reason for the shift is that the plasmon band is very sensitive to the interparticle distance as well as to the aggregate size (see Chapter 5).

Halas, West, and co-workers (Hirsch et al., 2003) used gold nanoshells to demonstrate a rapid immunoassay, capable of detecting analyte within a complex biological media such as blood. The plasmon resonances of these metallic nanoshells have been discussed in Chapter 5. They used nanoshell-conjugated antibodies. The principle utilized is that when antibody-conjugated particles are presented with a multivalent analyte, multiple particles bind to the analyte, forming particle dimers and higher-order aggregates. Aggregation of these metallic nanoparticles produces a red shift in the plasmon resonance, together with a decrease in the extinction coefficient. Hirsch et al. (2003) demonstrated the detection of immunoglobulins. Figure 13.2 shows the decrease in the extinction coefficient of the plasmon band at 720 nm upon aggregation of the gold nanoshells, resulting from the binding to the analyte (immunoglobulin).

Photonic Crystal Biosensors. Yet another approach utilizing nanophotonics for biosensing is based on the shift of stop-gap wavelength in a photonic crystal, created by binding of analyte. This method has already been discussed in Chapter 9, Section 9.8.

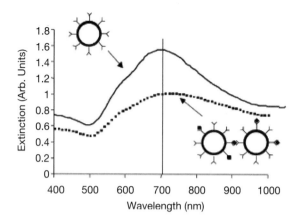

Figure 13.2. UV-vis spectrum of disperse nanoshells fabricated with a 96-nm-diameter core and a 22-nm-thick gold shell (———); spectrum of nanoshells/antibody conjugates following addition of analyte (- - -). Extinction reduction upon aggregation in the presence of analyte was monitored at 720 nm, as indicated. From Hirsch et al. (2003), reproduced with permission.

Porous Silicon Microcavity Biosensors. The porous silicon is a material that contains nanodomains of silicon and is highly luminescent. However, their emission profile is fairly broad (150 nm centered around 750 nm). Fauchet and co-workers (Chan et al., 2001a) used a porous silicon microcavity resonator configuration to narrow the fluorescence line width (~3 nm) and enhance its intensity, and they demonstrated its application for biosensing. The microcavity structure consists of a porous silicon medium between two distributed Bragg reflectors. They used the microcavity arrangement for DNA detection. The oxidized porous silicon surface was silanized and coupled with single-stranded DNA. The microcavity mode structure of fluorescence showed shift, when exposed to a complementary DNA. This result is shown in Figure 13.3. In contrast, no shift was observed when exposed to a noncomplementary DNA strand. Chan et al. (2001b) used this silicon microcavity sensor for identification of gram-negative bacteria. The results demonstrated the ability of a porous-silicon biosensor to distinguish between gram-negative and gram-positive bacteria.

PEBBLE Nanosensors for In Vitro Bioanalysis. A probe encapsulated by biologically localized embedding (PEBBLE), introduced by Kopelman and co-workers (Clark et al., 1999; Monsoon et al., 2003), enables optical measurement of changes in intracellular calcium levels, pH, and other biologically significant chemicals. It provides a major advancement in the field of nanoprobes and nanomedicine. PEBBLEs are nanoscale spherical devices, consisting of sensor molecules entrapped in a chemically inert matrix. Figure 13.4 shows a schematic diagram of a PEBBLE nanosensor that can provide many functions. The matrix materials used for production of PEBBLES are also shown in the figure. Examples of matrix media used for PEBBLE technology are polyacrylamide hydrogel, sol–gel silica, and

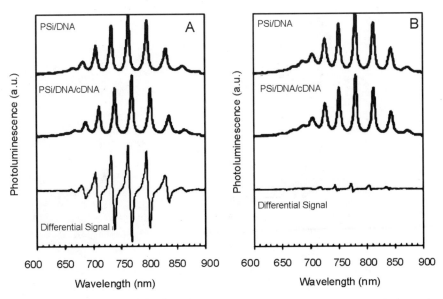

Figure 13.3. Mode structures for DNA attached porous silicon microcavity are shown as the top spectrum in both sets of plots. When complementary DNA is exposed to the DNA attached porous silicon, a 7-nm red shift is observed after binding (middle spectrum of part A). A large differential signal is obtained before and after binding. When a noncomplementary strand of DNA is exposed to the porous silicon sensor (middle spectrum of part B), no shifting of the luminescence peaks is observed and the differential signal is negligible. From Chan et al. (2001a), reproduced with permission.

cross-linked decyl methacrylate. These matrices have been used by Kopelman's group to fabricate sensors for H^+, Ca^{2+}, Na^+, Mg^{2+}, Zn^{2+}, Cl^-, NO_2^-, O_2, NO, and glucose. The PEBBLE size ranges from 30 to 600 nm. The matrix porosity allows entrapment and sensing of the analyte.

An example of a PEBBLE nanosensor is the calcium PEBBLE that utilizes calcium Green–1 and sulforhodamine dyes as sensing components. The calcium green fluorescence intensity increases with increasing calcium concentrations, while the sulforhodamine fluorescence intensity is unaffected. Thus, the ratio of the calcium green intensity to the sulforhodamine intensity can be used to measure cellular calcium levels.

According to Kopelman and co-workers, the PEBBLE technology offers the following benefits:

- It protects the cells from any toxicity associated with the sensing dye.
- It provides an opportunity to combine multiple sensing components (dyes, ionphores, etc.) and create complex sensing schemes.
- It insulates the indicator dyes from cellular interferences such as protein binding.

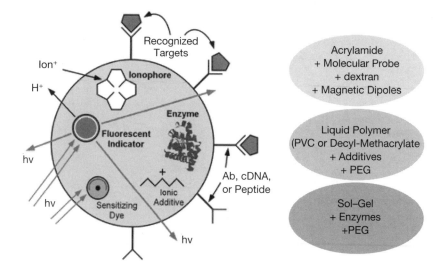

Figure 13.4. Schematics of a PEBBLE nanosensor, with various functions shown. Current matrix materials are presented on the right. From http://www.umich.edu/~koplab/research2/analytical/EnterPEBBLEs.html.

Dye-Doped Nanoparticle Bisensors. Tan and co-workers (Zhao et al., 2003) have developed dye-doped nanoparticles (NP) for DNA sensing. In this method, a number of luminescent molecules are embedded into the silica nanoparticles. These nanoparticles are highly luminescent, very resistant to photobleaching by the environment, and have been applied as pigments for cell staining and biorecognition. Tan and co-workers (Zhao et al., 2003) have also developed methodologies for modifying the silica surface of the nanoparticles for desired biochemical functionality, such as cell staining, enzymatic NPs, and DNA biosensors.

Using these nanoparticles, they developed useful biotechnologies for DNA/mRNA analysis and separation, as well as for single bacterium detection and cellular imaging. Sensitive DNA detection is extremely important in clinical diagnostics, gene therapy, and a variety of biomedical studies. The dye-doped silica nanoparticles, prepared using reverse micelle reaction method described in Chapter 7, are highly fluorescent, extremely photostable, and easy to bioconjugate for bioanalysis. The ultrasensitive DNA/mRNA analysis was realized using a sandwich assay, shown in Figure 13.5. There are three DNA strands in this assay: capture DNA1 (5′TAA CAA TAA TCC T-biotin 3′); probe DNA3 (5′ biotin-T ATC CTT ATC AAT ATT 3′) labeled with a NP (60 nm) to form a NP-DNA3 conjugate; and target DNA2 (5′GGA TTA TTG TTA AAT ATT GAT AAG GAT 3′). The combined sequences of DNA1 and DNA3 are complementary to that of the target DNA2. The biotinylated DNA1 is first immobilized on an avidin-coated glass substrate. DNA2 and NP-DNA3 are then added for hybridization. Detection of the captured DNA2 that hybridized to the DNA1 bound to the glass surface is accom-

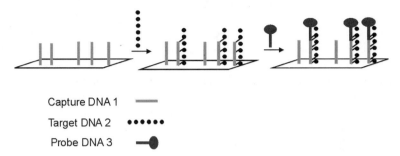

Figure 13.5. Schematic of a sandwich DNA assay based on NP. From Zhao et al. (2003), reproduced with permission.

plished by monitoring fluorescence signals of the NP-DNA3 conjugates left on the glass surface after thorough washing steps and with proper excitation. The nanoparticles exhibit an excellent signaling ability in the presence of trace amounts of DNA targets. With an effective surface modification, nonspecific binding and nanoparticle aggregation are minimized. In addition, the nanoparticle-based DNA bioanalysis assay can effectively discriminate one base mismatched DNA sequence.

Optical Nanofiber Sensors. These sensors utilize tapered optical fibers which are used for near field microscopy (Cullum and Vo-Dinh, 2003). The tapered fibers have the tip diameters ranging between 20 and 100 nm. These tapered fibers are also referred to as nanofibers. Like in near-field microscopy, these fibers are metal coated on the wall to confine light. The sensing biorecognition probe, which binds the analyte to be detected, is immobilized at the tip opening (the distal end of the nanofiber). The first optical nanosensors were demonstrated by Kopelman's group (Tan et al., 1992) for intracellular chemical sensing. Since then, several reports of measurements of pH, various ions, and other chemicals have appeared (Tan et al., 1995, Song et al., 1997; Koronczi et al., 1998; Bui et al., 1999; Vo-Dinh et al., 2000; Xu et al., 2001). An example is a nanobiosensor reported by Vo-Dinh and coworkers (Alarie and Vo-Dinh, 1996). In this work the fiber tip was silanized to allow for covalent attachment of antibody, using a reaction involving carbonyl dimidazole. The antibody employed in this sensor probe recognizes benzo[a]pyrene groups as a specific antigen which can detect benzo[a]pyrene tetrol (BPT), a DNA adduct of benzo[a]pyrene found in cells treated with this chemical carcinogen. This provides a convenient and simple method to rapidly detect cells that have be malignantly transformed with this chemical.

13.6 NANOCLINICS FOR OPTICAL DIAGNOSTICS AND TARGETED THERAPY

Our Institute for Lasers, Photonics, and Biophotonics has developed nanoparticles, called nanoclinics, which have a complex surface functionalized silica nanoshell containing various probes for diagnostics and drugs for targeted delivery (Levy et al., 2002). These "nanoclinics" (Figure 13.6) provide a new dimension to targeted diagnostics and therapy. They are produced using multistep nanochemistry in a reverse micelle nanoreactor (see Chapter 7 for this method), and surface functionalized with known biotargeting agents.

Nanoclinics are ~30-nm silica shells that can encapsulate various optical, magnetic, or electrical probes as well as externally activatable therapeutic agents (see Figure 13.6). The size of these nanoclincs is small enough for them to enter the cell, in order for them to function from within the cell. Through the development of nanoclinics (functionalized nanometer-sized particles that can serve as carriers), new therapeutic approaches to disease can be accomplished from within the cell. At our Institute, prototypic nanoclinics were produced through the integration of the ferrofluid, nanotechnology, and a peptide hormone analogue targeting agent (Levy et al., 2002). This multilayered nanosized structure consists of an iron oxide core, a two-photon optical probe, and a silica shell with a targeting peptide hormone analogue (luteinizing hormone-release hormone analog: LH-RH) covalently coupled to the surface of the shell. This protocol can produce nanoclinics with a tunable size from 5 to 40 nm in diameter. Their size enables them to diffuse into the tissue and enter the cells. These nanoparticles are large enough to respond to the applied DC magnetic field at 37°C. High-resolution transmission electron microscopy shows that the structure of the nanoparticles is composed of a crystalline core corresponding to Fe_2O_3 and an amorphous silica layer (bubble). The same crystalline/amorphous structure was obtained by electron diffraction of the particle and also confirmed by X-ray diffraction.

The selective interaction and internalization of these nanoclinics within cells was visualized using two-photon laser scanning microscopy, allowing for real-time observation of the uptake of nanoclinics (Bergey et al., 2002). The selective uptake of targeted nanoclinics was readily demonstrated by their uptake by (KB) oral epithelial carcinoma cells (LH-RH receptor positive), whereas a similar accumulation was not observed in LH-RH-negative nanoclinics studies or LH-RH positive nanoclinics incubated with receptor negative cells (UCI–107).

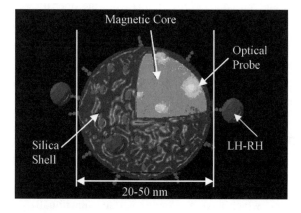

Figure 13.6. Illustrated representation of a nanoclinic.

The multifunctional nanoclinics, containing the magnetic Fe_2O_3 nanoparticles, also demonstrated the novel ability to selectively lyse targeted cells, when exposed to a DC magnetic field. Magnetic probes or particles have been investigated as a potential alternative treatment for cancer. Studies have demonstrated that the hyperthermic effect, generated by magnetic particles coupled to a high-frequency AC magnetic field (requiring tremendous power), could be used as an alternate or adjuvant to current therapeutic approaches for cancer treatment. This hyperthermic effect (heat produced by the relaxation of magnetic energy of the magnetic material) was shown to effectively destroy tumor tissue surrounding the probes or particles resulting in the reduction of tumor size. In contrast, our work demonstrated a new mechanism to mechanically disrupt the cellular structure, using a DC magnetic field at a strength typically achievable by magnetic resonance imaging (MRI) systems, for selectively destroying targeted cells.

13.7 NANOCLINIC GENE DELIVERY

Genetic manipulation of cells has become the foundation of what we know as molecular biology. There are many potential applications of gene therapy for treating human diseases, including renal diseases (polycystic kidney disease, renal cancer, chronic interstitial disease, and glomerulonephritides); cystic fibrosis; various immunodeficiency diseases (infectious and genetic); cancer (pancreatic, breast, prostrate, etc.); Parkinson's disease; multiple sclerosis, and so on (Anderson, 1998; Kay et al., 1997).

As we gain a deeper understanding of the genetic links to human diseases, the ability to prevent or treat using gene therapeutic approaches will become the new "medical wave of the future." However, the major blockage in effective gene therapy is the development of a safe vector for the in vivo delivery of therapeutic genes to the appropriate cells or tissues. Current in vivo technology has attempted to use attenuated viral strains (not capable of causing disease) as vectors for moving specific genetic material, but this technique has had limited success—sometimes with serious side effects (El-Aneed, 2004).

Because of the vast potential of gene therapy in the treatment of chronic human illnesses, there is considerable interest in the development of safe and effective transport vectors. The successful development of nonviral gene delivery nanoparticle platforms, using our nanoclinic concept, represents an important step toward enabling physicians to tailor therapy to be individualized for a given patient. Once the genetic defect for a particular disease has been identified, a gene-based therapy can be devised to repair the defect. Engineering of nonviral gene transfer/transfection vectors will facilitate the targeting of these therapeutic genes into the appropriate site in a tissue and cell.

For nonviral gene delivery, we have developed and tested a formulation of nanoparticles which consists of a stable aqueous dispersion of ultrafine organically modified silica-based nanoparticles (ORMOSIL) (average diameter 30 nm). ORMOSIL-based nanoparticles can be synthesized with varying surface charge and are stable (nonaggregating) in aqueous environments. The genetic payload (negatively charged DNA) is bound to the cationic groups present on the surface of the nanoparticles. The nonaggregating potential, coupled with the designed ability to bind and protect DNA, provides a second platform from which a nonviral gene transfer vector will be produced and tested. In addition, we have synthesized these particles to contain selective fluorescent dyes that we have used to optically track the delivery of DNA directly to the nucleus of cells. The nanoparticles for DNA delivery are functionalized with amino groups, which made it possible to electrostatically attach DNA molecules on the surface, as shown schematically in Figure 13.7.

Our studies used DNA molecules that were stained with (a) YOYO-1, a well-known nucleic acid stain with high binding efficiency (molecular probes), and (b) the cationic dye, And-10, as the dye encapsulated within the ORMOSIL shell. Thus, the nanoparticle-DNA complex has two dye molecules, one encapsulated inside the particle and the other on the DNA molecule adsorbed on the surface. The dyes were selected in such a way that the emission of And-10 overlaps to a great extent with the absorption of YOYO-1. At the same time, YOYO-1 has minimal absorption in the range of And-10 absorption. Thus, if the And-10→YOYO-1 fluorescence resonance energy transfer (FRET), a nanophotonic process described in Chapter 2, manifests itself under the excitation of And-10, one can conclude that this is because of close proximity of And-10 encapsulated in ORMOSIL nanoparticles and YOYO-1 intercalated into DNA molecules which are adsorbed on the surface of nanoparticles (see Figure 13.8).

In fact, in an ORMOSIL-containing buffer solution of DNA stained with the YOYO-1 we have observed the YOYO-1 emission under excitation at 380 nm (close

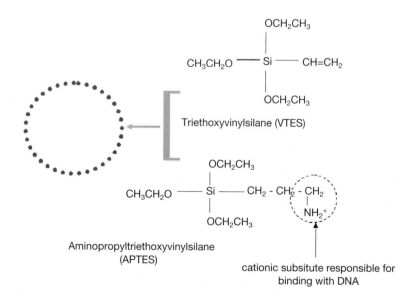

Figure 13.7. Schematics of the ORMOSIL nanoparticle composition.

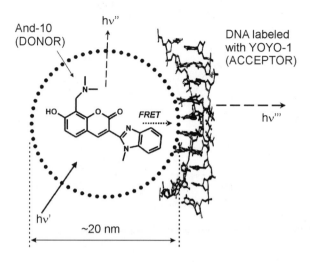

Figure 13.8. Schematics of the dye encapsulation and attachment of DNA on the ORMOSIL nanoparticle surface for FRET imaging.

to the And-10 absorption maximum) (Figure 13.9). The intensity of this emission increases significantly in comparison with the emission from a buffer solution of DNA-YOYO-1 without ORMOSIL nanoparticles. At the same time, the emission from the donor dye, And-10, dropped down in an ORMOSIL-containing buffer solution of DNA-YOYO-1, in comparison with a buffer solution of ORMOSIL nanoparticles without DNA. All other conditions (pH, temperature) remained the same, and the concentration of absorbing species was carefully controlled spectrophotometrically.

DNA release from ORMOSIL nanoparticles in cell cytoplasm can take place in response to changes in pH. We simulated these changes, *in vitro,* and found that a decrease in the And-10 → YOYO-1 FRET efficiency takes place with a decrease in pH, which we attribute to the detachment of the DNA molecules from nanoparticles. Since FRET occurs only when the donor and acceptor molecules are in close proximity (see Chapter 2), we can use this property to monitor the release of DNA from the nanoparticles inside live cells, and determine the conditions under which it occurs with better efficiency, without deterioration of the DNA delivery to the cells. This can have potential applications in gene delivery where the optically trackable nanoparticles can be used as nonviral vectors, to carry genetic material into cells and release them once inside the cytoplasm.

We have shown an expression of delivered genes in cells using the amino-functionalized ORMOSIL nanoparticles as nonviral vector. The plasmid, pGFP, encoding for the green fluorescent protein (see Chapter 12 for GFP), has been successfully transfected by the ORMOSIL nanoparticles in COS-1 cells, as seen by the confocal fluorescence images of individual cells (Figure 13.10). The origin of this cellular fluorescence from the green fluorescent protein (as opposed to cellular autofluorescence) has been confirmed by localized spectroscopy of the transfected cells.

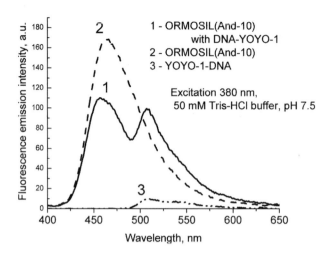

Figure 13.9. Fluorescence emission spectra indicating the existence of FRET between And-10, dye encapsulated in ORMOSIL nanoparticle, and YOYO-1, dye intercalated into DNA absorbed on the surface of the nanoparticle.

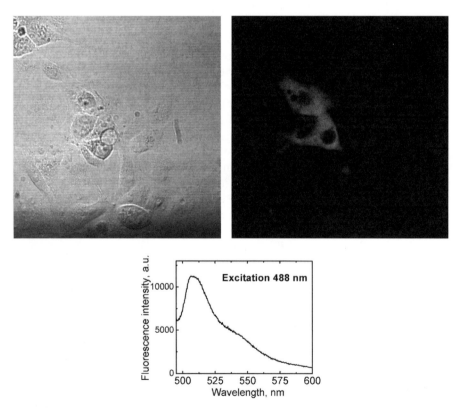

Figure 13.10. Cells transfected with eGFP (enhanced green fluorescent protein) vector delivered with ORMOSIL nanoparticles. Transmission (*left*) and fluorescence images (*right*) are shown. **Below:** eGFP Fluorescence spectra taken from cell cytoplasm.

13.8 NANOCLINICS FOR PHOTODYNAMIC THERAPY

Photodynamic therapy is an emerging modality for the treatment of a variety of oncological, cardiovascular, dermatological, and ophthalmic diseases. Photodynamic therapy is based on the concept that light-sensitive species or photosensitizers (PS) can be preferentially localized in tumor tissues upon systemic administration (Prasad, 2003). When such photosensitizers are irradiated with an appropriate wavelength of visible or near infrared (NIR) light, the excited molecules can transfer their energy to molecular oxygen (in its triplet gound state) present in the surrounding media. This energy transfer results in the formation of reactive oxygen species (ROS), like singlet oxygen ($^1O_2$) or free radicals (Prasad, 2003). ROS are responsible for oxidizing various cellular compartments, including plasma, mitochondria, lysosomal and nuclear membranes, and so on, resulting in irreversible damage of cells (Prasad, 2003). Therefore, under appropriate conditions, photodynamic therapy offers the advantage of an effective and selective method of destroying diseased tissues, without damaging adjacent healthy ones (Hasan et al., 2000).

However, most photosensitizing drugs (PS) are hydrophobic (i.e., poorly water-soluble); therefore, preparation of pharmaceutical formulations for parenteral administration is highly hampered (Konan et al., 2002). To overcome this difficulty, different strategies have evolved to enable a stable dispersion of these drugs into aqueous systems, often by means of a delivery vehicle. Upon systemic administration, such drug-doped carriers are preferentially taken up by tumor tissues by the virtue of the "enhanced permeability and retention effect" (Konan et al., 2002; Duncan, 1999), which is a property of such tissues to engulf and retain circulating macromolecules and particles owing to their 'leaky' vasculature. The carriers include oil dispersions (micelles), liposomes, low-density lipoproteins, polymeric micelles, and hydrophilic drug–polymer complexes. Oil-based drug formulations (micellar systems) using nonionic polyoxyethylated castor oils (e.g., Tween-80, Cremophor-EL, or CRM, etc.) have shown enhanced drug loading and improved tumor uptake over free drugs, presumably due to interaction with plasma lipoproteins in blood (Kongshaug et al., 1993). However, such emulsifying agents also have been reported to elicit acute hypersensitivity (anaphylactic) reactions *in vivo* (Michaud, 1997). Liposomes are concentric phospholipid bilayers encapsulating aqueous compartments, which can contain hydrophilic and lipophilic drugs (Konan et al., 2002). Although the tumor uptake of liposomal formulation of drugs is better than that of simple aqueous dispersions, many suffer from poor drug loading and increased self-aggregation of the drug in the entrapped state (Damoiseau et al., 2001). Liposomes are also prone to opsonization and subsequent capture by the major defense system of the body (reticuloendothelial system, or RES) (Damoiseau et al., 2001). Recently, drugs incorporated inside pH-sensitive polymeric micelles have shown improved tumor phototoxicity, compared to Cremophor-EL formulations *in vitro*. However, *in vivo* studies resulted in poor tumor regression and increased accumulation in normal tissues (Taillefer et al., 2000). Another major disadvantage, common to all of the above-mentioned delivery systems which are based on controlled-release of the photosensitive drugs, is the post-treatment accumulation of the free drugs in the skin and the eye, resulting in phototoxic side effects, which may last for weeks or even months (Dillon et al., 1988).

Hydrated ceramic-based nanoparticles, doped with photosensitive drugs, carry the promise of solving many of the problems associated with free as well as polymer-encapsulated drugs. Such ceramic particles have a number of advantages over organic polymeric particles. First, the preparative processes involved, which are quite similar to the well-known sol–gel process (Brinker and Schrer, 1990), require simple, ambient temperature conditions. These particles can be prepared with desired size, shape, and porosity and are extremely stable. Their ultra-small size (less than 50 nm) can help them evade capture by the reticuloendotheial system. In addition, there are no swelling or porosity changes with changes in pH, and they are not vulnerable to microbial attack. These particles also effectively protect doped molecules (enzymes, drugs, etc.) against denaturation, induced by extreme pH and temperature (Jain et al., 1998). Such particles, like silica, alumina and titania, are also known for their compatibility in biological systems (Jain et al., 1998). In addition, their surfaces can be easily modified with different functional groups (Lal et al., 2000). Therefore, they can be attached to a variety of monoclonal antibodies or other ligands to target them to desired sites *in vivo*.

Ceramic nanoparticles are highly stable and may not release any encapsulated biomolecules, even at extreme conditions of pH and temperature (Jain et al., 1998). Conventional drug delivery methods require the carrier vehicle to free the encapsulated drug to elicit the appropriate biological response (Hasan et al., 2000). However, this is not a prerequisite when macromolecular carrier molecules are used for the delivery of photosensitizing drugs in photodynamic therapy (Hasan, 1992; Hasan et al., 2000). Keeping this in view, we have developed ceramic-based nanoparticles as carriers of photosensitizing drugs for applications in photodynamic therapy. Although these particles do not release the entrapped drugs, their porous matrix is permeable to oxygen. Therefore, the desired photodestructive effect of the drug will be maintained even in the encapsulated form.

We have reported the synthesis of photosensitizer-doped organically modified silica-based nanoclinic nanoparticles (diameter ~30 nm), by controlled hydrolysis of triethoxyvinylsilane in micellar media (Roy et al., 2003). The drug/dye used is 2-devinyl-2-(1-hexyloxyethyl) pyropheophorbide (HPPH), an effective photosensitizer that is in Phase I/II clinical trials at Roswell Park Cancer Institute, Buffalo, New York (Henderson et al., 1997). The doped nanoparticles were shown to be spherical and highly monodispersed. Since ceramic matrices are generally porous, photosensitizing drugs entrapped within them can interact with molecular oxygen that has diffused through the pores. This can lead to the formation of singlet oxygen by energy transfer from the excited photosensitizer to oxygen, which can then diffuse out of the porous matrix to produce cytotoxic effect in tumor cells. The generation of singlet oxygen, after excitation of HPPH, was checked by singlet oxygen luminescence at 1270 nm. Figure 13.11 shows the singlet oxygen luminescence generated by HPPH, solubilized in micelles as well as entrapped in nanoparticles. Both spectra show similar intensity and peak positions (1270 nm), indicating similar efficiencies of singlet oxygen generation in both cases. A control spectrum, using nanoparticles without HPPH, shows no singlet oxygen luminescence. The doped nanoparticles are actively taken up by tu-

13.8 NANOCLINICS FOR PHOTODYNAMIC THERAPY 373

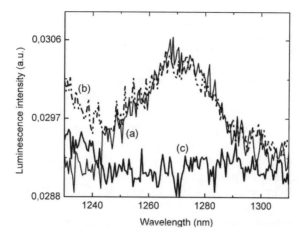

Figure 13.11. Emission spectra of singlet oxygen generated with (a) HPPH in AOT/BuOH/D_2O micelles (solid line), (b) HPPH doped in silica nanoparticles dispersed in D_2O, and (c) void silica nanoparticles in D_2O.

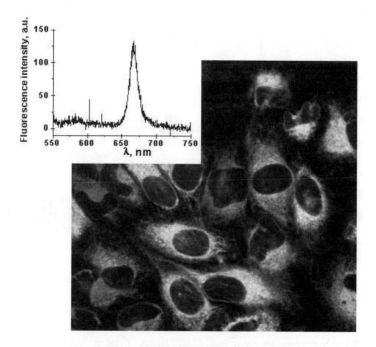

Figure 13.12. Confocal fluorescence image of tumor cells treated with HPPH-doped nanoparticles. **(Inset:)** HPPH fluorescence spectra taken from the cytoplasm of cell.

mor cells and irradiation with visible light results in irreversible destruction of such impregnated cells (Figure 13.12). These observations suggest the potential of ceramic-based particles as carriers for photodynamic drugs (Roy et al., 2003).

Another use of nanophotonics is multiphoton photodynamic therapy using up-converting nanoparticles. At our Institute for Lasers, Photonics, and Biophotonics, the silica-encapsulated rare-earth doped Y_2O_3 nanoparticles, discussed in Section 13.4, are also being investigated for multiphoton photodynamic therapy (Holm et al., 2002). The basic concept is similar to two-photon photodynamic therapy (Bhawalkar et al., 1997). However, here one utilizes the IR-to-visible up-conversion in these nanophores and not a two-photon active dye. The benefits are again greater penetration into a tissue, offered by the use of a near-IR (974-nm) excitation source. For this purpose, we again used the photosensitizer, HPPH, to test the ability of the nanophosphors to excite HPPH for photodynamic therapeutic action. The following study was performed. Sintered nanoparticles were dispersed in DMSO to obtain a translucent colloidal dispersion of nanophosphors. Equal volumes of the nanophosphor solution and 1 mM of HPPH in DMSO were mixed in individual cuvettes. Identical solutions containing only HPPH or the nanoparticles were also

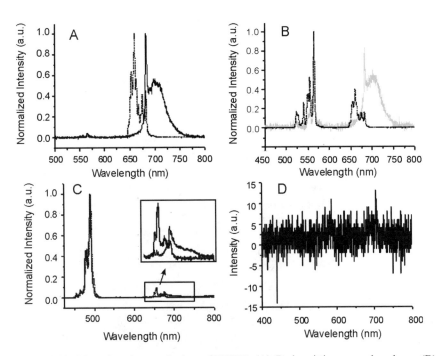

Figure 13.13. Nanophosphor excitation of HPPH; (A) Red-emitting nanophosphors, (B) green-emitting nanophosphors. (C) blue nanophosphors, and (D) HPPH excitation by 974 nm. Emission of the nanophosphor alone is represented by dotted line. Emission of HPPH is represented by solid lines.

placed into cuvettes. Each cuvette was pumped with a 974-nm CW diode laser, and the emission spectra were collected at 90° to the excitation laser with a fiber-coupled CCD spectroscope. The data were normalized to the maximum peak intensity and plotted with identical nanophosphor blank solutions.

It is clearly seen in Figure 13.13 that within the experimental parameters, both the green- and red-emitting nanophosphors are capable of exciting HPPH. Coupling of the emission of the nanophosphor with HPPH is shown by the loss of nanoparticle emission and appearance of the HPPH emission. The blue-emitting nanophosphors, however, did not demonstrate any significant coupling with HPPH. This lack of fluorescence from HPPH is due to absence of overlap of emission of the blue nanophosphors with the absorption band of the HPPH.

13.9 HIGHLIGHTS OF THE CHAPTER

- Nanophotonics is useful for optical diagnostics, as well as for light-guided (optically tracked) and light-activated therapies using nanomedicine.
- Nanomedicine is an emerging field that utilizes nanoparticles for targeted drug delivery with increased effectiveness of therapy.
- Near-field bioimaging allows one to probe structures and functions of microbes and biological assemblies that are significantly smaller than the wavelength of light.
- Nanoparticles of dimensions smaller than 50 nm provide a flexible structural platform for intracellular optical diagnostics and therapy, by offering the opportunity to tailor their interior, the surface, and the pores.
- Semiconductor quantum dots, with their narrow emission line width and tunability of emission wavelength by compositional and size variations, have emerged as an important group of biological markers for multispectral bioimaging
- Rare-earth-doped nanoparticles, which absorb IR light by multistep multiphoton absorption to produce up-converted emission in the visible, provide opportunity for background free bioimaging and deeper penetration in cells and tissues, compared to fluorescent markers excited by single-photon absorption.
- Nanophotonics provide several approaches for biosensing.
- Plasmonic biosensors utilize enhancement of fluorescence or a change in the plasmon resonance as a result of binding of the analyte, to be detected, on the surface of metallic nanostructures.
- Photonic crystal biosensors utilize a shift of stopgap wavelength in photonic crystal, created by binding of analyte.
- Porous silicon microcavity biosensors utilize analyte-induced shift in the mode structures of a microcavity formed by a nonporous silicon medium between two distributed Bragg reflectors.

- PEBBLE is an acronym for probe encapsulated by biologically localized embedding, and it refers to sensor molecules entrapped in an inert nanoparticle. These devices are advantageous, because cells and the indicator dyes are protected from each other. Also, multiple sensing mechanisms can be combined into one particle.
- Dye-doped nanoparticle biosensors utilize luminescent molecules embedded into a silicon particle where they are protected from photobleaching and environmental degradation.
- Optical nanofiber biosensors use the tapered optical fiber geometry of near-field microscopy, with the biorecognition element immobilized on the fiber tip opening.
- Nanoclinics are surface-functionalized silica nanoshells that encapsulate probes as well as externally activatable drugs or therapeutic agents.
- The nanoclinics provide a versatile nanomedicine platform for light-guided (optically tracked) and light-activated therapy.
- Magnetic nanoclinics are capable of destroying cancer cells by intracellular mechanical disruption, produced in the presence of a DC magnetic field.
- Optical tracking of nanoclinics has been used to establish effective gene delivery, providing the exciting prospects of using them for nonviral, safe gene therapy.
- Nanoclinics can be used to increase the effectiveness of light activated therapy, such as photodynamic therapy.
- The use of nanoclinics, encapsulating photodynamic therapeutic drugs (photosensitizers), allows these drugs, which are generally hydrophobic, to be more effectively distributed in biological fluids. Optical imaging can be used to track their distribution and localization in tumors.
- The porosity in the nanoclinic surface allows the diffusion of oxygen in to the interior of the nanoclinic, where singlet oxygen is formed by interaction with light-activated photodynamic therapeutic drug. The singlet oxygen produces the photodynamic therapy effect.
- Nanoclinics, containing up-converting nanoparticles, allow excitation of photodynamic therapeutic drugs in the IR, thus providing the opportunity for deeper penetration of light to treat deeper tumor, compared to that produced by visible light.

REFERENCES

Akerman, M. E., Chen, W. C. W., Laakkonen, P., Bhatia, S. N., and Ruoslalti, D., Nanocrystals Targeting *In Vivo, Proc. Natl. Acad. Sci.* **99,** 12617–12621 (2002).

Alarie, J. P., and Vo-Dinh, T., Antibody-Based Submicron Biosensor for Benzo[*a*]pyrene DNA Adduct, *Polycyclic Aromat. Compd.* **9,** 45–52 (1996).

Anderson, W. F., Human Gene Therapy, *Nature* **392 [Suppl.],** 25–30 (1998).

Bergey, E. J., Levy, L., Wang, X., Krebs, L. J., Lal, M., Kim, K.-S., Pakatchi, S., Liebow, C., and Prasad, P. N., Use of DC Magnetic Field to Induce Magnetocytolysis of Cancer Cells Targeted by LH-RH Magnetic Nanoparticles *In Vitro, Biomed. Microdevices* **4**, 293–299 (2002).

Betzig, E., Chichester, R. J., Lanni, F., and Taylor, D. L., Near-Field Fluorescence Imaging of Cytoskeletal Actin, *Bioimaging* **1**, 129–135 (1993).

Bhawalkar, J. D., Kumar, N. D., Zhao, C.-F., and Prasad, P. N., Two-Photon Photodynamic Therapy, *J. Clin. Laser Med. Surg.* **15**, 201–204 (1997).

Brinker, C. J., and Schrer, G., *Sol–Gel Science: The Physics and Chemistry of Sol–Gel Processing*, Academic Press: San Diego, 1990.

Bruchez, M., Jr., Moronne, M., Gin, P., Weiss, S., and Alivisatos, A. P., Semiconductor Nanocrystals as Fluorescent Biological Labels, *Science* **281**, 2013–2016 (1998).

Bui, J. D., Zelles, T., Lou, H. J., Gallion, V. L., Phillips, M. I., and Tan, W. H., Probing Intracellular Dynamics in Living Cells with Near-Field Optics, *J. Neurosci. Meth.* **89**, 9–15 (1999).

Chan, S., Horner, S. R., Fauchet, P. M., and Miller, B. L., Identification of Gram Negative Bacteria Using Nanoscale Silicon Microcavities, *J. Am. Chem. Soc.* **123**, 11797–11798 (2001a).

Chan, S., Li, Y., Rothberg, L. J., Miller, B. L., and Fauchet, P. M., Nanoscale Silicon Microcavities for Biosensing, *Mater. Sci. Eng. C* **15**, 277–282 (2001b).

Chan, W. C., and Nie, S., Quantum Dot Bioconjugates for Ultrasensitive Nonisotopic Detection, *Science* **281**, 2016–2018 (1998).

Chen, Y., Kalas, R. M., and Faris, W., Spectroscopic Properties of Up-converting Phosphor Reporters, *SPIE Proc.* **3600**, 151–154 (1999).

Clark, H. A., Hoyer, M., Philbert, M. A., and Kopelman, R., Optical Nanosensors for Chemical Analysis Inside Single Living Cells. 1. Fabrication, Characterization, and Methods for Intracellular Delivery of PEBBLE Sensors, *Anal. Chem.* **71**, 4831–4836 (1999).

Cullum, B. M., and Vo-Dinh, T., Nanosensors for Single-Cell Analyses, in *Biomedical Photonics Handbook*, T. Vo-Dinh, ed., CRC Press, Boca Raton, FL, 2003, pp. 60-1–60-20.

Damoiseau, X., Schuitmaker, H. J., Lagerberg, J. W. M., and Hoebeke, M., Increase of the photosensitizing efficiency of the Bacteriochlorin *a* by liposome-incorporation, *J. Photochem. Photobiol. B Biol.* **60**, 50–60 (2001).

Dillon, J., Kennedy, J. C., Pottier, R. H., and Roberts, J. E., *In Vitro* and *In Vivo* Protection Against Phototoxic Side Effects of Photodynamic Therapy by Radioprotective Agents WR-2721 and WR-77913, *Photochem. Photobiol.* **48**, 235–238 (1988).

Duncan, R., Polymer Conjugates for Tumour Targeting and Intracytoplasmic Delivery. The EPR Effect as a Common Gateway, *Pharm. Sci. Tech. Today* **2**, 441–449 (1999).

El-Aneed, A., An Overview of Current Delivery Systems in Cancer Gene Therapy, *J. Controlled Rel.* **94**, 1–14 (2004).

Hasan, T., *Photodynamic Therapy: Basic Principles and Clinical Applications*, Marcel Dekker, New York, 1992.

Hasan, T., Moor, A. C. E, and Ortel, B., Photodynamic Therapy of Cancer, in *Cancer Medicine, 5th ed.*, J. F. Holland, E. Frei, R. C. Bast, et al., eds., B. C. Decker, Hamilton, 2000, pp. 489–502.

Henderson, B. W., Bellnier, D. A., Graco, W. R., Sharma, A., Pandey, R. K., Vaughan, L., Weishaupt, K., and Dougherty, T. J., An *In Vivo* Quantitative Structure–Activity Relationship for a Congeneric Series of Pyropheophorbide Derivatives as Photosensitizers for Photodynamic Therapy, *Cancer Res.* **57,** 4000–4007 (1997).

Hirsch, L. R., Jackson, J. B., Lee, A., Halas, N. J., and West, J. L., A Whole Blood Immunoassay Using Gold Nanoshells, *Anal. Chem.* **75,** 2377–2381 (2003).

Holm, B. A., Bergey, E. J., De, T., Rodman, D. J., Kapoor, R., Levy, L., Friend, C. S., and Prasad, P. N., Nanotechnology in BioMedical Applications, *Mol. Cryst. Liq. Cryst.* **374,** 589–598 (2002).

Jain, T. K., Roy, I., De, T. K., and Maitra, A. N., Nanometer Silica Particles Encapsulating Active Compounds: A Novel Ceramic Drug Carrier, *J. Am. Chem. Soc.* **120,** 11092–11095 (1998).

Kapoor, R., Friend, C., Biswas, A., and Prasad, P. N., Highly Efficient Infrared-to-Visible Upconversion in $Er^{3+}:Y_2O_3$, *Opt. Lett.* **25,** 338–340 (2000).

Kay, M. A., Liu D., and Hoogerbrugge, P. M., Gene Therapy, *Proc. Natl. Acad. Sci* **94,** 12744–12746 (1997).

Konan, Y. N., Gruny, R., and Allemann, E., State of the Art in the Delivery of Photosensitizers for Photodynamic Therapy, *J. Photochem. Photobiol. B: Biology.* **66,** 89–106 (2002).

Kongshaug, M., Moan, J., Cheng, L. S., Garbo, G. M., Kolboe, S., Morgan, A. R., and Rimington, C., Binding of Drugs to Human Plasma Proteins, Exemplified by Sn(IV)-Etiopurpurin Dichloride Delivered in Cremophor and DMSO, *Int. J. Biochem.* **25,** 739–760 (1993).

Koronczi, I., Reichert, J., Heinzmann, G., and Ache, H. J., Development of a Submicron Optochemical Potassium Sensor with Enhanced Stability Due to Internal Reference, *Sensors Actuators B Chem.* **51,** 188–195 (1998).

Lal, M., Levy, L., Kim, K. S., He, G. S., Wang, X., Min, Y. H., Pakatchi, S., and Prasad, P. N., Silica Nanobubbles Containing an Organic Dye in a Multilayered Organic/Inorganic Heterostructure with Enhanced Luminescence, *Chem. Mater.* **12,** 2632–2639 (2000).

Levy, L., Sahoo, Y., Kim, K.-S., Bergey, E. J., and Paras, P. N., Nanochemistry: Synthesis and Characterization of Multifunctional Nanoclinics for Biological Applications, *Chem. Mater.* **14,** 3715–3721 (2002).

Malicka, J., Gryczynski, I., and Lakowicz, J. R., DNA Hybridization Assays Using Metal-Enhanced Fluorescence, *Biochem. Biomed. Res. Commun.* **306,** 213–218 (2003).

Mattoussi, H., Mauro, J. M., Goldman, E. R., Anderson, G. P., Sundor, V. C., Mikulec, F. V., and Bawendi, M. G., Self-Assembly of CdSe–ZnS Quantum Dot Bioconjugates Using an Engineered Recombinant Protein, *J. Am. Chem. Soc.* **122,** 12142–12150 (2000).

Michaud, L. B., Methods for Preventing Reactions Secondary to Cremophor EL., *Ann. Pharmacother.* **31,** 1402–1404 (1997).

Monsoon, E., Brasuel, M., Philbert, M. A., and Kopelman, R., PEBBLE Nanosensors for *In Vitro* Bioanalyses, in *Biomedical Photonics Handbook,* T. Vo-Dinh, ed., CRC Press, Boca Raton, FL, 2003, pp. 59-1–59-14.

Muramatsu, H., Chiba, N., Homma, K., Nakajima, K., Ataka, T., Ohta, S., Kusumi, A., and Fujihira, M., Near-Field Optical Microscopy in Liquids, *Appl. Phys. Lett.* **66,** 3245–3247 (1995).

Muramatsu, H., Homma, K., Yamamoto, N., Wang, J., Sakuta-Sogawa, K., and Shimamoto, N., Imaging of DNA Molecules by Scanning Near-Field Microscope, *Mater. Sci. Eng.* **12C,** 29–32 (2002).

Prasad, P. N., *Introduction to Biophotonics,* Wiley-Interscience, New York, 2003.

Roy, I., Ohulchanskyy, T., Pudavar, H. E., Bergey, E. J., Oseroff, A. R., Morgan, J., Dougherty, T. J., and Prasad, P. N., Ceramic-Based Nanoparticles Entrapping Water-Insoluble Photosensitizing Anticancer Drugs: A Novel Drug-Carrier System for Photodynamic Therapy, *J. Am. Chem. Soc.* **125,** 7860–7865 (2003).

Song, A., Parus, S., and Kopelman, R., High-Performance Fiber Optic pH Microsensors for Practical Physiological Measurements Using a Dual-Emission Sensitive Dye, *Anal. Chem.* **69,** 863–867 (1997).

Storhoff, J. J., and Mirkin, C. A., Programmed Materials Syntheses with DNA, *Chem. Rev.* **99,** 1849–1862 (1999).

Taillefer, J., Jones, M. C., Brasseur, N., Van Lier, J. E., and Leroux, J. C., Preparation and Characterization of pH-Responsive Polymeric Micelles for the Delivery Of Photosensitizing Anticancer Drugs, *J. Pharm. Sci.* **89,** 52–62 (2000).

Tan, W. H., Shi, Z. Y., Smith, S., Birnbaum, D., and Kopelman, R., Submicrometer Intracellular Chemical Optical Fiber Sensors, *Science* **258,** 778–781 (1992).

Tan, W. H., Shi, Z. Y., and Kopelman, R., Miniaturized Fiberoptic Chemical Sensors with Fluorescent Dye-Doped Polymers, *Sensors Actuators B Chem.* **28,** 157–161 (1995).

Vo-Dinh, T., Alarie, J. P., Cullum, B. M., and Griffin, G. D., Antibody-Based Nanoprobe for Measurement of a Fluorescent Analyte in a Single Cell, *Nat. Biotechnol.* **18,** 764–767 (2000).

Yanagida, T., Tamiya, E., Muramatsu, H., Degennar, P., Ishii, Y., Sako, Y., Saito, K., Ohta-Iino, S., Ogawa, S., Marriot, G., Kusumi, A., and Tatsumi, H., Near-Field Microscopy for Biomolecular Systems, in *Nano-Optics,* S. Kawata, M. Ohtsu and M. Irie, eds., Springer-Verlag, Berlin, 2002, pp. 191–236.

Zhao, X., Tapec-Dytioco, R., and Tan, W., Ultrasensitive DNA Detection Using Highly Fluorescent Bioconjugated Nanoparticles, *J. Am. Chem. Soc.* **125,** 11474–11475 (2003).

Xu, H., Aylott, J. W., Kopelman, R., Miller, T. J., and Philbert, M. A., A Real-Time Ratiometric Method for the Determination of Molecular Oxygen Inside Living Cells Using Sol–Gel-Based Spherical Optical Nanosensors with Applications to Rat C6 Glioma, *Anal. Chem.* **73,** 4124–4133 (2001).

CHAPTER 14

Nanophotonics and the Marketplace

This chapter analyzes various applications of nanophotonics, describes the current status of the commercial marketplace, and discusses its long-term future. As presented in Chapter 1, the fundamental aspects of light–matter interactions at a scale significantly smaller than the wavelength of light are most fascinating, and create challenges in the quest of knowledge. This chapter brings the perspective that scientific discoveries lead to technological inventions that can produce commercializable technologies for enriching our lives. Nanophotonics applications already exist, and more will be developed in the years ahead. There is little doubt that nanotechnology in general, and nanophotonics in particular, will play an important role in the development of products and their utilities in the decades ahead. However, any predictions made today are likely to be outdated rather quickly. While there will be the occasional nanophotonics breakthrough giving rise to exciting, new, and highly visible products, most will be made in relative obscurity, and yet will make significant advancements in technologies that will ultimately benefit society.

There are several measures of a new technology's economic impact. A quick search of the Internet clearly shows a rapidly growing number of companies being formed in the area of nanophotonics. Many are still at the research and development stage, but some have matured to the point where products are being offered. One must be careful to distinguish between the technology and the industry it enables. Readers should obviously use caution when encountering market projections because many of them are self-serving and/or exaggerated.

Some technological areas are well established, and the commercial potential of nanophotonics can perhaps be best appreciated by looking at the current market sizes and extrapolating what new nanotechnology can offer in the way of improved performance. Other technologies are very young, and it is much more difficult to predict the future market size. This chapter is an attempt to examine some of the most visible and important examples of how nanophotonics is being, and will be, commercialized.

Section 14.1 presents a general scope of the market and trends in nanotechnology, lasers, and photonics, the areas that together create nanophotonics. Section 14.2 describes the current commercial status of nanomaterials. The examples of materials included in this section are nanoparticles, photonic crystals, fluorescent quantum dots, and nanobarcodes. Section 14.3 covers quantum-confined lasers. Most semiconductor lasers being commercialized are based on quantum-confined struc-

Nanophotonics, by Paras N. Prasad
ISBN 0-471-64988-0 © 2004 John Wiley & Sons, Inc.

tures. Section 14.4 describes the applications and commercial status of near-field microscopy. Section 14.5 covers the applications and trend for nanolithography. Here, the major application, miniaturization of microprocessors, is analyzed in detail. Section 14.6 provides a subjective future outlook for nanophotonics, identifying some examples of future business opportunities. Section 14.7 gives the highlights of the chapter.

14.1 NANOTECHNOLOGY, LASERS, AND PHOTONICS

We first look at the general scope of the market and trends in the areas that combine to define nanophotonics. These are nanotechnology, lasers, and photonics. Despite the downturn of several technology sectors in and after 2000, nanotechnology continues to attract investments from venture capital groups. There are several sources for information on the market and projections in nanotechnology (Third European Report, 2003; Hwang, 2003; The Nanotech Report, 2003; Dunn and Whatmore, 2002). Sales figures for lasers are compiled annually by *Laser Focus World,* and they are reported in their January and February issues for nondiode and diode lasers, respectively. More generally, there are some figures and projections for the optoelectronic industry. The greater photonics marketplace is reviewed less quantitatively by *Photonics Spectra* in their January issues, wherein they discuss important growing industry developments.

Many important business decisions are made based upon expectations of growth in different market sectors. Market research companies attempt to satisfy this need by issuing reports that they claim have been rigorously researched. While the reports can cost many thousands of dollars in their full text versions, it is possible to purchase selected sections of interest for less. A cautionary note is that before placing much confidence in market forecasts of these reports, one must look at their track record in making past projections.

14.1.1 Nanotechnology

Nanotechnology, in general, is a rapidly growing field worldwide. In many countries, it is a national priority area, with the government committing major funding for research and development. The United States began to develop the National Nanotechnology Initiative (NNI) in 1996. Research funding provided by the NNI has significantly increased over this last decade as it has in other countries (global government funding was $432 million in 1997 and was $2154 million in 2002). The Engineering and Physical Sciences Research Council (EPSRC) of UK invests around $800 million in R&D each year; they have identified nanotechnology as a priority area (Gould, 2003). According to Hwang, the venture capital investment in the nanotechnology sector has been ~ $1 billion since 1999, with $386 million in 2002 (Hwang, 2003).

Few people are willing to make projections of the market potential. Dr. Roco of the National Science Foundation (Roco and Bainbridge, 2001) has predicted that the worldwide annual industrial production for this sector will exceed $1 trillion in 10–15 years. He has further broken this down as follows:

Market Size ($ billion)	Subcategory
340	Materials (nonchem)
100	Chemical (catalysts)
70	Aerospace
300	Electronics
180	Pharmaceutics
20	Tools

Source: http://www.ehr.nsf.gov/esie/programs/nsee/workshop/mihail_roco.pdf.

Nanophotonics, however defined, is a subset of these projections and is present in every category.

14.1.2 Worldwide Laser Sales

Lasers are powerful tools for many photonics-based technologies. The marketplace for lasers is well established and continues to grow, as more compact and energy efficient solid-state lasers, covering a wide spectral range, become readily available. Lasers have impacted not only the advanced technologies such as the telecommunications and healthcare, but also the consumer market in products such as CD players, laser printers, and so on. Nanophotonics has played a significant role in expanding the market for lasers. Most of the diode lasers use quantum-confined materials as the lasing media (quantum wells). The market analyses for lasers (Table 14.1) are taken from *Laser Focus World,* "Laser Marketplace 2003; Review and Forecast of the Laser Market; Part I: Non-diode Lasers," January 2003, pp. 73–96. Diode lasers (2003) are taken from *Laser Focus World,* "Laser Marketplace 2003; Review and Forecast of the Laser Market; Part II: Diode Lasers," February 2003, pp. 63–76. The periodical *Laser Focus World* provides this analysis annually in their January and February issues.

14.1.3 Photonics

Photonics and optoelectronics are terms that apply to many areas, including generating, modifying, steering, amplifying and detecting light. Clearly, these enabling technologies are applied to numerous fields of commercial interest, which include CD players, telecom equipment, medicine, manufacturing, and so on.

The market analysis for optoelectronic products is represented by Figures 14.1–14.4. These figures have been made available by the Optoelectronics Industry Development Association (OIDA www.oida.org)—a reputable industry group that publishes market data annually for its members.

The overall market size for optoelectronic products (according to OIDA) is shown in Figure 14.1. This market is large by any standard and has been, at the time

Table 14.1. Market Analysis for Lasers

Application	Nondiode Lasers		Diode Lasers	
	Units	Sales ($ million)	Units (K)	Sales ($ million)
Materials processing	30,278	1,316	1	4
Medical therapeutics	11,265	383	272	53
Instrumentation	34,298	52	0	0
Basic research	5,943	147	0	0
Telecommunications	0	0	2,521	884
Optical storage	130	.6	487,280	1,557
Entertainment	1,605	12.9	43,040	23
Image recording	7393	16	7,645	44
Inspection, measurement, and control	11,330	10	10,030	8
Barcode scanning	8,250	1	4,520	18
Sensing	705	15	1,365	7
Solid-state laser pumping	NA	NA	144	135
Other	665	15	9,580	80
Totals	112,506	1,995	566,398	2,814

Total Laser Sales: 112,500 nondiode lasers selling for approximately $2 billion.
566 million diode lasers (more than 90% are quantum confined lasers) selling for approximately $2.8 billion.

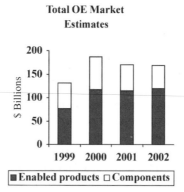

Figure 14.1. Total optoelectronics market measured in components and enabled products. Courtesy of Optoelectronics Industry Development Association.

14.1 NANOTECHNOLOGY, LASERS, AND PHOTONICS 385

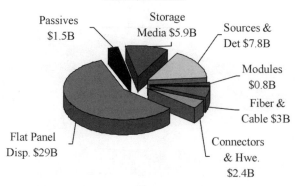

Figure 14.2. Photonic components market breakdown. Courtesy of Optoelectronics Industry Development Association.

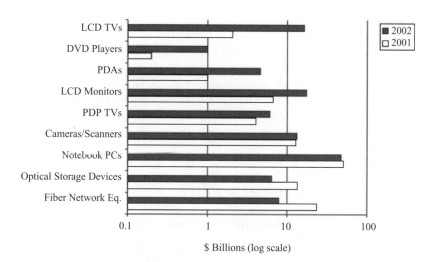

Figure 14.3. Optoelectronics-enabled markets for 2001 and 2002. Courtesy of Optoelectronics Industry Development Association.

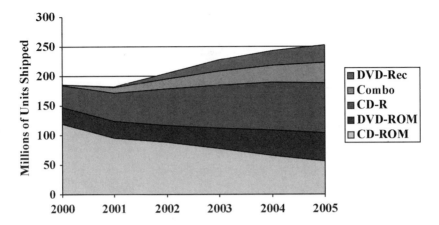

Figure 14.4. Technology trends for personal computer optical drive. Courtesy of Optoelectronics Industry Development Association.

of writing this book, relatively flat in part due to the effect of a slow economy. Figure 14.2 provides the market breakdown on the basis of components. Figure 14.3 provides a glimpse of the market trend by comparing optoelectronic markets for 2001 and 2002, for which data are available. Figure 14.4 provides the technology trend for the specific case of optical data storage.

14.1.4 Nanophotonics

While it is not possible to identify all or most of the specific market opportunities at this time, nanophotonics can be expected to impact further development of all components of photonics described in Figure 14.2. One may ask why the R&D innovations of today will be important to industry tomorrow. Nanophotonics products will be smaller, and smaller devices generally are preferred. However, if no other performance feature is added, then one must compare the value of the smaller product against the costs of developing and manufacturing it. In the following sections we examine the market and trends for the various areas of nanophotonics, covered in this book.

14.2 OPTICAL NANOMATERIALS

Nanomaterials in general and nanoparticles in particular are at the forefront of nanotechnology (Pitkethly, 2003). Most of the nanotechnology start-up companies attracting venture capital funding are in the area of nanomaterials (Cheetham and Grubstein, 2003). Estimates are that the size of this market for nanomaterials will be $900 million in 2005 and $11 billion by 2010 (The Nanotech Report, 2003).

Among all optical nanomaterials, the most well established markets for nanoscale optical materials at the current time are for very low technology applications, such as optical coatings and sunscreen lotions. Some high-tech commercial applications have recently emerged. These include photonic crystals for complex optical circuitry and signal processing, and nanomaterial based sensors to detect and respond to chemical and biological threats. However, it is too early to tell how significant the business opportunities are for these applications. There are some estimates of over 300 companies worldwide producing nanomaterials in various forms, of which about 200 are quoted as nanoparticle producers. Many of these are start-ups that have originated in universities or government laboratories or that have been established by entrepreneurs. Table 14.2 provides some examples of companies (websites) that offer products in the area of nanomaterials.

Titanium dioxide, TiO_2, and zinc oxide, ZnO, nanoparticles are used as UV blockers. Sunscreen lotions have zinc oxide or titania nanoparticles as UV blockers. While these nanoparticles absorb UV, they are transparent in the visible wavelength and are too small, relative to the wavelength of light, to scatter light. In addition to scattering less, they also absorb UV light more efficiently. Titania nanoparticles have also been investigated for more efficient solar cell applications. There are a number of start-up companies using this nanoparticle concept to develop next-generation solar cells, but they are still in the developmental stage. Nanoparticles also show promise for drug delivery as well as for optical diagnostics and light-activated therapies.

Photonic crystals and holey fibers show promise for optical communications. One-dimensional photonic crystals containing liquid crystal nanodroplets can be used for producing dynamically switchable gratings. This topic has been discussed in Chapter 11. Dynamically switchable gratings have already been introduced in the marketplace (SBG Lab Inc.). Photonic crystal components have yet to find a place in the commercial market, but photonic crystal fibers are now being commercialized (Blazephotonics; Crystal Fibres A/S). There are still many practical challenges to be met.

Here, a detailed analysis of some selected examples is presented.

14.2.1 Nanoparticle Coatings

Very few materials are used without some type of surface coating. The business of applying these coatings is very large. Table 14.3 lists U.S. production and sales for coatings in general. Successful coatings often permit less expensive materials to be considered for demanding applications. The last 10–20 years have seen the emergence of nanoparticles used to achieve various effects. As the size of the component particles decreases, the surface area to volume ratio of the particles grows quickly and places more importance on the particle–host matrix interface. The electrical, mechanical, and optical behavior of these smaller particles is often quite different from their microscopic counterparts.

"Ceramiclear" is the first nanotechnology-based automotive clearcoat and is a product of PPG. The coating was developed in cooperation with Mercedes Benz and was placed on the first automobiles in August 2002. The approximately 40-nm-thick layer is primarily composed of silica nanoparticles, 10–20 nm in diameter.

Table 14.2. Example of Some Companies that Manufacture Nanomaterials and Some Devices Utilizing Them

Material Class	Application	Company Website
Metal nanoparticles		
Au	Biosensors	www.nanospheric.com
Au	Biosensors	www.nanoprobes.com
Au/Ag nanorods	Security bar codes	www.nanoplextech.com
Si	Displays	www.ultradot.com
Oxide nanoparticles		
Various oxides	Particle production	www.nanophase.com
		www.nanosonic.com
	Coatings	www.ppg.com
	Polymer composites	www.tritonsys.com
	Displays and batteries	www.ntera.com
TiO_2	Particle production	www.altairinc.com
	Sunscreens	www.oxonica.com
		www.granula.com
	Photovoltaic cells	www.konarka.com
	Self-cleaning glass	www.pilkington.com
		www.afgglass.com
		www.ppg.com
		www.saint-gobain.com
Ln-doped oxides	Phosphors	www.nanocrystals.com
Other inorganic nanoparticles		
Ln phosphates	Security printing and marking	www.nano-solutions.de
II–IV quantum dots and wires		
CdSe quantum dots	Production	www.evidenttech.com
Quantum dots	Biosensors; fluorescent markers	www.qdots.com
Quantum dots/quantum wires	Photovoltaics; nanolasers	www.nanosysinc.com
Organics		
Nanocrystals	Drug delivery systems	www.nanocrystal.com
Dendrimers	Drug delivery systems	www.dnanotech.com
Nanoemulsions	Drug delivery systems	www.nanobio.com
Lipid vesicles	Drug delivery systems	www.imarx.com
Polymer nanocomposites	Holographic data storage	www.lptinc.com
Polymer dispersed liquid crystals	Switchable gratings	www.sbglab.com
Organic–inorganic hybrids		
Nanoclinics	Nanomedicine	www.nanobiotix.com
Photonics crystals and photonics fibers	Waveguiding, Supercontinuum generation	www.blazephotonics.com www.crystal-fibre.com

Table 14.3. U.S. 2002 Production and Sales for Coatings in General

Category	Volume (millions of gallons)	Value ($ million)
Architectural	719	1,023
OEM products	412	5,538
Special purpose	183	3,352
Misc.	149	1,188
TOTAL	1,463	17,210
Automobile finishes	48	1,069

Source: U.S. Census (http://www.census.gov/industry/1/ma325f02.pdf).

The coating is advertised to have superior resistance to repeated washings in comparison to previous coatings.

As one of its developers stated (conversation with Kurt Olson of PPG), there were several challenges to be overcome. The coatings needed to be compatible with existing equipment, have satisfactory aesthetic appeal, be cost effective, and have a shelf life of more than six months. Agglomerization of the nanoparticles (especially during the cure phase) tended to result in an unacceptable haze. The problem was overcome, in part, by developing a proprietary inorganic/organic binder that optimized the crosslink density and still permitted a densely packed, silica-rich air-coating interface that is in part responsible for its improved performance. The clearcoat (*Source:* http://corporate.ppg.com/PPG/Corporate/AboutUs/ Newsroom/Corporate/pr_Ceramiclear.htm) is applied to paint finishes using existing equipment and demonstrates superior protection against mechanical action stemming from repeated washings.

14.2.2 Sunscreen Nanoparticles

The opportunity to use nanoscale-size titania and zinc oxide particles in sunscreen formulations has improved UV protection, increased transparency (virtually eliminated the chalky appearance), and introduced an antibacterial function to using sunscreens. TiO_2 and zinc oxide are both candidates, but Nanophase is pursuing (with BASF) zinc oxide due to its superior UV absorbing properties and its added antimicrobial properties (*Source:* Conversation with Bo Kowalczyk at Nanophase). Although the market demand for nanoscopic particles is approximately 25 million pounds, this is small compared to the 8,000-million-pound market for TiO_2 as illustrated in Table 14.4.

14.2.3 Self-Cleaning Glass

A long-standing dream of plate glass manufacturers has been the development of a glass window that cleans itself. Such an innovation would have obvious value in downtown Manhattan, residential skylights, and a host of other glass installations, including those for aerospace industries.

In the last few years, virtually all major glass manufacturers have developed

Table 14.4. Analysis of Market for TiO$_2$ Nanoparticles

Nanoparticle Type	Diameter	Cost per Pound	Annual Market (million pounds)
Global TiO$_2$ (60% for paints, etc.)	~200 nm	$0.90	8000
Nanoparticles for sunscreens	10–20 nm	$3–10	25
TiO$_2$ for catalysts (including supports)	Various	Various	20
Photovoltaic	—	—	1

Source: Marc S. Reisch, Essential Minerals, *Chemical and Engineering News,* Vol. 81, pp. 13–14, March 31, 2003. Jim Fisher, President of International Business Management Associates (IBMA), predicts that the sunscreen crystals have a potential market size of 25 million pounds per year.

plate glass with coatings that claim to be self-cleaning. The principle behind this innovation is the photocatalytic decomposition attained with titanium dioxide nanocrystals, which have the added effect of making the surface hydrophilic (literally, "water loving"), thereby accounting for the sheeting effect of water on this surface. In fact, clean glass is itself hydrophilic, which is almost never seen, because glass soils easily.

The following manufacturers have developed these coatings and have marketed them under the following tradenames:

Company	Trade Name
PPG	SunClean
Saint Gobain Glass (SGG)	Aquaclean
Pilkington	Activ
Ashai Glass Company (AFG)	Radiance Ti

The secondary requirements for this coating are that it not diminish the optical clarity of the window (no haze, tint, etc.), that it has reasonable toughness to withstand ordinary environments and care, and that its cost is on par with its value added.

14.2.4 Fluorescent Quantum Dots

The merits of quantum dots for fluorescence based imaging have been discussed in Chapter 13. Quantum Dot Corporation, located in Hayward California, is a company commercializing quantum dot-based fluorescent markers. The chief advantages of using these products are the much narrower emission spectra, and their robustness toward photobleaching.

Another known fluorescent quantum dots application is security marking. Nanosolutions, Germany (http://www.nano-solutions.de/en/index.html), is the manufacturer of security pigments REN®-X red and REN®-X green, which contain nanoparticles dispersed in an ink. The resulting dispersion is colorless and completely transparent. When illuminated using UV radiation, the image printed with this ink fluoresces.

14.2.5 Nanobarcodes

Nanobarcodes utilize another interesting concept. In this case, one has little segments of metals in the form of nanosized metallic rods connected to each other. Different rods are used, which could be silver, gold, and so on. These metallic nanorods exhibit specific surface plasmon resonances that depend on the length of the nanorod, as well as the nature of the metal material (see Chapter 5). Light reflected through these different segments are collected, which produce different colors for color coding. Nanoplex Technologies Inc. (www.nanoplextech.com) launched its first product, a kit containing Nanobarcodes™ particles with six different stripping patterns, at the Pittsburgh Conference held in Orlando, Florida, March 4–14, 2003.

14.2.6 Photonic Crystals

The property of photonic crystals to steer light at sharp angles (Chapter 9) has led to great expectations that a large variety of compact integrated optical devices, unattainable with classical waveguide technologies, can be manufactured. Current work is targeting materials and manufacturing methods that are compatible with semiconductor production equipment and methodology. The most popular material is silicon-on-insulator (SOI), where the hole patterns are etched in a thin layer of silicon on top of an insulating silica substrate; this material has good compatibility with established manufacturing practices and is being pioneered by such companies as Galian, Luxtera, and Clarendon Photonics. Among other approaches being investigated, NanoOpto is developing a proprietary molding process, imprinting the circuit design into a polymer resist layer; the pattern is then created in the silicon/silica layers with anisotropic reactive-ion etching. NeoPhotonics is creating photonic structures in polymers, while Micro Managed Photons utilizes patterns in thin gold films on a glass substrate. In spite of the all the work being performed on the component level, many industry experts believe that the availability of these products is still some way off. Jaymin Amin, the director of optical systems development at Corning, cautiously states: "I don't think we'll see photonic-crystal components in the next five years" (Mills, 2002).

14.2.7 Photonic Crystal Fibers

There are widespread R&D efforts to realize some of the laboratory demonstrations of photonic crystal fiber applications into practical products. Using modified fiber drawing procedures, fiber manufacturing methods were developed and optimized throughout the 1990s to the point where PCFs have become commercially available at the writing of this book. One of the earliest companies to do so is Blaze Photonics, which is a UK-based company with a substantial number of fiber types, with zero group velocity dispersions at a specified wavelength, commercially available for between $20 and $400 per meter; these fibers are available in lengths up to kilometers. Clearly, the price of these fibers will drop with product acceptance and scale-up. Products available already include hollow core fibers (for visible, 800, 1060, and

1550 nm), high nonlinearity fibers (for supercontinuum generation), polarization maintaining fibers, and endlessly single mode fibers (www.blazephotonics.com). These applications have been discussed in Chapter 9. Crystal Fibre A/S (www.crystal-fibre.com) of Denmark is another company commercializing photonic crystal fibers. This company produces pure or doped (e.g., rare-earth ion-doped) silica-based PCFs.

14.3 QUANTUM-CONFINED LASERS

Quantum well, quantum wire, and quantum dot (Qdot) lasers are nanostructured devices that have been used to produce efficient semiconductor lasers. As mentioned in Section 14.1, most semiconductor lasers, currently sold, utilize quantum well structures. Hence, there is already a well-established market for this area of nanophotonics, and it is growing, as miniaturization of optical devices continues to be an area of demand. As indicated in Chapter 4, there are certain advantages of Qdot lasers, which continue to be an area of commercial interest. Zia Laser Inc. of Albuquerque already lists Qdot lasers among its products.

Another laser, being commercialized and finding important application in chemical and biological sensing, is a Quantum Cascade laser (QCL). Alpes Lasers (Switzerland) and Applied Optoelectronics, Inc. (United States) are manufacturers of the Quantum Cascade lasers. Distributed Feedback single-mode QCLs from Alpes Lasers, operating in pulsed mode, are marketed for detection of contaminations in semiconductors, foods, medical diagnostics, and for explosive detection. A promising development in QCL is room-temperature operation at 9.1 μm, produced by U.S.-based Maxima Corporation, for communication applications. It is believed that this much longer wavelength could greatly enhance transmission through atmosphere in free-space communication (http://optics.org/articles/ole/7/6/5/1).

14.4 NEAR-FIELD MICROSCOPY

Near-field microscopy is increasingly being recognized as a valuable tool for nanoscopic imaging. It has provided new research tools to probe nanostructures and dynamics as well as to perform nondestructive evaluation of nanostructured devices. Nanoscale imaging using near-field microscopy has a variety of applications such as in the semiconductor industry for nondestructive evaluation of materials. In biomedical research, one can image bacteria and viruses. Medical imaging is an area that is now receiving a great deal of attention from the point of view of being able to look at bacteria and other very small objects at the micron and submicron sale. Near-field biosensors using near-field probes and nanoscale structures are also receiving a great deal of attention. However, the near-field technique is still in the research stage and is far away from any routine application.

14.5 NANOLITHOGRAPHY

This, again, is an area that has generated interest because of advancements in the techniques as described in Chapter 11. However, the application of most techniques, presented in Chapter 11 to produce commercial nanostructures, is still far into the future.

A major application of nanolithography is in semiconductor industries. The semiconductor industry has distinguished itself by the rapid introduction of technological improvements in its products. Exponentially decreasing the minimum feature sizes used to fabricate integrated circuits (according to the famous Moore's Law, the number of components per chip doubles every 18 months) has resulted in decreasing the cost per function, which translates into improvements of productivity and quality of life as evidenced through the proliferation of computers, electronic communication, and consumer electronics. The typical feature size on a DRAM was 5 μm in the 1960s; it is now a quarter micron, and the technology aims at reaching 50 nm by 2012. A very thorough resource for further reading on this topic is the International Technology Roadmap for Semiconductors (ITRS) documents (http://public.itrs.net).

Optical lithography for the patterning of silicon CMOS devices continues to push forward, or downward, creating smaller and smaller transistors. While fabrication facilities that produce silicon chips with the "enormous" minimum MOSFET gate length of 1.6 μm still exist, the cutting edge microprocessors are into the deep-submicron sizes. The newest Intel Pentium–4 processor is fabricated using a 0.13-μm technology; that is, the smallest gate length of the MOSFET is no larger than 0.13 μm or 130 nm (actual gate lengths for this process are under 70 nm). The original Pentium 4 was fabricated with a 0.18-μm technology. It is a $100 billion market (*Source*: Intel website).

Research into smaller technology sizes is ongoing, with an operational 90-nm technology process, according to some, expected in late 2003 (Thompson et al., 2002). Beyond this, 30-nm gate length devices have been demonstrated using standard optical lithographical techniques (248-nm wavelength) (Chau et al., 2000). However, in order to have production-level processes at 30-nm technology, new lithographic techniques must be used. Current thinking is that extreme ultraviolet lithography (EUVL) with wavelengths on the order of 13 nm is needed for this step (Garner, 2003). EUVL is currently the focus of much research (see, for example, Stulen and Sweeney, 1999; Sweeney, 2000; Kurz et al., 2000). After the end of this decade, it is unclear what the next technology will be for smaller devices, although there are many possibilities such as molecular electronics.

In addition, wafer sizes are now as large as 300 mm in diameter, requiring mask development and UV light sources that can produce a uniform intensity across the entire wafer.

A nanolithography technique, being advertised at the time of writing this book, is dip-pen lithography, discussed in Chapter 11. Nanoink, Inc. of Chicago, IL (http://www.nanoink.net/) has announced a product NSCRIPTOR™ DPNWriter™ system based on a fully functional commercial scanning probe microscope (SPM) system, environmental chamber, pens, inkwells, substrates, substrate holders, and accessories for DPN experiments.

Another manufacturer of nanolithography tools is Molecular Imprints Inc., Austin, TX (http://www.molecularimprints.com/), which produces a series of systems, Imprio 50, Imprio 55, and Imprio 100, based on the unique Step and Flash Imprint Lithography technology called S-FIL™. The Imprio 100 provides sub-50-nm lithography.

Nanonex Inc., Monmouth Junction, NJ (http://www.nanonex.com/) offers nanoimprint lithography tools, resists, and masks.

14.6 FUTURE OUTLOOK FOR NANOPHOTONICS

The scientific quest for knowledge, along with society's insatiable thirst for compact, energy-efficient, and multimodal technologies, ensures a bright future for nanophotonics. Like in most technology areas, market-driven inventions will create economic opportunities. However, recognizing that nanophotonics is an emerging area, new scientific discoveries will also play a dominant role in the development of new technologies. Not all discoveries and resulting laboratory demonstrations of technology result in commercialized products. The competitive edge of a particular technology, performance reliability, production scalability, and cost-effectiveness are some important measures to be met by a commercially viable product. While research scientists are good at producing innovations, most are not well-suited to transitioning an innovation to commercial opportunities. This is where the university–industry–investment partnership can play a vital role in transitioning scientific discoveries to commercial products.

Even though there is inherent risk in making predictions about future developments and in identifying areas of future growths, it may still be useful to provide perspectives for the future outlook. Four major thrust areas are projected as those that could significantly benefit from breakthroughs in nanophotonics. Therefore, these thrusts are presented with some selected examples of economic opportunities.

14.6.1 Power Generation and Conversion

As we strive for cleaner and efficient sources of energy, solar energy conversion becomes a priority area. A nanophotonic approach utilizing inorganic:organic hybrid nanostructures and nanocomposites can produce broadband harvesting of solar energy while using flexible low-cost, large-area roll-to-roll plastic solar panels and solar tents. Some promising features of this nanocomposite approach have been discussed in Chapter 10. Other power conversion sources can involve rare-earth-doped nanoparticle up-converters and quantum cutters, described in Chapter 6. These photon converters can be utilized to harvest solar photons at the edges of solar spectrum, specifically in the IR and in the deep UV. A major direction for basic research, needed to mature this technology, is an understanding and subsequent control of dynamic processes at the nanoscale. Much emphasis has been placed on the nanoscopic structure–function relationship. However, the dynamics in nanostructures is equally important in controlling many photonic functions, such as in the case of photon conversion.

Another major area of opportunity is the utilization of quantum-cutter nanoparticles for lighting applications. As described in Chapter 6, there is a strong push to produce mercury-free, efficient lighting sources. Efficient quantum cutters, which can even be spray-coated, will provide a means to realize this goal. Some other applications of photon up-converting nanoparticles are for display and security marking.

14.6.2 Information Technology

Despite its slow down in early 2000, the IT market can be expected to grow, as society has to deal with the rapid increase of information and the need to store, display, and disseminate it. Hence, increased processing speed, increased bandwidth (more channels to transmit information), high-density storage, and high-resolution, flexible thin displays are going to demand new technological innovations. In addition, as we get more accustomed to wireless communications, coupling of photonics to RF/microwave will play a major role in future information technology. Nanophotonics can be expected to create a major impact in all these areas. Photonic crystal-based integrated photonic circuits, as well as hybrid nanocomposite-based display devices and RF/photonic links, are some specific examples of opportunities. A major challenge lies in formulating processing methods that meet the needs of reliable device performance, together with batch processing and cost effectiveness.

14.6.3 Sensor Technology

There is an ever-increasing need to enhance the capability of sensor technology for health, structural, and environmental monitoring. One area of great concern is new strains of microbial organisms and the spread of infectious diseases that require rapid detection and identification. This requires point detection as well as environmental monitoring. Another area of major concern, worldwide, is the threat of chemical and biological terrorism. The detection here is not only for the danger posed to health, through chemical and biological agents, but also for structural damage (to bridges, monuments, etc., through explosives). Nanophotonics-based sensors utilizing nanostructured multiple probes provide the ability for simultaneous detection of many threats, as well as the ability for remote sensing where necessary. A useful future approach can utilize nanoscale optoelectronics with hybrid detection methods involving both photonics and electronics.

14.6.4 Nanomedicine

As an increasingly aging world population presents unique healthcare problems, new approaches for early detection and treatment of diseases will be required.

Nanomedicine utilizing light-guided and light-activated therapy, with the ability to monitor real-time drug action, will lead to new approaches for more effective and personalized molecular-based therapy. Nanomedicine is thus, in the opinion of this author, an area of tremendous future opportunity. However, it should be noted that due to the long process of clearance for any therapy, nanomedicine cannot be expected to be available in the marketplace any time soon. A major concern already raised by many is any long-term adverse health effects (such as toxicity, accumulation in vital organs, obstruction of circulatory system, etc.) produced by nanoparticles. Another version of the nanomedicine approach can be useful for cosmetic industry.

In summary, there is strong evidence that there will be numerous business opportunities for nanophotonics and that nanophotonic innovations will have both evolutionary and revolutionary impacts in a wide range of markets.

14.7 HIGHLIGHTS OF THE CHAPTER

- Nanophotonics provides numerous business opportunities, many of which already have a presence in the marketplace.
- Nanotechnology, lasers, and photonics, the three areas that are fused to define nanophotonics, have existing established markets worth billions of dollars.
- A major impetus for growth of nanotechnology, including nanophotonics, is provided by national priority funding from government in many countries.
- In the nanotechnology sector, most venture capital funding has gone to the area of nanomaterials.
- Optical nanomaterials have well-established markets, which are mostly focused on low-technology applications, such as in sunscreen lotions and optical coatings.
- Fluorescent quantum dots and nanobarcodes are other examples of optical nanostructures being currently commercialized.
- Another nanophotonic material, recently introduced in the market, is photonic crystal fibers.
- Most semiconductor lasers utilize quantum well structures. More recently, quantum dot lasers have also been introduced in the market.
- Quantum cascade lasers are yet another example of quantum-confined lasers, being marketed for detection of contaminants in semiconductors, foods, and medical diagnostics and for explosive detection.
- Near-field microscopy has much to offer in nondestructive evaluation of nanostructured devices and in bioimaging.
- A major application of nanolithography is in semiconductor industries for producing smaller microprocessors with more capability.
- Dip-pen lithography and imprint lithography are some new nanolithographic techniques being commercialized.
- Some examples of general areas with future prospects and representing eco-

nomic opportunities are (i) power generation and conversion, (ii) information technology, (iii) sensor technology, and (iv) nanomedicine.

REFERENCES

Chau, R., Kavalieros, J., Roberds, B., Schenker, R., Lionberger, D.; Barlage, D., Doyle, B., Arghavani, R., Murthy, A., and Dewey, G., 30 nm Physical Gate Length CMOS Transistors with 1.0 ps n-MOS and 1.7 ps p-MOS Gate Delays, presented at Electron Devices Meeting, 2000, IEDM Technical Digest International, 2000.

Cheetham, A. K., and Grubstein, P. S. H., Nanomaterial and Venture Capital, *Nanotoday* **December,** 16–19 (2003).

Dunn, S., and Whatmore, R. W., Nanotechnology Advances in Europe, working paper STOA 108 EN, European Commission, Brussels, 2002.

Garner, C. M., Nano-materials and Silicon Nanotechnology, presented at NanoElectronics & Photonics Forum Conference, Mountain View, VA, 2003.

Gould, P., UK Invests in the Nanoworld, *Nanotoday* **December,** 28–34 (2003).

Hwang, V., Presented at Nanorepublic Conference, Los Angeles, 2003 (www.larta.org/nanorepublic).

Kurz, P., Mann, H.-J., Antoni, M., Singer, W., Muhlbeyer, M., Melzer, F., Dinger, U., Weiser, M., Stacklies, S., Seitz, G., Haidl, F., Sohmen, E., and Kaiser, W., Optics for EUV Lithography, presented at Microprocesses and Nanotechnology Conference, 2000 International, 2000.

Mills, J., Photonic Crystals Head Toward the Marketplace, *Opto and Laser Europe* **November,** 2002; http://optics.org/articles/ole/7/11/1/1.

Pitkethly, M., Nanoparticles As Building Blocks?, *Nanotoday* **December,** 36–42 (2003).

Roco, M. C., and Bainbridge, W. S., eds., *Societal Implications of Nanoscience and Nanotechnology,* Kluwer Academic Publishers, Hingham, MA, 2001.

Stulen, R. H., and Sweeney, D. W., Extreme Ultraviolet Lithography, *IEEE J. Quantum Electron.* **35,** 694–699 (1999).

Sweeney, D., Current Status of EUV Optics and Future Advancements in Optical Components, presented at Microprocesses and Nanotechnology Conference, 2000 International, 2000.

The Nanotech Report 2003, *Investment Overview and Market Research for Nanotechnology,* Vol. 2, Lux Capital, New York, 2003.

Third European Report on Science of Technology Indicators, EUR 2002, European Commission, Brussels, 2003.

Thompson, S., Anand, N., Armstrong, M., Auth, C., Arcot, B., Alavi, M., Bai, P., Bielefeld, J., Bigwood, R., Brandenburg, J., Buehler, M., Cea, S., Chikarmane, V., Choi, C., Frankovic, R., Ghani, T., Glass, G., Han, W., Hoffmann, T., Hussein, M., Jacob, P., Jain, A., Jan, C., Joshi, S., Kenyon, C., Klaus, J., Klopcic, S., Luce, J., Ma, Z., Mcintyre, B., Mistry, K., Murthy, A., Nguyen, P., Pearson, H., Sandford, T., Schweinfurth, R., Shaheed, R., Sivakumar, S., Taylor, M., Tufts, B., Wallace, C., Wang, P., Weber, C., and Bohr, M., A 90 nm Logic Technology Featuring 50 nm Strained Silicon Channel Transistors, 7 Layers of Cu Interconnects, Low k ILD, and 1 μm^2 SRAM Cell, Proc. Electron Devices Meeting, 2002. IEDM '02, pp. 61–64.

Index

Absorption coefficient, 97
Absorption profile, 311
Acoustic phonons, disappearing, 157
AF-281 two-photon photoinitiator, 312
Aggregates, formation of, 34
AlGaAs/GaAs quantum well, 82
"Allowed light," 45, 46, 74
All-*trans*-retinal, 339
"All"-wet-chemical approach, 232
Anderson-type photon localization, 279
Angular multiplexing, 340
Anisotropic etching process, 330
Anisotropy mapping, 61
Anomalous dispersion, 271
Anomalous group velocity dispersion, 270–271
  in photonic crystals, 247–249
Anomalous refractive index dispersion, in photonic crystals, 249–250
Anti-Stokes Raman process, 156. *See also* Raman spectroscopy
Aperture-controlled near-field optics, 42–43
Apertureless near-field microscopy, 51, 62–65
Apertureless NSOM, 41, 74
  approaches to, 63
Aperture NSOM, 74
Apparent quantum yield, 143, 144
APSS, 349–350
Arene–arene interactions, 211
Asymmetric quantum well, 99
Atomic force microscope (AFM), 310, 328
Atomic force microscopy (AFM), 50, 200, 202–204, 206
Attenuated total reflection (ATR), 30–31
Au nanoparticles, 187, 188. *See also* Gold nanoparticles

Autofluorescence, 359
Axial nanoscopic localization, 29–32

Back-scattered electrons, 196
Bacteria, as biosynthesizers, 347–350
Bacteria bioreactors, 337
Bacteriorhodopsin (bR), 337, 339–341
Bandgap
  in photonic crystals, 246, 272
  photon and electron, 21–24
"Bandgap engineering," 80
Bandgap renormalization, 96
Band splitting, 24
Band structure, photonic crystal, 270
Band-to-band optical transition, 84
Barcodes. *See* Nanobarcodes
BBO crystal, 69, 70
Biexcitons, 26, 28, 34, 38, 121
  formation of, 96
Bioanalysis. *See In vitro* bioanalysis
Biocolloids, 337, 343–344
Bioderived materials, 337, 350
Bioimaging, 94
  FRET and, 36
  near-field, 356–357, 376
  semiconductor quantum dots for, 358–359
  up-converting nanophores for, 359–360
Bioinspired materials, 209, 337, 344–346, 350
Biology, technology development in, 6
Biomaterials, 337–350. *See also* Bioinspired materials
  bacteria and, 347–350
  biotemplates and, 346–347
  naturally occurring, 338–344, 350
Biomedical research, technology development in, 6

Nanophotonics, by Paras N. Prasad
ISBN 0-471-64988-0 © 2004 John Wiley & Sons, Inc.

Biomimicry, 337
Bioobjects, 343–344
Bioscience, single-molecule detection and, 54
Biosensing, 356, 360–365, 376
Biosensors
  dye-doped nanoparticle, 364–365, 376–377
  optical nanofiber, 377
  photonic crystal, 361–362, 376
  plasmonic, 360–361, 376
  porous silicon microcavity, 362, 376
Biosynthesizers, bacteria as, 347–350
Biotechnology, nanophotonics for, 355–377
Biotemplates, 337, 338, 346–347, 350
Bleached state, 96
Block copolymers, 118
  multifunctionality of, 213–215
Blue transparent organic polymers, 232
Bohr radius, 27–28
Bonding, molecular orbital theory of, 115
Bosons, 12
Bragg grating, 241, 257
Bragg scattering, 21–22
Bragg's equation, 190
Bragg stack, 244–246
Bridger DNA, 347
Brillouin zone, 243
Broadband continuum generation, 266
Broadband solar cells, 297
Bruggeman geometry, 285, 303
Bulk GaAs band-edge absorption, 91

Cadmium. See CdS entries; CdSe entries
Calcium PEBBLE, 363
Carbon nanotubes, 118
Carrier generation, study of, 72
Carrier-induced photoluminescence quenching, 72
Catenanes, 216
Cathode ray tube (CRT), 196
Cavity enhancement factor, 251
Cavity optical response, 251
CdSe nanoparticles, 93
CdSe quantum dots, 103
CdSe quantum rods, 94
CdSe–ZnS core-shell quantum dots, 104–105
CdS/HgS/CdS core-shell structure, 106

CdS nanoparticles, DNA as a template for, 346
CdS quantum dots, 347
Cell fabrication method, 252–253
Ceramic-based nanoparticles, hydrated, 372
"Ceramiclear," 387–389
Channel waveguides, 140
  guiding in, 15–16
Charge-transfer bands, 38
Charge-transfer complexes, 34–35
Chemical beam epitaxy (CBE), 183, 205
Chemical/biological sensor, 269
Chemical composition information, EDS, 199
Chemical etching method, 50
Chemical vapor deposition (CVD) polymerization, 118
Chemistry. See also Microemulsion chemistry; Nanochemistry; Reverse micelle chemistry; Sol–gel chemistry; Wet chemistry
  supramolecular, 225–226
  technology development in, 5
Chromosomes, bioimaging of, 356
Coated spherical particles, 135
Coatings, nanoparticle, 387–389
Collection mode NSOM, 74
Collimated light propagation, 249
Colloidal crystal arrays (CCAs), 267–268
Colloidal crystal mask, 326
Colloidal crystals, 241, 271
Colloidal photonic crystal self-assembly, 252–255
Colloidal quantum dots, close-packing of, 102–103
Colloidal self-assembly, 271
Colloidal synthesis, 186–189, 205
Communications. See Optical communications
Computer optical drive, technology trends for, 386
"Concentration quenching," 161
Conduction band, 22
Confinement, of photons and electrons, 14–19
Confinement effect, 80. See also Quantum confinement effect
  dielectric, 99–100

INDEX **401**

Conjugated structures, with delocalized electrons, 115–116
Convective self-assembly method, 254
Convergent approach, 234
  to dendrimer production, 221–225
Cooperative effects, 38
  photon and electron, 10, 24–28
Cooperative emission, 36–37, 38, 172
Cooperative transitions, new, 34–35
Cooper pairs, 26
  formation of, 27
Core-shell nanoparticles, 135, 153–154
Core-shell quantum dots, 104–106, 122
Core-shell structures
  fabricating, 136–137
  wet-chemical approach to, 105–106
Coulomb interaction, enhancement of, 99–100
Cowpea mosaic virus (CPMV) particles, 347
Critical angle, 29
Critical temperature, 26
Cross-linking, 282
Cross-relaxation, 158, 165, 166, 168, 172
Crystal field effect, 155
Crystals, electronic and photonic, 21
Cut-off frequency, 157
Cutting, quantum, 159–160, 166–170, 173. *See also* Quantum cutting process

Damped-harmonic dipoles, 72
DEANST crystals, 60–61
De Broglie equation, 10–11, 194
Debye model, 155
Defect modes, in bandgap region, 250, 251
Defect structures, 259, 260
Degeneracy, 86
Degree of confinement, 87–88
Delocalization energy, 116
Delocalized electrons, conjugated structures with, 115–116
Dendrimers, 209, 217–225
  convergent approach to producing, 221–225
  divergent approach to producing, 219–221
  light-harvesting, 224, 350
  synthesis of, 234

Dendritic architecture, interaction modulation by, 224–225
Dendritic boxes, 222
Dendritic fluorescence sensors, 225
Dendritic structure, 221
  core isolation by, 223–224
  multifunctional, 221–223
Dendrons, 221
  synthesis of, 220
Dense wavelength division multiplexing (DWDM), 272, 302
Density of states, 84–85, 86, 87, 121
  quantum confinement and, 89
Diagnostics, optical, 357–358, 365–367
Diblock copolymers, 213–214, 234
Dielectric confinement effect, 99–100
Dielectric constant, 11–12, 37, 133, 284
Dielectric contrast model, 92–93
Dielectric modification, 153
Dielectric sensitivity, of plasmon resonance, 142
Dielectric stack, 269
Diffraction, in a superprism, 249
Diffraction efficiency, 299–300
Diffraction techniques, 189–204
Digital alloys, 180–181
Dimer formation, excited-state, 35
Diode lasers, 383
Dipolar structures, organic molecular, 224
Dipole–dipole interactions, 224
Dip-pen nanolithography, 310, 328–329, 333, 393–394, 396
Dipping process, 183
Direct bandgap, 22
Direct process, 156
Direct-Space Integral Equation Method (DSIEM), 44
Discotic molecules, 226
Discotic structures, 225
Discrete dipole approximation (DDA), 134–135, 137–138, 148
Disease, personalized approaches to, 355
Dispersion
  anomalous group velocity, 247–249
  anomalous refractive index, 249–250
Dispersion relation, 13
Distributed Bragg reflectors (DBR), 112
Divergent approach, 234
  to dendrimer production, 219–221

DNA. *See also* Bridger DNA; Native DNA
  as a biotemplate, 346
  detection of, 364–365
  naturally occurring, 341–343, 350
  Watson–Crick model of, 211
DNA–lipid film, 341
DNA periodic arrays, 2-D and 3-D, 347
DNA/polycation complex
    Langmuir–Blodgett films, 347
Domains, nanoscopic, 277
Double heterostructure semiconductor
    (DHS) laser, 109
Down-conversion process, 159–160, 166
  mechanisms of, 167–168
Drude model, 14, 133
Drug-doped carriers, 371
Dry etching, 257
Dye, photobleaching of, 318
Dye-doped nanoparticle biosensors,
    364–365, 376–377
Dynamic AFM, 203

Edge-emitting laser, 112
Education, multidisciplinary, 3–4
Effective bandgap, 120–121
Effective refractive index, 261, 299
  manipulation of, 279
Effusion cells, 179
Eigenfunctions, 11
Eigenmodes, 16
Eigenvalue, 11
Eigenvalue wave equations, 37
  photon and electron, 10
Elastomeric phase masks, 322, 324
Electric field effects, 97–99
Electric field enhancement, 277
Electric force microscopy (EFM), 200
Electro-absorption, 99
Electrochemical etching, 257
Electroluminescence, 164
  in PPV, 289
  quantum dot:polymer nanocomposites
    and, 297
Electroluminescent polymer, 214
Electromagnetic field, nanoscale
    localization of, 28
Electron beam (E-Beam) lithography,
    255–256
Electron beam techniques, 197–199

Electronic crystal, 21
Electronic energy transfer, nanoscale,
    35–36
Electronic excitation. *See also* Excitation
    dynamics
  dephasing of, 172
  processes in, 156
Electronic interactions, nanoscale
    confinement of, 33–37
Electronic transition dephasing, 156
Electron microscopy, 194–197, 205
  techniques in, 189
Electron–phonon coupling, 155–158
Electron–phonon (lattice vibration)
    interactions, 26, 160
Electrons
  bandgap and, 21–24
  characteristics of, 10
  confinement of, 14–19
  cooperative effects for, 24–28
  delocalization of, 115–116
  free-space propagation of, 12–14
  leakage of, 19–20
  tunneling of, 19–20
  versus photons, 10–28
Electron spectroscopy for chemical analysis
    (ESCA), 192
Electron tunneling, 20, 102
Electro-optic effect, 95
Electro-optic modulators, 99, 224
Electrostatic donor-acceptor interactions,
    226, 227
Electrostatic force microscopy (EFM), 203
Electrostatic interactions, 211
Embedded nanoparticles, 333
Emission bands, size dependence of, 160
Emission efficiency, 163
  enhancing, 159
Emissions, types of, 153
Emission spectra, up-converted, 163
Energy barrier, 16–17
Energy-dispersive spectroscopy (EDS),
    199, 206
Energy down-conversion, 158
Energy funnels, 224
Energy harvesting antenna approach,
    344–345
Energy level structure, control of, 168
Energy up-conversion, 158

Engineering, technology development in, 5–6
Enhanced green fluorescent protein (eGFP), 370
"Enhanced permeability and retention effect," 371
Enhancement, nanoscale, 65–69
E-O modulators, high-frequency, 349
Epitaxial growth, 178, 179, 204
  methods of, 179–183
Erbium-doped fiber amplifiers (EDFA), 160
Etching
  for photonic crystal fabrication, 256–257
  wet-chemical, 329
Evanescent waves, 20, 29–30, 37
  excitation of, 142
Excimers, 35, 38
Exciplexes, 35, 38
Excitation dynamics, 153–173
  control of, 287
  nanoscopic interactions for, 154–158
  nanostructure and excited states, 154–158
  photon avalanche, 165–166
  quantum cutting, 166–170
  rare-earth doped nanostructures, 158–161
  site isolating nanoparticles, 171
  up-converting nanophores, 161–165
Excitation energy, 155
  transfer of, 303
Excited-state absorption (ESA), 165
Excited-state dimer formation, 35
Excited-state dynamics, 171
Excited states, relaxation of, 156
Exciton binding, 90
Exciton Bohr radius, 27, 28
Excitonic peaks, 97, 98
Excitonic transitions, 84, 91, 121
Exciton Rydberg energy, 27
Excitons, 26, 95
Extinction coefficient, 133
Extreme ultraviolet lithography (EUVL), 393
Extrinsic size effect, 133

Face-centered cubic (FCC) crystal structure, 261
Femtosecond laser pulses, 320
Fermions, 12
Fermi's Golden Rule, 143

Fiber geometry, 50
Field distribution, modeling of, 266–267
Field enhancement, 148
Field-enhancing apertureless NSOM, 63
Finite-Difference Time Domain (FDTD) method, 44, 243
Fl-DPPE dye-labeled lipid, 145–146
Fluorescence
  directional emission of, 146–147
  increasing lifetime of, 171
  metal-induced enhancement of, 144
Fluorescence-based optical sensors, 142
Fluorescence detection
  sensitivity of, 142
  single-molecule, 54
Fluorescence quantum yield, 148
Fluorescence resonance energy transfer (FRET), 36, 38, 73, 338, 368, 369, 370
Fluorescence sensors, dendritic, 225
Fluorescent dyes, 341
Fluorescent lamps, mercury-free, 167
Fluorescent quantum dots, 390, 396
Fluorophores, 30
Focused ion beam (FIB) system, 138–139
Foldmers, 226–228
Forbidden light, 44, 45, 46, 74
"Forbidden zone," 44
Förster-type energy transfer, 103, 345
Franz–Keldysh effect, 97
Free carrier absorption, 89
Free-space propagation, photon and electron, 10, 12–14
Frenkel exciton, 26
Frequency conversion, 25
Frequency-domain techniques, 242, 270
Fresnel reflection formulas, 32
Fundamental wave (FW), 45
  electric field intensity of, 47

GaAs (gallium arsenide), 23
Gain curve, 108
GaMnSb digital-layer alloy, 181
Gap semiconductor, indirect, 90
G band, 64
Gene delivery, by nanoclinics, 367–370
Generations, dendrimer, 217
Gene therapy, 367
Glancing incidence exposure, 325

Glass, self-cleaning, 389–390
Glass–polymer composite, 301–302, 304
Glucose sensing, 268–269
Gold dot nanoarrays, 329
Gold nanoparticles, optical absorption spectra of, 131. *See also* Au nanoparticles
Gold nanoshells, rapid immunoassay and, 361
Gravity sedimentation, 252
Green fluorescent protein (GFP), 338. *See also* Enhanced green fluorescent protein (eGFP)
Ground-state absorption (GSA), 165
Group velocity dispersion, in photonic crystals, 247–249, 265
Group V precursors, 188, 189

Hamiltonian operator, 12
Harmonics, generation of, 261
Heating-and-pulling method, 50
Heavy hole (hh), 83
Helical foldmers, 228
Helicates, 226
Hermite–Gaussian laser beams, 63
Heterocircuit catenane, 216
Heterostructured nanoparticles, 135
Hexagonal photonic lattice, 2D, 258
High-quality factor (Q), 251
High-resolution, near-field Raman microscopy, 64
Hole (positive vacancy), 23. *See also* Heavy hole (hh); Light hole (lh)
Hole-burning experiment, 156
Hole carriers, enhanced mobility of, 296–297
Hole transfer, 294
Hole transporting polymer, 303
Holey fibers, 264, 272, 387
Holographic data storage, 339–341
Holographic grating, 293–294
Holographic methods, 271
 for photonic crystal fabrication, 257–259
Holographic two-photon-induced polymerization (H-TPIP), 316
Holography, two-photon, 333
Homocircuit catenane, 216
Homogeneous line broadening, 156
Hot emission bands, 157

HPPH, 372
Hückel theory, 22, 116
Hybrid optical materials, 279
Hybrid optical nanocomposites, 281, 303
Hydrogel, 268
Hydrogen bond, 210–211
 molecular pair-specific, 226
Hydrophobic interactions, 211
Hyperbranched polymers, 217
Hyperthermic effect, 367

Illumination mode NSOM, 74
Immersion spectroscopy, 134
Impurity pair interactions, 158
Incoherent nonlinearity, 95
Indirect bandgap, 22
Indirect gap semiconductor, transition probability in, 90
Indirect semiconductors, 121
Infiltration techniques, 254–255
Information technology, nanophotonics and, 395
Inorganic quantum dots, 278
Inorganic semiconductors, 79, 80–88
In-plane laser, 112
Institute for Lasers, Photonics, and Biophotonics, 161, 184, 365, 374
Intensity-dependent refractive index, 285
Interaction dynamics, nanoscopic, 34
Interaction potential, photon and electron, 10
Interactions, noncovalent, 210–212
Interband transitions, 84, 88, 116, 133
Interchain coupling, 289
Interlocked structures, 215
Intermolecular interactions
 dendritic architecture and, 224–225
 modulation of, 224–225
International Technology Roadmap for Semiconductors (ITRS), 393
Intraband transitions, 84, 89–90, 113
Intrinsic optical bistability, 142
Intrinsic size effect, 133
Introduction to Biophotonics (Prasad), 26
Inverse opal structure, 271
In vitro bioanalysis, PEBBLE nanosensors for, 362–364
Ionic interactions, 211
Ion–ion interactions, 158

Ion-pair interactions, 158
  optimization of, 168
Iridovirus, 343–344
Isomerism, positional, 216

J-aggregates, 34, 35, 38, 64

Kerr effect (optical Kerr effect), 26, 265
Kerr nonlinearity, 246
Knudsen cells, 179
Korriga–Kohn–Rostoker (KKR) method, 242
Kramer–Kronig relation, 95, 96, 339
Kretschmann configuration of ATR, 30–31
Kretschmann geometry, 65–66, 138

Langmuir–Blodgett films, 31, 32, 229–232
  multilayer, 235
  octadecanoic acid (ODA), 145
Langmuir–Blodgett techniques, 119, 120, 229, 235, 260
Langmuir–Blodgett Y-type deposition process, 229, 230
Laser-assisted direct imprint (LADI) nanolithography, 330
Laser-assisted molecular beam deposition (LAMBD), 183
Laser-assisted vapor deposition (LAVD), 177, 183–185, 204, 205
Laser beams, 63
Laser cavities, 108
  design of, 112
Laser paints, 283–284
Lasers
  market analysis for, 382, 383, 384
  quantum-confined, 392
  quantum-confined semiconductors and, 80
Laser technology, 107–108
Lasing media, quantum-confined structures as, 106–115
Lateral force microscopy (LFM), 203
Lateral nanoscopic localization, 32–33
Lattice phonons, 89–90
Lenses, tunable, 300
LH-RH (luteinizing hormone-release hormone) nanoclinics, 366
Lifetime broadening, 156
Ligand field effect, 155

Ligands, 211
Light, resonance coupling of, 147
Light-emitting "nanodiode" arrays, 310
Light-emitting diodes (LEDs), 117–118
  dendrimer-based organic, 223
Light harvesting, energy funnels for, 224
Light-harvesting dendrimers, 344–345, 350
Light hole (lh), 83
Light/matter interaction, 9
Light scattering, 283
Linear conjugated structures, 115
Linear response theory, 155
Linear susceptibility, 25
Line broadening, 156, 160, 191
Liposomes, 371
Liquid crystals
  assembly processes for, 227
  supramolecular, 226
Liquid-phase epitaxy (LPE), 177, 183, 204, 205
Liquid-phase methods, 204
Lithography. *See also* Nanolithography
  electron beam (E-Beam), 255–256
  nanoimprint, 330–331, 333
  nanosphere, 325–328
  near-field, 317–322, 333
  near-field phase-mask soft, 322–324
  phase-mask soft, 333
  two-photon, 255, 271
Local field enhancement, 100, 130, 137–138, 143, 279, 284–286
  metallic nanoparticles and, 141–142
  in photonic crystals, 246–247, 270
Local field interactions, excited states and, 155
Local interaction dynamics, control of, 159
Localization
  of electromagnetic field, 28
  photon and electron, 10
Localized phonons, 156
Localized surface plasmons, 130
Longitudinal cavity modes, 108
Longitudinal field effect, 97
Low-frequency acoustic phonons, 155
Low-frequency phonon host nanocrystals, 162
Luminance value, up-conversion, 163

Luminescence intensity, up-conversion, 162
Luminescence photophysics, in quantum dots, 93
Luminescent silicon nanoparticles, 90

Macrocycle, 217
Magnetic force microscopy (MFM), 200, 203
Magnetic nanoclinics, 377
Magneto-optic materials, 322
Marketplace, nanophotonics and, 381–397
Mass-spectrometric techniques, 180
Master equation, 243
Materials. *See* Biomaterials; Nanomaterials; Quantum-confined materials
Matter, nanoscale confinement of, 2–3
Maxwell Garnett geometry, 284–285, 303
Maxwell's equations, 11, 243
MBE growth chamber, 179, 180. *See also* Molecular beam epitaxy (MBE)
Medical imaging, 392. *See also* Nanomedicine
Mesophases, 277, 284
Mesoporous foldmers, 225, 235
Metabolically engineered materials, 338, 350
Metabolic engineering, 348–349
Metal coordination, 211
Metal–dipole interaction, 143
Metal-enhanced fluorescence, 361
Metal ions, sensing, 268
Metallic nanoarrays, 326
Metallic nanoparticles, 130–135, 186
  features of, 131
  nonspherically shaped, 134
  optical properties of, 132–133
Metallic nanoshells, 135–137, 147–148
Metallic nanostructures, 129
  applications of, 141–142
  fluorophore interactions with, 143
  interaction with light, 147
  local field enhancement and, 137–138
  plasmonic guiding and, 140–141
  plasmon printing and, 324–325
Metallic particles, radiative decay rate enhancement for, 144. *See also* Metallic nanoparticles
Metal–ligand coordination, 226
Metal-organic chemical vapor deposition (MOCVD), 177, 178, 181–183, 204–205
Metal-organic vapor-phase epitaxy (MOVPE), 182
Metal-to-insulator transition, 131–132
Metal-to-ligand charge transfer (MLCT), 34
Metamaterials, 278, 280
  nanocomposite, 301–302, 304
Microcavity effects, 271
  in photonic crystals, 250–252
Microelectromechanical systems (MEMS), 311
Microemulsion chemistry, 205
Microemulsion nanoreactor, 185
Microfabrication, 313
Micro-patterns, 316–317
Microscopic techniques, 189–204
Microscopy
  apertureless near-field, 62–65
  atomic force, 202–204
  electron, 194–197
  near-field, 48–51, 392
  scanning probe, 199–204
  scanning tunneling, 200–202
Microsphere, 14
Minibands, formation of, 101–102
Modeling
  of near-field nanoscopic interactions, 44–47
  of photonic crystals, 242–246
Molecular architectures, 233
  nanostructured, 209–235
Molecular assemblies, monolayer and multilayer, 210, 229–233, 235
Molecular beam epitaxy (MBE), 177, 178, 179–181, 204. *See also* MBE growth chamber
Molecular biology, 367
Molecular engineering, 209
Molecular machines, 209, 215–217, 234
Molecular motor, light-fueled, 218
Molecular orbital theory of bonding, 115
Molecular recognition, 268
Molecules
  conjugated, 115
  electronic transitions for, 154
  nanoscale behavior of, 53–55
Monodispersed nanosize bioobjects, 343
Monolayer molecular assemblies, 229–233

Monomers, photopolymerization, 311–312
M state, 340–341
Multifunctionality, nanostructure control and, 280
Multilayer molecular assemblies, 229–233
Multiphasic nanocomposites, 277, 286–290
  optical, 303
Multiphonon relaxation process, 156, 157, 172
Multiple multipole (MMP) model, 44–45, 74
Multiple quantum wells (MQW), 80
  laser, 110
Multiple scattering formula, 244

Nanoarrays, photonically aligned, 331–332
Nanobarcodes, 391
Nanochemistry, 177, 178, 185–189, 204, 205
Nanoclinics, 377
  for gene delivery, 367–370
  for optical diagnostics and targeted therapy, 365–367
  for photodynamic therapy, 371–376
Nanocomposite metamaterials, 301–302, 304
Nanocomposites, 2–3, 277–304. *See also* Multiphasic nanocomposites
  laser paints, 283–284
  local field enhancement, 284–286
  for optoelectronics, 290–297
  as photonic media, 278–280
  polymer-dispersed liquid crystals, 297–301
Nanocomposite waveguides, 280–283
Nanocontrol, of excitation dynamics, 153–173
Nanocrystal environment, 34
Nanocrystals
  Group III–V, 188
  re-precipitation method for, 118
  second harmonic imaging of, 59
Nanocrystal superlattices, 102
"Nanodiode" arrays, 326
Nanodomains, 70
  optical communications between, 279–280
"Nanoenvironment," 54

Nanoimprint lithography, 310, 330–331, 333
Nanolithography, 309–333, 393–394, 396
  dip-pen, 328–329, 333
  nanosphere lithography, 325–328
  near-field lithography, 317–322
  near-field phase-mask soft lithography, 322–324
  photonically aligned nanoarrays, 331–332
  plasmon printing, 324–325
  two-photon lithography, 311–317, 333
"Nano-mania," 1
Nanomaterials, 4, 79, 177–178
  characterization of, 189–204
  companies that manufacture, 388
  growth methods for, 178–189
  imaging of, 194
  optical, 386–392
  quantum-confined, 79–122
Nanomedicine, 3, 376
  future of, 395–396
  nanophotonics for, 355–377
Nanomers, 2, 115, 116
Nano-OLED fabrication, 327
Nanoparticle biosensors, dye-doped, 364–365. *See also* Biosensors
Nanoparticle coatings, 387–389
Nanoparticles, 2
  embedded, 333
  metallic, 130–135
  for optical diagnostics and targeted therapy, 357–358
  preparation of, 185
  rare-earth, 158–161
  rare-earth-doped, 376
  self-assembling of, 102
  site isolating, 171
  sunscreen, 389
  up-converting, 359
Nano-PDLCs, 297, 299
Nanophores, up-converting, 161–165. *See also* Up-converting nanophores
Nanophotonics, 1–3. *See also* Biomaterials
  for biotechnology and nanomedicine, 355–377
  foundations for, 9–38
  future outlook for, 394–396
  marketplace and, 381–397

Nanophotonics *(continued)*
  products, 386
  SPIE short course on, 1
  technology thrust areas for, 5–6
Nanorods
  metallic, 130–135
  organic, 118
Nanoscale confinement, of electronic interactions, 33–37
Nanoscale control, 173
Nanoscale dynamics, time- and space-resolved studies of, 69–73
Nanoscale electronic energy transfer, 35–36
Nanoscale enhancement, of optical interactions, 65–69
Nanoscale letter writing, 328
Nanoscale manipulation, advantages of, 166
Nanoscale optical interactions, 28–33
Nanoscopic aperture (fiber tip), 43
Nanoscopic domains, 277
Nanoscopic interaction dynamics, 34
Nanoscopic interactions, near-field, 44–47
Nanoscopic localization
  axial, 29–32
  lateral, 32–33
Nanosensors, PEBBLE, 362–364
Nanoshelled particles, 135
Nanoshells, metallic, 135
Nanosphere lithography, 310, 325–328, 333. *See also* Nanolithography
Nanostructure. *See also* Nanostructures
  control of, 160–161, 168, 209
  excited states and, 154–158
  manipulation of, 156–157, 171
  processing of, 280
Nanostructured molecular architectures, 209–235
  dendrimers, 217–225
  molecular machines, 215–217
  monolayer and multilayer molecular assemblies, 229–233
  nanostructured polymeric media, 212–215
  noncovalent interactions in, 210–212
  supramolecular structures, 225–228
Nanostructures. *See also* Nanostructure
  dendritic, 217–225
  hierarchical, 104
  metallic, 141–142

rare-earth-doped, 158–161
three-dimensional real-space images of, 199–204
Nanotechnology, 382–383, 396
Nanotubes, organic, 118
Nanowires
  GaN, 93
  single-crystal InP, 92–93
Narrow-band filtering, 267, 272
Native DNA, 337
Naturally occurring biomaterials, 338–344, 350
Nature-inspired materials. *See* Bioinspired materials
n-doping, 22
Near-field bioimaging, 356–357, 376
Near-field excitation, 54
Near-field geometry, 33
Near-field interaction/microscopy, 41–75
Near-field lithography, 309, 317–322, 333
Near-field microscopes, sources for, 73
Near-field microscopy, 48–51, 355, 392, 396
  instrumentation used for, 49
Near-field nanoscopic interactions, theoretical modeling of, 44–47
Near-field optics, 42–44
Near-field phase-mask soft lithography, 322–324
Near-field scanning optical microscope (NSOM), 33, 41, 48, 73, 200, 356
Near-field spectroscopy, apertureless, 62–65
Near-field studies, 51–62
Neuron imaging, 357
New cooperative transitions, 34–35
New technologies, research and development of, 5–6
Noncovalent interactions, 210–212, 233–234
  interlocked structures and, 215–216
"Nonfluorescent" molecules, 144
Nonlinear coupled vector wave equations, 45
Nonlinear optical effects, 121, 271–272
  enhancement of, 285–286
Nonlinear optical interactions, 25, 45
Nonlinear optical processes, 74, 229–231, 332

studies of, 55–62
Nonlinear optical properties, 95–96
Nonlinear photonic crystals, 239, 260–264
Nonlinearity
  dynamic, 95
  in solid-core photonic crystal fiber, 265–266
Nonradiative decay probability, 159
Nonradiative decay rate, 144
Nonspherical nanoparticles, field enhancement for, 138
Normal polymer (N), 231
Normal refractive index dispersion, 249
NPP nanocrystals, 57–58

Octadecanoic acid (ODA) Langmuir–Blodgett films, 145
Oligoamides, 228
Oligomers, 115, 116. *See also* Foldmers
One-dimensional electron gas (1DEG), 85
One-phonon process, 156
One-photon absorption, 318, 319
Onsager model, 294
Opal structure, 242, 244, 245, 271
Optical amplification, 158, 171
Optical circuitry, photonic crystal, 259–260
Optical communications, 272
  control of, 279–280
  photonic crystals and, 266–267
Optical correlation, 340
Optical diagnostics
  nanoclinics for, 365–367
  nanoparticles for, 357–358
Optical feedback, 283
Optical fiber geometries, 51
Optical interactions
  nanoscale, 28–33
  nanoscale enhancement of, 65–69
Optical lithography, 393
Optically induced refractive index change, 95
Optical materials, nanoscale, 79
Optical microelectromechanical systems (optical MEMS), 309
Optical nanocomposite media, 303
Optical nanocomposites, 277, 279, 287, 289–290, 302–303

Optical nanofiber biosensors, 365, 377
Optical nanomaterials, 396
  markets for, 386–392
Optical phonons, 155–156
Optical processes, nonlinear, 55–62
Optical propagation loss, 282
Optical properties, 88–90
  nonlinear, 95–96
  size-dependent, 147
Optical saturation, 70
Optical science and technology, nanoscale, 2
Optical signal processing, 95
Optical tracking, of nanoclinics, 377
Optical transitions, 84–85, 89, 121
  linear and nonlinear, 142
Optical transmission
  enhanced, 148
  through subwavelength hole arrays, 138
Optical waveguides, 303
  materials for, 280
Optics
  near-field, 42–44
  nonlinear, 260–264
Optoelectronic products, market analysis for, 383
Optoelectronics, nanocomposites for, 290–297
Optoelectronics-enabled markets, 385
Organically modified silica-based nanoparticles (ORMOSILs), 367–370. *See also* ORMOCER entries
Organic conjugated structure, 122
Organic crystals, surface modification of, 320–321
Organic molecules, bifunctional, 187
Organic nanomers, as quantum wires, 116–118
Organic nanotubes, 118
Organic quantum-confined structures, 115–120
Organic quantum dots, 118–119
Organic quantum wells, 119–120
Organic structures, integration with inorganic structures, 4
Organometallic structures, 34
Orientational alignment, 234

ORMOCER-1, 315
ORMOCERs, 281. *See also* Organically modified silica-based nanoparticles (ORMOSILs)
Oscillator strength, 85, 86
   enhancement of, 89
Oxidation state, control of, 159
Oxide nanoparticles, 187

Parametric mixing (generation), 25
Passivating reagent, 186
Pauli exclusion principle, 96
PDLC gratings, 332. *See also* Polymer-dispersed liquid crystals (PDLCs)
PEBBLE nanosensors, for *in vitro* bioanalysis, 362–364
Penetration depth, 19
Periodic potential, photon and electron localization under, 21–24
Perturbation theory, 101
Phase separation, limiting, 301
Phase-mask soft lithography, 310, 333
Phase-matching, 250, 261, 271–272
Phase-space filling, 95–96
Phenylene vinylene, polymerization of, 116. *See also* Poly-*p*-phenylene vinylene (PPV)
Phonons, 155
   interactions of, 172
   low-frequency acoustic, 155
   optical, 155–156
Phonon spectrum, 172
   modification of, 160
Photobleaching, 69, 75, 309, 318
   abrupt irreversible, 55
   of dyes, 146
Photocharge generation, 304
   enhanced, 294–295
Photochemical processes, 317
Photodynamic therapy, nanoclinics for, 371–376
Photofabrication, 66
Photogenerated nanostructures, 333
Photoinitiators, 312–313
Photoluminescence
   of CdSe-ZnS core-shell quantum dots, 104–105
   in silicon nanocrystals, 94
   studies of, 53

Photomasks, 320, 324
Photon avalanche, 165–166, 172–173
Photon cascade emission, 169
Photonically aligned nanoarrays, 331–332
Photonic applications, bacteriorhodopsin and, 339
Photonic bandgap materials, 241, 270
Photonic bandgap structure, 242, 247
Photonic components market, 385
Photonic crystal biosensors, 361–362, 371
Photonic crystal fibers (PCF), 264–266, 272, 391–392, 396
   solid-core, 265–266
Photonic crystals, 21, 22, 37, 239–272, 387, 391. *See also* Nonlinear photonic crystals
   bioobjects and, 343
   dip-pen nanolithography and, 329
   fabrication methods for, 252–259, 271
   features of, 246–252
   microcavity effect in, 250–252
   optical circuitry in, 259–26C
   optical communications and, 266–267
   structures of, 140
   theoretical modeling of, 242–246
   transmission and reflection spectra of, 240
   two-dimensional, 241
Photonic crystal sensors, 267–269, 272
Photonic media
   DNA as, 341
   nanocomposites as, 278–280
Photon localization, 270, 279
   Anderson-type, 279
Photon mode density, 143, 144
Photon recycling, 251, 271
Photons
   bandgap and, 21–24
   characteristics of, 10
   confinement of, 14–19
   cooperative effects for, 24–28
   free-space propagation of, 12–14
   leakage of, 19–20
   propagation as waves, 11
   tunneling of, 19–20, 37
   versus electrons, 10–28
Photon scanning tunneling microscopy (PSTM), 49, 74
Photophysical bleaching, 96

Photopolymerization, 309
  two-photon-induced, 311
Photoprocesses, nanoscale, 3
Photorefractive effect, 280, 293
Photorefractive hologram, 292, 293
Photorefractive holographic grating, 339
Photorefractive medium, 290
Photorefractive phase shift, 293
Photorefractivity, 277–278, 303
  steps in producing, 292
Photoresist, 323. *See also* Resist
Photosensitization, selective, 294
Photosensitizers (PS), 290, 371
Physics, technology development in, 5
  bonds, 115–116
  electrons, delocalized, 115–116
Plane Wave Expansion Method (PWEM), 242, 243, 246, 270
Plasmon bands, 130
Plasmon-guided evanescent waves, 310. *See also* Plasmonic guiding
Plasmonic biosensors, 360–361, 376
Plasmonic effects, types of, 148
Plasmonic-enhanced fluorescence, versus SERS, 145
Plasmonic guiding, 148. *See also* Plasmon-guided evanescent waves; Plasmonic wave guiding
Plasmonic printing, 142
Plasmonics, 2, 129–149
  local field enhancement, 137–138
  metallic nanoparticles and nanorods, 130–135
  metallic nanoshells, 135–137
  metallic nanostructure applications, 141–142
  radiative decay engineering, 142–147
  subwavelength aperture, 138–139
Plasmonic wave guiding, 139–141
Plasmon oscillation, 132, 133, 141
Plasmon printing, 324–325, 333
Plasmon resonance frequencies, metallic nanoshell, 135
Plasmon resonances, 147–148
  dielectric sensitivity of, 142
Plasmons, 129
Pockels effect, 26
Polarization anisotropy, 92, 93
Polarization dependence, 58, 59

Polarons, charged, 72
Polydiacetylene, 118
  absorption spectra of, 120
  structure of, 119
Polydimethylsiloxane (PDMS) film prepolymer, 323
Polyhydroxyalkanoates (PHA), 338, 348, 350)
Polymer-dispersed liquid crystals (PDLCs), 278, 297–301, 304, 331–332
Polymer electro-optic modulators, 224
Polymeric materials, 234, 281
Polymeric media, 209
  nanostructured, 212–215
Polymeric nanoparticles, 117
Polymeric structure, multifunctionality in, 234
Polymerization, multiphoton, 259
Polymerized crystalline colloidal array (PCCA), 268–269
Polymer light-emitting diode (PLED), 213
Polymer photonics, applications of, 212
Polymer photo-oxidation, reduction of, 137
Polymers
  crescent-backboned, 228
  crosslinking of, 331
  optical nanocomposites of, 290
  supramolecular, 225, 235
Polypeptides, as biotemplates, 346
Poly-*p*-phenylene vinylene (PPV), 116, 117, 118, 214, 287–289, 303, 304
Poly(propylene imine) dendrimer, 222
Polystyrene sphere packing, 242
Polystyrene sphere structure, band diagram for, 244
Poly(vinylpyrrolidone) (PVP), 282
Population inversion, 107
Porous silicon microcavity biosensors, 362, 376
Powder X-ray diffraction, 190, 205
  patterns in, 191
Power generation, nanophotonics and, 394–395
PPV:glass nanocomposite, 301–302, 304
PPV polymer, 301
PPV–silica nanocomposite, 289, 291
Prism gratings, 300–301
Prism–metal film–air configurations, 66, 67
PRL-701 dye, 67, 68, 69

Probe encapsulated by biologically localized embedding (PEBBLE), 362, 376
Probe–sample distance, 71, 72
  maintaining constant, 50
Protein–protein interactions, 36
Purcell effect, 251, 271

Q-CdS:PVK nanocomposite, 295
Quantum Cascade lasers (QCLs), 90, 106, 113, 122, 392, 396
  schematics of, 114
Quantum-confined lasers, 392
Quantum-confined materials, 79–122
  core-shell quantum dots and quantum dot–quantum wells, 104–106
  dielectric confinement effect in, 99–100
  inorganic semiconductors, 80–88
  as lasing media, 106–115
  manifestations of quantum confinement in, 88–99
  superlattices, 100–103
Quantum-confined semiconductors, 80, 106
  optics of, 89
Quantum-confined Stark effect, 96–99, 122
Quantum-confined structures, 122
  as lasing media, 106–115
  organic, 115–120
Quantum confinement effect, 17, 34, 79, 120
  examples of, 91–95
  exciton binding and, 90
  manifestations of, 88–99
  oscillator strength and, 89
Quantum cutters, 2, 167, 395
Quantum cutting processes, 153, 158, 166–170, 173
  example of, 169
  generation of, 159–160
Quantum dot emissions, 358
Quantum dot lasers, 111–112, 396
Quantum dot:polymer nanocomposites, 297
Quantum dot–quantum wells (QDQWs), 104–106
Quantum dots, 80, 86–88, 121, 280, 303–304. *See also* Semiconductor quantum dots
  as photosensitizers, 294, 303–304
  core-shell, 104–106

fluorescent, 390
organic, 118–119
preparation of, 93
studies of, 51–53
superlattices of, 102
Quantum dot solids, 102
Quantum numbers, 16
Quantum rings, 79, 88, 121
Quantum rods, 86, 94
Quantum states, wave functions for, 18
Quantum well lasers, 106, 109–110
Quantum wells, 80, 81–85, 97, 120
  effective bandgap for, 84
  fabrication of, 180
  multiple, 100–102
  organic, 119–120
Quantum wire lasers, 111
Quantum wires, 85–86, 91, 121
  organic nanomers as, 116–118

Radiation, nanoscale confinement of, 2
Radiative decay engineering, 142–147, 148
Radiative rate enhancement, 143–147
Radiative transition, microscopic theory of, 143
Raman microscopy, high-resolution, near-field, 64
Raman process, 156
Raman spectroscopy, surface-enhanced, 137
Random lasers, 283–284
  tunable systems of, 284
Rare-earth-doped nanoparticles, 376
Rare-earth-doped nanostructures, 158–161
Rare-earth-doped yttria nanoparticles, 360
Rare-earth ion pairs, 35
Rare-earth ions, 171, 359
  cooperative emission and, 36–37
  electronic transitions for, 154
  up-conversion processes in, 161, 162, 172
Reactive-ion etching (RIE), 257
Reactive oxygen species (ROS), 371
Reciprocal-Space Perturbative Method (RSPM), 44
Reflection high-energy electron diffraction (RHEED) technique, 177, 180, 197–198, 206
Reflectivity minimum, 32

Refractive index, 11, 12, 21, 25, 37
 of liquid crystals, 297–298
 local field enhancement and, 100
 manipulation of, 282
 nonlinear, 97
Refractive index contrast, 270
Refractive index dispersion, in photonic crystals, 249–250
Refractive index grating, 339
Regular structure photonic crystals, 242
Research, multidisciplinary, 3–4. *See also* Studies; Technology
Research and development, opportunities for, 5–6
Resist, 256. *See also* Photoresist
Resonance frequency, 134
Reverse micelle cavity nanoreactor, 116
Reverse micelle chemistry, 360
Reverse micelle synthesis, 185–186, 205
Rhodamine 6G-doped film, 341–343
Roswell Park Cancer Institute, 372–376
Rotaxanes, 216
Rotaxane shuttle, hydrogen-bond-assembled, 217
Rydberg energy, 28

Saturable absorber, 96
Scanning electron microscopy (SEM), 194–195, 195–197, 206
Scanning near-field optical microscope (SNOM), 33, 48
Scanning near-field photolithography (SNP), 319
Scanning probe microscopy (SPM), 199–204, 206
Scanning tunneling microscopes (STMs), 33
Scanning tunneling microscopy (STM), 189–190, 200–202
Scattering random media, 279, 297, 302
Scattering-type nanocomposites, 277
Scherrer's formula, 191
Schrödinger equation, 12, 17, 18
Second harmonic generation (SHG) process, 25, 55, 56, 58
 Langmuir–Blodgett films and, 231
Second harmonic (SH) wave, 45
Security marking, 390
Sedimentation, 2525

Self-assembled monolayer method, 235
Self-assembled monolayers (SAM), 229, 232–233
 near-field lithography and, 319–320
Self-assembled superlattices, 233
Self-assembly fabrication methods for 3D photonic crystals, 252–255
Self-cleaning glass, 389–390
Self-collimation, 249, 271
Semiconductor industries, nanolithography and, 393
Semiconductor lasers, 396
Semiconductor quantum dots, 376
 for bioimaging, 358–359
Semiconductor quantum well laser, 109
Semiconductors, 79
 direct-gap and indirect-gap, 23
 inorganic, 80–88
 material parameters for, 81
 quantum confined, 89, 106
 wide bandgap, 94
Sensors. *See also* Biosensors
 optical nanofiber, 365
 photonic crystal, 267–269, 272
Sensor technology, 395
Shear-force feedback technique, 50
Side-chain polymer, 215
Signal-to-noise ratio (SNR), 54
Silicon nanoparticles, 94, 359
 luminescent, 90
Silicon-on-insulator (SOI), 391
Single-molecule fluorescence detection, 54
Single-molecule spectroscopy, 53–55
 requirements for, 54
Single quantum dot spectroscopy, 52–53
Single quantum well (SQW) laser, 109, 110
Site encapsulation, 171, 173
Site isolating nanoparticles, 171
"Smart glue," 346
Soft lithography, 310, 322, 333. *See also* Dip-pen nanolithography (DPN)
Sol–gel chemistry, 171, 280–282, 286
Solid-core photonic crystal fiber, enhanced nonlinearity in, 265–266
Solution phase lasing, 288
Space-charge field, 292–293
Space-resolved studies, of nanoscale dynamics, 69–73

Spectroscopic methods, 189–204
Spectroscopy. *See also* Surface-enhanced Raman spectroscopy (SERS)
  apertureless near-field, 62–65
  energy-dispersive, 199
  single-molecule, 53–55
  X-ray photoelectron, 192
Spin coating, 326
Stark effect, 95, 122
  quantum-confined, 96–99, 122
  measurement, 55
Stop-gaps, 246, 263
Strained-layer lasers, 110
Stranski–Krastanov (SK) growth, 180
Studies. *See also* Research entries; Technology
  nanoscale dynamic, 69–73
  near-field, 51–62
Sub-bands, 83
  in quantum-confined structures, 90
Sub-band transitions, 121
Subdiffraction structures, 315
Subwavelength aperture plasmonics, 138–139
Subwavelength guiding structures, 141
Sulfur hexafluoride, 184
Sunscreen nanoparticles, 389
Superlattices, 79–80, 100–103, 122
  self-assembled, 232
Superprism phenomenon, 248
Supramolecular architecture/structures, 209–210, 225–228, 235
Supramolecular liquid crystals, 226
Surface coatings, 387–389
Surface emitting laser (SEL), 112
Surface-enhanced Raman spectroscopy (SERS), 137, 141, 144–145, 148
Surface phonon modes, 158
Surface plasmon coupling angles, 66, 69
Surface plasmon enhancement, 65, 67–69
Surface plasmonic coupling, 149. *See also* Surface plasmon coupling angles
Surface plasmon resonance (SPR), 74, 129. *See also* Plasmonic biosensors
  coupling angle in, 66
  size dependence of, 133
  technique of, 30–32
Surface plasmon resonance enhanced two-photon excited emission, 67

Surface plasmon resonance frequency, 142
Surface plasmon waves, 38, 130

"Tapping mode," 202
Targeted nanoclinics, 366
Targeted therapy
  nanoclinics for, 365–367
  nanoparticles for, 357–358
Technology. *See also* Biotechnology; Information technology; Laser technology; Nanotechnology; New technologies; Optical science and technology; Research entries; Sensor technology; Studies
  economic impact of, 381
  photonic-based, 3
TE modes, electric field distribution for, 16
Theoretical modeling, of photonic crystals, 242–246
Therapies. *See also* Photodynamic therapy
  light-guided and light-activated, 355
  nanoclinics for, 365–367
  nanoparticles for, 357–358
Third harmonic generation (THG), 26, 56, 59–62, 261–262
Three-dimensional confinement, 86
Three-dimensional structures, 233
Three-dimensional topography, STM and, 202
3D optical waveguide structures, 314
Time-correlated single-photon detection, 54
Time-domain techniques, 243
Time-resolved collection-mode near-field microscopy, 70
Time-resolved studies, of nanoscale dynamics, 69–73
Tip fabrication, 50–51
Tipping process, 183
Total internal reflection (TIR), 29–30
Trace gas analysis, 114–115
Training, multidisciplinary, 3–4
Transfer Matrix Method (TMM), 243–244, 245, 246, 270, 242
Transient transmittance, 71, 72
Transition probability, in indirect gap semiconductor, 90
Translight package, 244, 245
Transmission electron microscopy (TEM), 194, 195, 206

Transmission probability, 20
Transparent optical nanocomposites, 302–303
Transport vectors, for gene therapy, 367
Tunneling, photon and electron, 19–20
Tunneling microscopy, 200–202
Two-beam asymmetric coupling, 293–294
Two-dimensional confinement, 85–87
Two-dimensional electron gas (2DEG), 82
Two-photon absorbing organic chromophores, 312
Two-photon absorption, 26, 59
Two-photon excitation (TPE), 56, 59–62, 67, 142, 311, 314, 318, 333
Two-photon excited efficient light harvesting, 345
Two-photon excited emission, 262–264, 272
Two-photon excited fluorescence, 56
  image via, 64, 65
Two-photon holography, 309, 316, 333
Two-photon lithography, 255, 271, 309. See also Nanolithography
  growth of, 314–315
Two-photon near-field microscopy, 64
Two-photon optical memory, 311
Two-photon polymerization, 259–260, 315
Two-photon 3D microfabrication, 313
Type I/II quantum wells, 81

Ultra-diffraction, 249
Ultrafast nanostructure processes, dynamics of, 69
Ultra-refraction, 271
Up-conversion processes, 153
  generation of, 159
Up-converted fluorescence, 69
Up-converting nanophores, 161–165
  for bioimaging, 359–360
Up-converting transparent glass, 163, 165
UV (ultraviolet) blockers, 387

Vacuum UV photons, 160
Vacuum UV radiation, 167
Valence band, 22
Van der Waals forces, 211, 229
Vapor deposition
  laser-assisted, 183–185

metal-organic chemical, 181–183
Vapor-phase epitaxy, 181–183
Vapor-phase methods, 204
Vertical cavity surface emitting laser (VCSEL), 112
Vertical deposition method, 254
Viruses, as biotemplates, 347
Voxels, 315–316

Wannier exciton, 27
Wave excitation, evanescent, 142
Waveguides. *See also* Waveguiding
  evanescent wave-coupled, 29
  nanocomposite, 280–283
Waveguiding. *See also* Waveguides
  photonic crystal fiber, 264–265
  plasmonic, 139–141
Wavelength, photon and electron, 10
Wavelength division multiplexing (WDM), 267
Wavelength multiplexing, 340
Well superlattice, electron tunneling and, 102
Wet-chemical etching, 329. *See also* Wet etching
Wet chemistry, 177
Wet etching, 257. *See also* Wet-chemical etching
Whole-blood immunoassay, as a metal nanoshell application, 137
Wide bandgap semiconductors, 94

Xe (xenon) discharge, 167
X-ray diffraction, 190–192
X-ray diffractogram, 191
X-ray nanomaterial characterization, 190–194
X-ray photoelectron spectroscopy (XPS), 192, 205

YOYO-1 dye, 356, 368–369
Yttria nanoparticles
  rare-earth-doped, 360
  up-converted emission spectra in, 163

Zero-dimensional electron gas (0DEG), 87
ZnSe, growth of, 198
ZnSe/ZnCdSe superlattice, 180, 181